Solving Problems in Geometry

Insights and Strategies for Mathematical
Olympiad and Competitions

Mathematical Olympiad Series

ISSN: 1793-8570

Series Editors: Lee Peng Yee *(Nanyang Technological University, Singapore)*
Xiong Bin *(East China Normal University, China)*

Published

The complete list of the published volumes in the series can be found at
http://www.worldscientific.com/series/mos

Vol. 10 | Mathematical Olympiad Series

Solving Problems in Geometry

Insights and Strategies for Mathematical Olympiad and Competitions

Kim Hoo Hang
Institute of Advanced Studies, Nanyang Technological University, Singapore

Haibin Wang
NUS High School of Mathematics and Science, Singapore

World Scientific

NEW JERSEY · LONDON · SINGAPORE · BEIJING · SHANGHAI · HONG KONG · TAIPEI · CHENNAI · TOKYO

Published by

World Scientific Publishing Co. Pte. Ltd.

5 Toh Tuck Link, Singapore 596224

USA office: 27 Warren Street, Suite 401-402, Hackensack, NJ 07601

UK office: 57 Shelton Street, Covent Garden, London WC2H 9HE

Library of Congress Cataloging-in-Publication Data
Names: Hang, Kim Hoo, author. | Wang, Haibin, author.
Title: Solving problems in geometry : insights and strategies for mathematical Olympiad
    and competitions / by Kim Hoo Hang (NTU, Singapore),
    Haibin Wang (NUS High School of Mathematics and Science, Singapore).
Description: New Jersey : World Scientific, [2017] | Series: Mathematical olympiad series ;
    volume 10
Identifiers: LCCN 2017021633| ISBN 9789814590723 (hardcover : alk. paper) |
    ISBN 9789814583749 (pbk : alk. paper)
Subjects: LCSH: Geometry--Problems, exercises, etc.
Classification: LCC QA459 .H25 2017 | DDC 516.076--dc23
LC record available at https://lccn.loc.gov/2017021633

British Library Cataloguing-in-Publication Data
A catalogue record for this book is available from the British Library.

Printed in Singapore

Preface

Elementary geometry is a foundational and important topic not only in Mathematics competitions, but also in mainstream pre-university Mathematics education. Indeed, this is the first axiomatic system most learners encounter: definitions, theorems, proofs and counterexamples. While beginners find the basic theorems and illustrations intuitive, they may encounter difficulties and frequently become clueless when solving problems. For example, the concept of congruent triangles is the most straightforward and easy to understand, but many beginners find it difficult to identify congruent triangles in a diagram, not mentioning constructing congruent triangles intentionally to solve the problem. In particular, drawing auxiliary lines is perceived by many learners as a mysterious skill.

Geometry problems which appear at higher level Mathematics competitions are of course more challenging and require deeper skills. Even the most experienced contestant may spend an hour or so to solve one such problem – while the final solution may be elegantly written down in half a page. In this case, a beginner cannot learn much from merely reading the solution. Such obstacles, with insufficient scaffolding and the lack of guidance, hinder many learners when studying problem-solving in geometry.

In this book, we focus on showing the readers **how** to seek clues and acquire the geometric insight. One may find a few paragraphs named *"Insight"* for almost every problem, where we illustrate how to start tackling the problem, which clues could be found, and how to link the clues leading to the conclusion. Note that such a process is inevitably a lengthy one, during which the reader could attempt a number of strategies and *fail* repeatedly before reaching the final conclusion. A formal proof, usually much shorter, will be presented after we obtain the insight. Occasionally, if sufficient clues have been revealed, we will leave it to the reader to complete the proof.

In the first few chapters, we introduce the basic properties of triangles, quadrilaterals and circles. Proofs and explanatory notes are written down so that the learners will gain the geometric insight of those results, instead of memorizing the literal expression of the theorems. Examples, which range from easy and straightforward to difficult, are used to elaborate how these properties are applied in problem-solving.

In the later chapters, we give a list of commonly used facts, useful skills and problem-solving strategies which could help readers tackle challenging geometry problems at high-level Mathematics competitions. Such a collection of facts, skills and strategies are seldom found in any mainstream textbooks as these are not standard theorems. They essentially focus on ideas and methodology. We illustrate these skills and strategies using geometry problems from recent-year competitions. The following is a list of these competitions.

1.   International, Regional and Invitational Competitions

- IMO   International Mathematical Olympiad
              (including shortlist problems)
- APMO   Asia Pacific Mathematical Olympiad
- EGMO   European Girls' Mathematical Olympiad
- CMO   China Mathematical Olympiad
- CGMO   China Girls' Mathematical Olympiad
- CWMO   China West Mathematical Olympiad (Invitation)
- CZE-SVK   Czech and Slovak Mathematical Olympiad
- IWYMIC   Invitational World Youth Mathematics Intercity
              Competition

2.　　National Competitions and Selection Tests

- AUT　　Austria
- BLR　　Belarus
- BRA　　Brazil
- BGR　　Bulgaria
- CAN　　Canada
- CHN　　China
- HRV　　Croatia
- GER　　Germany
- HEL　　Greece
- HUN　　Hungary
- IND　　India
- IRN　　Iran
- ITA　　Italy
- JPN　　Japan
- ROU　　Romania
- RUS　　Russia
- SVN　　Slovenia
- TUR　　Turkey
- UKR　　Ukraine
- USA　　U.S.A.
- VNM　　Vietnam

Elementary geometry is a beautiful area of mathematics. Upon the mastery of the basic knowledge and skills, one will always find solving a geometry problem an exciting experience. We wish the readers a pleasant experience with the time spent on this book. Enjoy Mathematics and enjoy problem-solving!

Contents

Chapter 1

Congruent Triangles

We assume the reader knows the following basic geometric concepts, which we will not define:

- Points, lines, rays, line segments and lengths
- Angles, right angles, acute angles, obtuse angles, parallel lines $(/\!/)$ and perpendicular lines $(\perp)$
- Triangles, isosceles triangles, equilateral triangles, quadrilaterals, polygons
- Height (altitudes) of a triangle, area of a triangle
- Circles, radii, diameters, chords, arcs, minor arcs and major arcs

1.1 Preliminaries

We assume the reader is familiar with the fundamental results in geometry, especially the following, the illustration of which can be found in any reasonable secondary school textbook.

(1) For any two fixed points, there exists a unique straight line passing through them (and hence, if two straight lines intersect more than once, they must coincide).

(2) For any given straight line ℓ and point P, there exists a unique line passing through P and parallel to ℓ.

(3) Opposing angles are equal to each other. (Refer to the diagram on the right. $\angle 1$ and $\angle 2$ are opposing angles. We have $\angle 1 = 180° - \angle 3 = \angle 2$.)

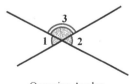

Opposing Angles

(4) In an isosceles triangle, the angles which correspond to equal sides are equal. (Refer to the diagram on the right.)

The inverse is also true: if two angles in a triangle are the same, then they correspond to the sides which are equal.

Equal angles correspond to equal sides.

(5) Triangle Inequality: In any triangle $\triangle ABC$, $AB + BC > AC$.

(A straight line segment gives the shortest path between two points.)

(6) If two parallel lines intersect with a third, we have:
  • The corresponding angles are the same.
  • The alternate angles are the same.
  • The interior angles are *supplementary* (i.e., their sum is $180°$).
    (Refer to the diagrams below.)

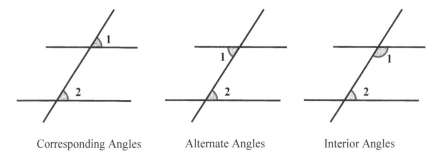

Corresponding Angles   Alternate Angles   Interior Angles

Its inverse also holds: equal corresponding angles, equal alternate angles or supplementary interior angles imply parallel lines.

One may use (6) to prove the following well-known results.

Theorem 1.1.1 *The sum of the interior angles of a triangle is* $180°$.

Proof. Refer to the diagram on the right. Draw a line passing through A which is parallel to BC. We have $\angle B = \angle 1$ and $\angle C = \angle 2$.

Hence, $\angle A + \angle B + \angle C = \angle A + \angle 1 + \angle 2 = 180°$. □

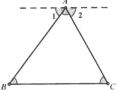

An immediate and widely applicable corollary is that an *exterior* angle of a triangle equals the sum of two non-neighboring interior angles. Refer to the diagram on the right. We have $\angle 1 = 180° - \angle C = \angle A + \angle B$.

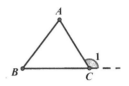

It is also widely known that the sum of the interior angles of a quadrilateral is $360°$. Notice that a quadrilateral could be divided into two triangles. Refer to the diagram on the right.

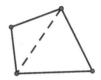

One sees that similar arguments apply to a general n-sided (convex) polygon: the sum of the interior angles is $180° \times (n-2)$.

Example 1.1.2 Find $\angle A + \angle B + \angle C + \angle D + \angle E + \angle F + \angle G$ in the left diagram below.

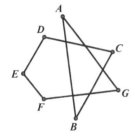
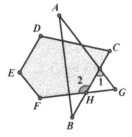

Ans. Refer to the right diagram above. Let BC and FG intersect at H. Notice that $\angle A + \angle B = \angle 1$ and $\angle 1 + \angle G = \angle 2$.
Now $\angle 2 + \angle C + \angle D + \angle E + \angle F = 540°$, as this is the sum of the interior angles of the convex pentagon (i.e., a 5-sided polygon) $CDEFH$.
In conclusion, $\angle A + \angle B + \angle C + \angle D + \angle E + \angle F + \angle G = 540°$. □

Note: Using the exterior angles of a triangle is an effective method to solve this type of questions. Refer to the following diagrams. Can you see $\angle A + \angle B + \angle C + \angle D + \angle E + \angle F + \angle G = 540°$ in both cases?

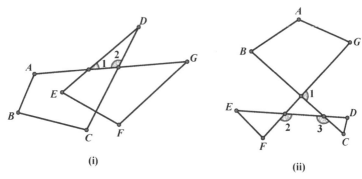

(i)   (ii)

Hint:

(i)   Connect *EG*. Can you see $\angle E + \angle F + \angle G = 180° + \angle 1$? A similar argument applies to $\angle A + \angle B + \angle C$.

(ii)   Connect *BG*. Can you see $\angle A + \angle B + \angle G = 180° + \angle 1$? Can you see $\angle E + \angle F = \angle 2$? Can you find $\angle 1 + \angle 2 + \angle 3$? (Consider their supplementary angles.)

Example 1.1.3   Refer to the diagram on the right. $\triangle ABC$ is an isosceles triangle where $AB = AC$. *D* is a point on *BC* such that $AB = CD$. Draw $DE \perp AB$ at *E*. Show that $2\angle ADE = 3\angle B$.

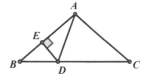

Insight.   We are not given the exact value of $\angle BAC$ or $\angle B$, but if we know either of them, then the positions of *D* and *E* are uniquely determined, according to the construction of the diagram. Let $\angle B = x$. Can you express $\angle ADE$ in term of *x*?

Proof.   Let $\angle B = \angle C = x$. We have $\angle BAC = 180° - 2x$. Notice that $\triangle CAD$ is an isosceles triangle, where $AC = AB = CD$. It follows that

$$\angle ADC = \angle CAD = \frac{1}{2}\left(180° - x\right) = 90° - \frac{x}{2}.$$

Now $\angle BAD = \angle BAC - \angle CAD = \left(180° - 2x\right) - \left(90° - \frac{x}{2}\right) = 90° - \frac{3}{2}x.$

Hence, $\angle ADE = \dfrac{3}{2}x$ (because $\angle BAD + \angle ADE = 90°$ in the right angled triangle $\triangle AED$). The conclusion follows. □

Example 1.1.4 Given a quadrilateral $ABCD$, E is a point on AD. F is a point inside $ABCD$ such that CF, EF bisects $\angle ACB$ and $\angle BED$ respectively. Show that $\angle CFE = 90° + \dfrac{1}{2}(\angle CAD + \angle CBE)$. (**Note:** an *angle bisector* divides the angle into two equal halves.)

Insight. Refer to the diagram on the right. One sees that $\angle CAD$ and $\angle CBE$ are NOT related. For example, if $\angle CAD$ is given, one may move E along AD and $\angle CBE$ will vary. On the other hand, if $\angle CBE$ is given, one may choose A' on DA extended so that $\angle CA'D$ is smaller than $\angle CAD$.

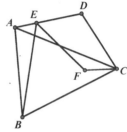

Hence, if we let $\angle CAD = \alpha$, we cannot express $\angle CBE$ in α (and vice versa). How about letting $\angle CBE = \beta$? We **should** be able to express $\angle CFE$ in α and β.

Notice that $\angle CFE$ is constructed via angle bisectors EF and CF. Let $\angle BED = 2x$ and $\angle ACB = 2y$. Refer to the diagram on the right. Let AC and EF intersect at G. In $\triangle CFG$, one sees that $\angle CFE = 180° - y - \angle CGF$, where $\angle CGF = \angle AGE = \angle DEF - \angle EAG = x - \alpha$. Hence, $\angle CFE = 180° - x - y + \alpha$. (1)

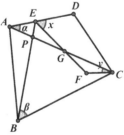

We are to show $\angle CFE = 90° + \dfrac{1}{2}(\angle CAD + \angle CBE) = 90° + \dfrac{1}{2}(\alpha + \beta)$.

How are x, y related to α, β? Let AC and BE intersect at P. Consider $\triangle AEP$ and $\triangle BCP$. One sees that $\angle PAE + \angle PEA = 180° - \angle APE$ $= 180° - \angle BPC = \angle PBC + \angle PCB$.

Hence, $\alpha + (180° - 2x) = \beta + 2y$, which implies $x + y = 90° + \frac{1}{2}(\alpha - \beta)$.

Now (1) gives $\angle CFE = 180° - \left(90° + \frac{1}{2}(\alpha - \beta)\right) + \alpha = 90° + \frac{1}{2}(\alpha + \beta)$.

Note: This is not an easy problem, but it could be solved by elementary knowledge. When solving problems purely about angles, it is a useful technique to set an unknown angle as a variable and apply algebraic manipulations. If one variable is not enough (to express the other angles), one may set more variables, but remember to work towards cancelling out those variables, simply because they should **not** appear in the conclusion. In order to cancel out the variables, one should seek for equalities among angles. Useful clues include right angles, isosceles triangles, exterior angles and angle bisectors.

The following examples give standard results which are frequently used in problem-solving. One should be very familiar with these results.

Example 1.1.5 In $\triangle ABC$, $\angle A = 90°$ and $AD \perp BC$ at D. Show that $\angle BAD = \angle C$ and $\angle CAD = \angle B$.

Proof. Refer to the diagram on the right. We have $\angle BAD = 90° - \angle B = \angle C$ and similarly, $\angle CAD = 90° - \angle C = \angle B$.   □

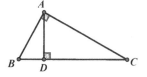

Example 1.1.6 Refer to the diagram on the right. AB and CD intersect at E. If $\angle B = \angle D$, show that $\angle A = \angle C$.

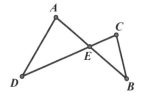

Proof. We have $\angle A = 180° - \angle D - \angle AED$ and $\angle C = 180° - \angle B - \angle BEC$.
Since $\angle B = \angle D$ and $\angle AED = \angle BEC$, it follows that $\angle A = \angle C$.   □

Notice that $\angle A + \angle D = \angle B + \angle C$ always holds even if we do not have $\angle B = \angle D$. We used this fact in Example 1.1.4.

Example 1.1.7 In an acute angled triangle $\triangle ABC$, BD, CE are heights. Show that $\angle ABD = \angle ACE$.

Proof. Refer to the diagram on the right. We have $\angle ABD = 90° - \angle A = \angle ACE$. □

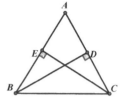

One may also see this as a special case of Example 1.1.6, where $\angle BEC = \angle BDC = 90°$.

Example 1.1.8 In $\triangle ABC$, M is the midpoint of BC. Show that if $AM = \dfrac{1}{2} BC$, then $\angle A = 90°$.

Proof. Refer to the diagram on the right. Since $AM = BM = CM$, we have $\angle 1 = \angle B$ and $\angle 2 = \angle C$, i.e., $\angle A = \angle B + \angle C$. Since $\angle A + \angle B + \angle C = 180°$, $\angle A = 90°$. □

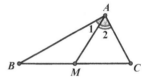

Example 1.1.9 In $\triangle ABC$, D is on BC. Show that the angle bisectors of $\angle ADB$ and $\angle ADC$ are perpendicular to each other.

Proof. Refer to the diagram on the right. Let DE, DF be the angle bisectors of $\angle ADB$ and $\angle ADC$ respectively.

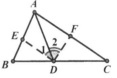

Since $\angle 1 = \dfrac{1}{2} \angle ADB$, $\angle 2 = \dfrac{1}{2} \angle ADC$ and $\angle ADB + \angle ADC = 180°$, we have $\angle 1 + \angle 2 = 90°$, and hence the conclusion. □

Example 1.1.10 Refer to the diagram on the right. Let AD bisect $\angle A$. If $BD /\!/ AC$, show that $AB = BD$.

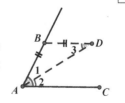

Proof. We are given $\angle 1 = \angle 2$. Since $BD /\!/ AC$, $\angle 2 = \angle 3$. Now $\angle 1 = \angle 3$ and it follows that $AB = BD$. □

Note:

(1) It is a commonly used technique to construct an isosceles triangle from an angle bisector and parallel lines. Besides giving equal angles, angle bisectors have many other useful properties, which we will see in later chapters.

(2) Notice that the inverse also holds:
   - If we are given that $AB = BD$ and AD bisects $\angle A$, then we must have $BD \,/\!/\, AC$.
   - If we are given that $AB = BD$ and $BD \,/\!/\, AC$, then AD must be the angle bisector of $\angle A$.

Example 1.1.11   Given lines $\ell_1 \,/\!/\, \ell_2$ and a point P, draw $PA \perp \ell_1$ at A and $PB \perp \ell_2$ at B, then P, A, B are collinear (i.e., the three points lie on the same line).

Proof.   Refer to the diagrams below. Suppose otherwise that P, A, B are not collinear. Let AP extended intersect ℓ_2 at C. Now $\angle PCB = 90°$ and $\triangle PBC$ has two $90°$ interior angles. This is absurd.

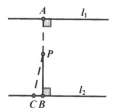   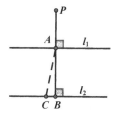

Notice that the argument holds even if ℓ_1, ℓ_2 are on the same side of P. Refer to the diagram above on the right.   □

1.2   Congruent Triangles

Congruent triangles are the cornerstones of elementary geometry. We say two triangles $\triangle ABC$ and $\triangle A'B'C'$ are congruent if they are exactly the same: $AB = A'B'$, $AC = A'C'$, $BC = B'C'$, $\angle A = \angle A'$, $\angle B = \angle B'$ and $\angle C = \angle C'$. We denote this by $\triangle ABC \cong \triangle A'B'C'$.

Moreover, if $\triangle ABC \cong \triangle A'B'C'$, **all** the corresponding line segments and angles are identical. Refer to the diagrams below for an example: Given $\triangle ABC \cong \triangle A'B'C'$, let AH be the height of $\triangle ABC$ on the side BC and $A'H'$ be the height of $\triangle A'B'C'$ on the side $B'C'$. Let M, M' be the midpoints of $AH, A'H'$ respectively. We must have $BM = B'M'$ and $\angle BMH = \angle B'M'H'$.

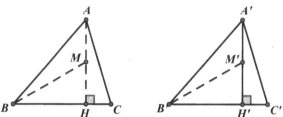

Applying the definition directly could verify a pair of congruent triangles. However, in most of the cases, this is unnecessary. It is taught in most secondary education that one can verify congruent triangles using one of the following criteria:

- S.A.S.: If two pairs of corresponding sides and the angles **between** them are identical, then the two triangles are congruent, i.e., if $AB = A'B'$, $AC = A'C'$ and $\angle A = \angle A'$, then $\triangle ABC \cong \triangle A'B'C'$.
- A.A.S.: If one pair of corresponding sides and any two pairs of corresponding angles are identical, then the two triangles are congruent.
- S.S.S.: If all the corresponding sides are identical, then the two triangles are congruent.

Note:

(1) S.A.S. applies only when two pairs of corresponding sides and the angles **between** them are identical. Otherwise, we **cannot** use this criterion. Refer to the following counter example:

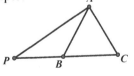

Let $\triangle ABC$ be an isosceles triangle where $AB = AC$. P is a point on CB extended. Refer to the diagram on the right. Consider $\triangle PAC$ and $\triangle PAB$.

We have $AB = AC$, $\angle P$ is a common angle and AP is a common side. However, one sees clearly that $\triangle PAC \ncong \triangle PAB$ because $\angle PBA > 90° > \angle PCA$.

(2) One may also write A.A.S. as A.S.A. In fact, it does not matter whether the corresponding sides are between the two pairs of corresponding angles, simply because two pairs of equal angles automatically gives the third pair of equal angles: the sum of the interior angles of a triangle is always $180°$.

(3) H.L.: If $\triangle ABC$ and $\triangle A'B'C'$ are right angled triangles, then they are congruent if their hypotenuses and one pair of corresponding legs are identical, i.e., if $\angle A = \angle A' = 90°$, $AB = A'B'$ and $BC = B'C'$, then $\triangle ABC \cong \triangle A'B'C'$.

Indeed, one may place the two right angled triangles together and form an isosceles triangle. Refer to the diagram on the right. $BC = B'C'$ immediately gives $\angle C = \angle C'$ and hence, we have $\triangle ABC \cong \triangle A'B'C'$ (A.A.S.).

One immediate application of congruent triangles on isosceles triangles is that the angle bisector of the vertex angle, the median on the base and the height on the base of an isosceles triangle coincide.

Definition 1.2.1 In $\triangle ABC$, let M be the midpoint of BC such that $BM = CM$, then AM is called the median on the side BC. (Refer to the diagram on the right.)

Theorem 1.2.2 *Let $\triangle ABC$ be an isosceles triangle such that $AB = AC$. Let M be the midpoint of BC. We have:*
(1) *$AM \perp BC$.*
(2) *AM bisects $\angle A$, i.e., $\angle BAM = \angle CAM$.*

Proof. The conclusion follows from $\triangle ABM \cong \triangle ACM$ (S.S.S.).   □

Notice that in the theorem above, any point P on the line AM gives an isosceles triangle $\triangle PBC$. Refer to the diagram on the right. Indeed, AM is the perpendicular bisector of the line segment BC.

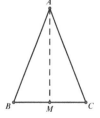

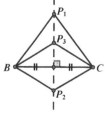

Definition 1.2.3 The perpendicular bisector of a line segment *AB* is a straight line which passes through the midpoint of *AB* and is perpendicular to *AB* .

Perpendicular bisector of *AB*

Theorem 1.2.4 *Given a line segment AB and a point P. We have PA = PB if and only if P lies on the perpendicular bisector of AB. In particular, if P,Q are two points such that PA = PB and QA = QB, then the line PQ is the perpendicular bisector of AB.*

One may show the conclusion easily by using congruent triangles. We leave it to the reader.

Notice that Theorem 1.2.2 states that in an isosceles triangle $\triangle ABC$ where $AB = AC$, the angle bisector of $\angle A$, the median on *BC*, and the height on *BC* coincide. Moreover, one could show by congruent triangles that the inverse is also true: if any two among these three lines coincide (for example, *AD* bisects $\angle A$ where *D* is the midpoint of *BC*), then $\triangle ABC$ is an isosceles triangle with $AB = AC$. This is an elementary property of isosceles triangles, but it may apply in a subtle manner in problem-solving, which confuses beginners.

Example 1.2.5   Given $\triangle ABC$ where $\angle A = 90°$ and $AB = AC$, *D* is a point on *AC* such that *BD* bisects $\angle ABC$. Draw $CE \perp BD$, intersecting *BD* extended at *E*. Show that $BD = 2CE$.

Insight.   Apparently, the conclusion does not give us any clue because *BD* and *CE* are not directly related. Perhaps we should seek clues from the conditions.

It is given that *BE* bisects $\angle ABC$ and we see that *BE* is *almost* a height: not a height of any given triangle, but $BE \perp CE$. If we *fill up* the triangle by extending *BA* and *CE*, intersecting each other at *F*, then *BE* is the height of $\triangle BCF$. Refer to the diagram on the right.

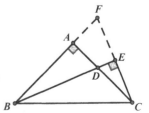

Can you see $\triangle BCF$ is an isosceles triangle? Moreover, E must be the midpoint of CF as well, which implies $CF = 2CE$. Hence, it suffices to show $BD = CF$.

How are BD and CF related? If it is not clear to you, seek clues from the conditions again! Which conditions have we not used yet? We are given $AB = AC$ and $\angle BAC = 90°$. How are they related to BD and CF? We **should** have $\triangle ABD \cong \triangle ACF$ if $BD = CF$. How can we show $\triangle ABD \cong \triangle ACF$? We have a pair of equal sides $AB = AC$ and a pair of right angles. Showing $AD = AF$ may not be easy because we do not know the position of A on BF. Can we find another pair of equal angles?

Proof. Let BA extended and CE extended intersect at F. Since BE bisects $\angle ABC$ and $BE \perp CF$, we have $\triangle BEC \cong \triangle BEF$ (A.A.S.) and hence, $CF = 2CE$. It is easy to see $\angle ABD = \angle DCE$ (Example 1.1.6). Since $AB = AC$ and $\angle BAD = 90° = \angle CAF$, we have $\triangle ABD \cong \triangle ACF$ (A.A.S.). It follows that $BD = CF = 2CE$.   □

Note:

(1) One may derive a few conclusions from the proof above. For example, can you see $\angle ADB = \angle BCE$ and $BC = AB + AD$?

(2) How did we see the auxiliary lines? Notice that we basically *reflected* $\triangle BCE$ along the angle bisector BE. This is an effective technique which utilizes the symmetry property of the angle bisector.

Recognizing congruent triangles is one of the most fundamental but useful methods in showing equal line segments or angles. In particular, one may seek congruent triangles via the following clues:

- Equilateral triangles and isosceles triangles
- Right angled triangles with the height on the hypotenuse (which gives equal angles, Example 1.1.5)
- Common sides or angles shared by triangles
- Parallel lines
- Medians and angle bisectors
- Opposite angles (Example 1.1.6)

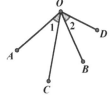

- Equal angles sharing the common vertex: Refer
  to the diagram on the right. If $\angle 1 = \angle 2$, then
  $\angle AOB = \angle COD$. Notice that the inverse also holds, i.e., if
  $\angle AOB = \angle COD$, then $\angle 1 = \angle 2$.

Example 1.2.6 Refer to the diagram on
the right. In $\triangle ABC$, draw equilateral
triangles $\triangle ABF$ and $\triangle ACE$ outwards
from AB, CA respectively. Show that
$BE = CF$.

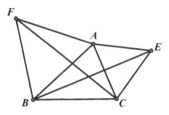

Proof. We have equal sides $AF = AB$
and $AC = AE$ due to the equilateral
triangles.
Notice that we also have
$\angle 1 = \angle 2 = 60°$.
Hence, $\angle 1 + \angle BAC = \angle 2 + \angle BAC$,
i.e., $\angle BAE = \angle CAF$.

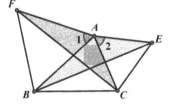

It follows that $\triangle BAE \cong \triangle FAC$ (S.A.S.), which leads to the conclusion
that $BE = CF$. □

Example 1.2.7 In an acute angled triangle $\triangle ABC$, $\angle A = 45°$. AD, BE
are heights. If AD, BE intersect at H, show that $AH = BC$.

Insight. Can we find a pair of congruent triangles where AH, BC are
corresponding sides? It is given $\angle A = 45°$ and $BE \perp AC$. Hence, it is
easy to see that $AE = BE$.
Refer to the diagram on the right. It *seems* that $\triangle AEH$ and $\triangle BEC$ are
congruent.
Since $\angle AEH = 90° = \angle BEC$, we only need one more
condition. Shall we prove $CE = EH$, or find another
pair of equal angles? Can you see $\angle CBE = \angle CAH$
(Example 1.1.6)?

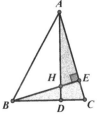

We leave it to the reader to complete the proof.

Example 1.2.8   Refer to the diagram on the right. In $\triangle ABC$, $\angle A = 90°$. P is a point outside $\triangle ABC$ such that $PB \perp BC$ and $PB = BC$. D is a point on PA extended such that $CD \perp PA$. E is a point on CD extended such that $BE \perp AB$. Show that AE bisects $\angle BAC$.

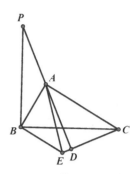

Insight.   We are given $PB = PC$ and one can easily see that $\angle ABP = \angle EBC$. Are there any congruent triangles? It *seems* from the diagram that $\triangle ABP \cong \triangle EBC$. Is it true? We are to show AE bisects $\angle BAC$, i.e., $\angle BAE = 45°$. Hence, we **should** have $\triangle ABE$ a right angled isosceles triangle where $AB = BE$, i.e., $\triangle ABP$ and $\triangle EBC$ **should** be congruent. Now can we find another pair of equal sides or angles?

Proof.   Notice that $\angle ABP = 90° - \angle ABC = \angle EBC$. We also have $\angle APB = \angle BCE$ (Example 1.1.6, BC intersecting PD). Since $PB = BC$, we conclude that $\triangle ABP \cong \triangle EBC$. Hence, $AB = BE$, which implies $\triangle ABE$ is a right angled isosceles triangle.

Now $\angle BAE = 45° = \dfrac{1}{2} \angle BAC$, which implies AE bisects $\angle BAC$.   □

Example 1.2.9   Refer to the diagram on the right. $\triangle ABC$ is an equilateral triangle with $AB = 10$ cm. D is a point outside $\triangle ABC$ such that $BD = CD$ and $\angle BDC = 120°$. M, N are on AB, AC respectively such that $\angle MDN = 60°$. Find the perimeter of $\triangle AMN$.

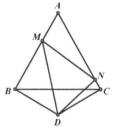

Insight. The difficulty is that M is arbitrary, i.e., it can be any point on AB. Even though we know $\angle MDN = \dfrac{1}{2} \angle BDC$, it is hard to apply this condition directly.

What if we choose a special point *M*, say when $\triangle DMN$ is an equilateral triangle? Refer to the left diagram below. Now $\triangle AMN$ is also an equilateral triangle. One may show (by studying the property of the right angled triangle $\triangle BDM$) that $AM = \dfrac{2}{3} AB$. Hence, the perimeter of $\triangle AMN = 2AB = 20$ cm.

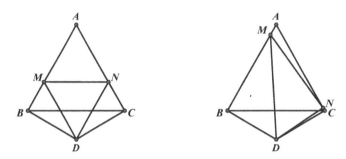

What if we choose *M* to be very close to *A*? Refer to the diagram above on the right. $\triangle AMN$ seems to become very narrow. *AM* is approaching to zero length while *AN* and *MN* are very close to *AC*. In this case, we may expect the perimeter of $\triangle AMN$ to be $0 + AC + AC = 2AC = 20$ cm.

It seems that we shall prove $AM + AN + MN = AB + AC$, i.e., $MN = BM + CN$. However, it may not be easy to show this directly as *BM* and *CN* are far apart. Notice that we encounter the same difficulty: given that $\angle MDN = \dfrac{1}{2} \angle BDC$, how to handle the remaining portions of $\angle BDC$? If we can put those portions together, an equal angle of $\angle MDN$ would appear. How can we put $\angle BDM$ and $\angle CDN$, as well as *BM* and *CN* together? Cut and paste!

Ans. Extend *AC* to *E* such that $CE = BM$. Connect *DE*. Notice that $\angle DBC = \angle DCB = \dfrac{1}{2}(180° - 120°) = 30°$. Hence, $\angle DBM = \angle DCE = 90°$ and we have $\triangle DBM \cong \triangle DCE$ (S.A.S.). This implies $\angle BDM = \angle CDE$ and $DM = DE$. Refer to the following left diagram.

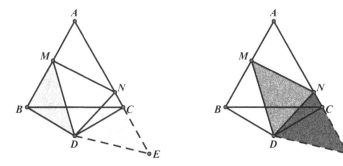

In order to show $MN = BM + CN = CE + CN = NE$, it suffices to show $\triangle DNM \cong \triangle DNE$. Since $\angle MDN = 60°$, $\angle BDM + \angle CDN = 60°$.
Hence, $\angle EDN = \angle CDE + \angle CDN = 60° = \angle MDN$. Since $DM = DE$, it follows that $\triangle DNM \cong \triangle DNE$ (S.A.S.).
In conclusion, $AM + MN + AN = AB + AC = 20\,\text{cm}$.   □

One may apply congruent triangles to prove the following useful properties. These are not the standard theorems, but one familiar with these results could have a better understanding of the basic geometrical facts and seek clues during problem-solving more effectively.

Example 1.2.10   Given a line segment AB and two points P, Q such that line PQ intersects AB at C, if $\angle APC = \angle BPC$ and $\angle AQC = \angle BQC$, then PQ is the perpendicular bisector of AB.

Proof.   Case I: P, Q are on the same side of AB.
Refer to the diagram on the right. We have
$\angle 1 = \angle AQC - \angle APC = \angle BQC - \angle BPC = \angle 2$.

It follows that $\triangle APQ \cong \triangle BPQ$ (A.A.S.), which implies $AP = BP$.
Since PC is the angle bisector of the isosceles triangle $\triangle PAB$, it is also the perpendicular bisector of AB.

Case II: P, Q are on different sides of AB. Refer to the diagram on the right. It is easy to see that $\triangle APQ \cong \triangle BPQ$ (A.A.S.). Hence, $PA = PB$ and $QA = QB$. The conclusion follows by Theorem 1.2.4. □

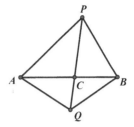

Example 1.2.11 Given $\triangle ABC$, draw squares $ABDE$ and $CAFG$ outwards based on AB, CA respectively. Let M be the midpoint of BC. Show that $AM = \dfrac{1}{2} EF$.

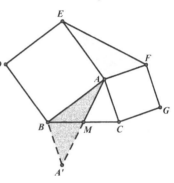

Insight. We see that $\triangle ABC$ and $\triangle AEF$ have two equal pairs of sides: $AB = AE$ and $AC = AF$. However, it is clear that $\triangle ABC \ncong \triangle AEF$ because $\angle BAC \neq \angle EAF$. In fact, $\angle BAC$ and $\angle EAF$ are supplementary. (Can you see it?)

Since M is the midpoint of BC, a commonly used technique is to double AM. Refer to the diagram above on the right. If we extend AM to A' such that $AM = A'M$, can you see that $\triangle BAA' \cong \triangle AEF$?

Proof. Extend AM to A' such that $AM = A'M$. Since $BM = CM$ and $\angle A'MB = \angle AMC$, we have $\triangle A'MB \cong \triangle AMC$ (S.A.S.), which implies $\angle CAM = \angle BA'M$ and hence, $AC \parallel A'B$. It follows that $\angle ABA' = 180° - \angle BAC$. Since $\angle EAF + \angle BAC = 360° - 90° - 90° = 180°$, we must have $\angle EAF = 180° - \angle BAC = \angle ABA'$.

Since $AB = AE$, $BA' = AC = AF$, we conclude that $\triangle BAA' \cong \triangle AEF$ (S.A.S.). It follows that $EF = AA' = 2AM$. □

Note: It is an important technique to extend and double the median of a triangle because this immediately gives congruent triangles.

Refer to the diagram on the right where AD is a median of $\triangle ABC$ and we have $\triangle ACD \cong \triangle A'BD$.

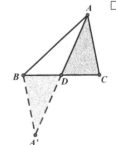

After this *rotation* of $\triangle ACD$, we may put together lengths and sides which are previously far apart and perhaps obtain useful conclusions.

Example 1.2.12 In $\triangle ABC$, D is the midpoint of *BC*. *E* is a point on *AC* such that *BE* intersects *AD* at *P* and *BP = AC*. Show that *AE = PE*.

Insight.   We are given $BP = AC$, which should be an important condition. However, *BP* and *AC* are far apart and it seems not clear how one could use this condition. How about the median *AD* doubled?

Refer to the diagram on the right. If we extend
AD and take $A'D = AD$, one sees immediately
that $\triangle ACD \cong \triangle A'BD$. In particular, we have
$AC = A'B$.

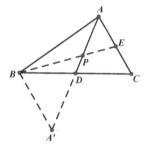

In fact, we rotated $\triangle ACD$ and hence, moved
AC to $A'B$. Now $A'B$ and *BP* are connected:
we can apply the condition $BP = AC$.

Proof.   Extend *AD* to *A'* such that $AD = A'D$. It is easy to see that
$\triangle ACD \cong \triangle A'BD$ (S.A.S.). Hence, $\angle A' = \angle CAD$ and $BP = AC = A'B$,
i.e., $\triangle BA'P$ is an isosceles triangle.
Now we have $\angle APE = \angle BPA' = \angle A' = \angle CAD$, which implies $AE = PE$.
$\square$

1.3   Circumcenter and Incenter of a Triangle

Given a triangle, there are many interesting points in it.

Recall the definition of the perpendicular bisector of a line segment. Since each triangle has three sides, one may draw three perpendicular bisectors. Note that these perpendicular bisectors are *concurrent*, i.e., they pass through the same point. Refer to the following diagrams.

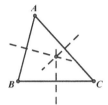 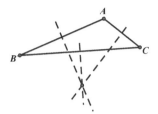

This particular point is called the **circumcenter** of the triangle. Notice that each triangle has exactly one circumcenter and it could be outside the triangle. Refer to the right diagram above.

Now we use congruent triangles to show the existence of the circumcenter of a triangle.

Theorem 1.3.1 *The perpendicular bisectors of a triangle are concurrent.*

Proof. Refer to the left diagram below. Let the perpendicular bisectors of AB, BC intersect at O. We are to show that the perpendicular bisector of AC passes through O as well.

Since O lies on the perpendicular bisector of AB, we have $AO = BO$ (Theorem 1.2.4). Similarly, $BO = CO$. It follows that $AO = CO$. Hence, O lies on the perpendicular bisector of AC (Theorem 1.2.4). □

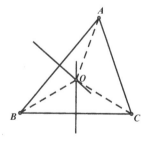 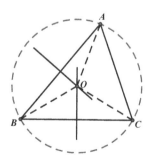

Note:

(1) This common point of intersection is called the circumcenter as it is the center of the circumcircle of $\triangle ABC$. Refer to the right diagram

above. A circle centered at O with radius OA passes through A, B and C, since $OA = OB = OC$.

(2) In the proof above, we assume the two perpendicular bisectors intersect at O and show that this point lies on the third perpendicular bisector. This is a common method to show three lines passing through the same point.

Theorem 1.3.2   *The angle bisectors of a triangle are concurrent.*

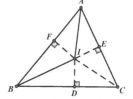

Proof. Refer to the diagram on the right. Let the angle bisector of $\angle A$ and $\angle B$ intersect at I. We show that the angle bisector of $\angle C$ passes through I as well, i.e., $\angle ACI = \angle BCI$.

Draw $ID \perp BC$ at D, $IE \perp AC$ at E and $IF \perp AB$ at F. Since AI is the angle bisector of $\angle A$, it is easy to see $\triangle AIF \cong \triangle AIE$ (A.A.S.). Hence, $IF = IE$. Similarly, $ID = IF$. It follows that $ID = IE$.

Now it is easy to see that $\triangle CID \cong \triangle CIE$ (H.L.), which leads to the conclusion that $\angle ACI = \angle ABI$.   □

Note:
(1) I is called the **incenter** of $\triangle ABC$.
(2)   Since $\triangle AIF \cong \triangle AIE$, one sees that $AE = AF$, i.e., A (and similarly I) lie on the perpendicular bisector of EF. Hence, AI is the perpendicular bisector of EF. A similar argument applies for BI and CI as well.

Theorem 1.3.3   *Let I be the incenter of $\triangle ABC$.* $\angle BIC = 90° + \dfrac{1}{2}\angle A$.

Proof.   Refer to the diagram below. Since I is the incenter of $\triangle ABC$, AI, BI, CI are angle bisectors. Since $2(\angle 1 + \angle 2 + \angle 3) = 180°$, we must have $\angle 1 + \angle 2 + \angle 3 = 90°$.

Now $\angle BIC = 180° - \angle 2 - \angle 3 = 180° - (90° - \angle 1)$

$= 90° + \angle 1 = 90° + \dfrac{1}{2}\angle A.$ □

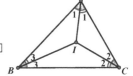

Example 1.3.4   Given $\triangle ABC$ where $\angle A = 60°$, D, E are on AC, AB respectively such that BD, CE bisects $\angle B, \angle C$ respectively. If BD and CE intersect at I, show that $DI = EI$.

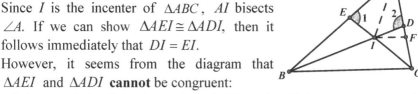

Insight.   Refer to the diagram on the right. Since I is the incenter of $\triangle ABC$, AI bisects $\angle A$. If we can show $\triangle AEI \cong \triangle ADI$, then it follows immediately that $DI = EI$.

However, it seems from the diagram that $\triangle AEI$ and $\triangle ADI$ **cannot** be congruent: $\angle 1$ is acute but $\angle 2$ is obtuse, i.e., $\triangle AEI$ and $\triangle ADI$ are **not** symmetric about AI. Why not reflect $\triangle AEI$ about AI and construct congruent triangles? Let us choose F on AC such that $AF = AE$. Now $\triangle AEI \cong \triangle AFI$ and we have $EI = FI$. Can we show $DI = FI$? Notice that $\triangle IDF$ **should** be an isosceles triangle. How can we show it? Since $\angle AFI = \angle 1$, it suffices to show that $\angle 1 = 180° - \angle 2$, or equivalently, $\angle 1 + \angle 2 = 180°$. This may not be difficult because both $\angle 1$ and $\angle 2$ can be expressed using $\angle B$ and $\angle C$ (using exterior angles) and we know $\angle B + \angle C = 180° - \angle A = 120°$!

Proof.   Choose F on AC such that $AE = AF$. Notice that I is the incenter of $\triangle ABC$, i.e., AI bisects $\angle A$. Now we have $\triangle AEI \cong \triangle AFI$ (S.A.S.) and hence, $EI = FI$.

Since $\angle 1 = \angle B + \dfrac{1}{2} \angle C$ and $\angle 2 = \angle C + \dfrac{1}{2} \angle B$, we have

$$\angle 1 + \angle 2 = \frac{3}{2}(\angle B + \angle C) = \frac{3}{2}(180° - \angle A) = 180°, \text{ i.e., } \angle 1 = 180° - \angle 2.$$

It follows that $\angle DFI = \angle 1 = \angle FDI$. Now $DI = FI = EI$. $\square$

1.4 Quadrilaterals

A quadrilateral is a polygon with four sides. In this book, we focus on convex quadrilaterals only. Refer to the following diagrams for examples.

Convex quadrilateral   Concave quadrilateral

There are two important types of quadrilaterals: parallelograms (including rectangles, rhombus and squares) and trapeziums. We will study their properties in this section.

Definition 1.4.1 A parallelogram is a quadrilateral with both pairs of opposing sides parallel to each other.

We give a list of equivalent ways to define a parallelogram.

(1) A parallelogram is a quadrilateral with two pairs of equal opposite sides.

(2) A parallelogram is a quadrilateral with a pair of opposite sides equal and parallel to each other.

(3) A parallelogram is a quadrilateral with both pairs of opposite angles equal.

(4) A parallelogram is a quadrilateral with two diagonals bisecting each other.

One may show that all these definitions are equivalent by the techniques of congruent triangles.

Note that these definitions also describe the properties of a parallelogram. One may pay particular attention to (4), which is less frequently mentioned in textbooks, but widely applicable in problem-solving.

Example 1.4.2 Given a parallelogram $ABCD$, draw equilateral triangles $\triangle ABE$ and $\triangle BCF$ outwards from AB, BC respectively. Show that $\triangle DEF$ is an equilateral triangle.

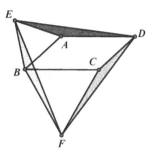

Insight. Refer to the diagram on the right. Given a parallelogram and equilateral triangles, one shall seek congruent triangles. Apparently, $\triangle ADE, \triangle CFD, \triangle BFE$ **should** be congruent. It is easy to show equal sides, while a bit of calculation might be needed to show equal angles.

Proof. We have $AE = AB = CD$ and $AD = BC = CF$. Notice that $\angle DAE = 360° - \angle BAD - \angle BAE$ and $\angle FCD = 360° - \angle BCF - \angle BCD$. Since $\angle BAD = \angle BCD$ and $\angle BCF = 60° = \angle BAE$, we have $\angle DAE = \angle FCD$. Hence, $\triangle ADE \cong \triangle CFD$ (S.A.S.) and $DE = DF$. Similarly, $BE = AB = CD$ and $BF = CF$. Notice that $\angle EBF = \angle ABE + \angle CBF + \angle ABC = 60° + 60° + 180° - \angle BCD = 300° - \angle BCD = 360° - \angle BCF - \angle BCD = \angle FCD$. Now $\triangle BEF \cong \triangle CFD$ (S.A.S.) and hence, $DF = EF$. This completes the proof. $\qquad\square$

Notice that the techniques for solving problems on quadrilaterals are still mainly through congruent triangles.

Example 1.4.3 Let $ABCD$ be a quadrilateral such that $\angle B = \angle D$. AC and BD intersect at P. If $AP = CP$, show that $ABCD$ is a parallelogram.

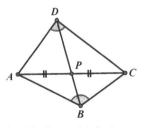

Insight. It is not easy to show the conclusion using congruent triangles directly. Although there are pairs of equal angles and identical lengths, they do not form congruent triangles. Refer to the diagram on the right.

Since DP is the median on AC, the median doubled could help to construct congruent triangles.

Moreover, among all the criteria to determine a quadrilateral, we may use (4): two diagonals bisecting each other. This is because we are given $AP = CP$ and we only need to show $DP = BP$. Bingo! This coincides with our strategy to double the median DP.

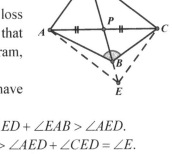

Proof. We claim that $BP = DP$, which leads to the conclusion immediately.

Suppose otherwise, say $BP < DP$, without loss of generality. We extend PB to E such that $DP = EP$. Now $AECD$ is a parallelogram, which implies $\angle D = \angle E$.

However, $\angle B = \angle D$ and we must have $\angle B = \angle E$. This is impossible! Notice that $\angle B = \angle ABD + \angle CBD$, where $\angle ABD = \angle AED + \angle EAB > \angle AED$. Similarly, $\angle CBD > \angle CED$. We have $\angle B > \angle AED + \angle CED = \angle E$.

In conclusion, we must have $BP = DP$ and hence, $ABCD$ must be a parallelogram. □

Example 1.4.4   Given an isosceles triangle $\triangle ABC$ where $AB = AC$, M is the midpoint of BC. P is a point on BA extended and $PD \perp BC$ at D. If PD intersects AC at E, show that $PD + DE = 2AM$.

Insight. AM is a median of $\triangle ABC$ and we need $2AM$ in the conclusion. Hence, it is natural to extend and double the median AM. Refer to the diagram below. Extend AM to A' such that $AM = A'M$. We are to show $PE + PD = AA'$. Can you see a line segment equal to DE?

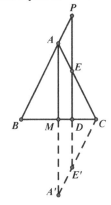

Proof.   Extend AM to A' such that $AM = A'M$. Since $AB = AC$, one sees that $\triangle ABM \cong \triangle A'CM \cong \triangle ACM$. Let PD extended intersect $A'C$ at E'. We have $\triangle CDE \cong \triangle CDE'$ (A.A.S.) and hence, $DE = DE'$. We also conclude that $PD \, /\!/ \, AM$ and $AP \, /\!/ \, A'C$. Now $AA'E'P$ is a parallelogram and $AA' = PE'$. It follows that
$PD + DE = PD + DE' = PE' = AA' = 2AM.$ □

Note: One may also draw $AN \perp PD$ at N and show that N is the midpoint of PE. Refer to the diagram on the right. Since $AMDN$ is a parallelogram (and in fact, a rectangle), we have $AM = DN$. Now it suffices to show $PD + DE = 2DN$. Note that this is equivalent to $PD - DN = DN - DE$, or $PN = EN$.

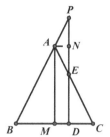

It is easy to see that N is the midpoint of PE because $\triangle APE$ is an isosceles triangle where $AP = AE$. (Can you show it?)

Definition 1.4.5 A rectangle is a quadrilateral with four right angles.

We give the following equivalent ways to define a rectangle.
(1) A rectangle is a parallelogram with a right angle.
(2) A rectangle is a parallelogram with equal diagonals.

One may show that all these definitions are equivalent by the techniques of congruent triangles.

Note that (2) is an important property of rectangles. In particular, in a rectangle $ABCD$ where AC, BD intersect at O, we have $\angle OAD = \angle ODA$. Refer to the diagram on the right.

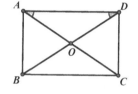

Given two parallel lines $\ell_1 // \ell_2$, the perpendicular distance from an arbitrary point on one line to the other line is a constant. Refer to the diagram on the right.

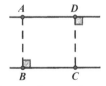

One could easily see that $ABCD$ is a rectangle and we always have $AB = CD$. This length is defined as the distance between ℓ_1 and ℓ_2.

Theorem 1.4.6 *In a right angled triangle $\triangle ABC$ where $\angle A = 90°$ and M is the midpoint of BC, we have $AM = \dfrac{1}{2} BC$.*

Observe the fact that the right angled triangle is half of a rectangle. Refer to the diagram on the right. One may show the conclusion easily by congruent triangles.

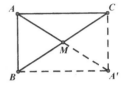

This is a simple but useful result. However, even the experienced contestants fail to recognize it occasionally, especially when the problem is complicated.

Note that Example 1.1.8 is the inverse of Theorem 1.4.6. In summary, given $\triangle ABC$ where M is the midpoint of BC, $\angle A = 90°$ if and only if $AM = \dfrac{1}{2}BC$.

Example 1.4.7   In an acute angled triangle $\triangle ABC$, BE, CF are heights on AC, AB respectively. Let D be the midpoint of BC. Show that $DE = DF$.

Proof.   This is an immediate application of Theorem 1.4.6. In the right angled triangle $\triangle BEC$, we have $DE = \dfrac{1}{2}BC$. Similarly, $DF = \dfrac{1}{2}BC$. The conclusion follows.   □

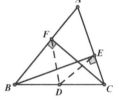

Example 1.4.8   In a right angled triangle $\triangle ABC$ where $\angle A = 90°$ and $\angle C = 30°$, $AB = \dfrac{1}{2}BC$.

Proof.   Refer to the diagram on the right. Let M be the midpoint of BC. By Theorem 1.4.6, $AM = BM$. We see that $\triangle ABM$ is an isosceles triangle where $\angle B = 60°$, and hence, an equilateral triangle.

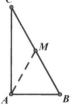

It follows that $AB = BM = \dfrac{1}{2}BC$.   □

Note:

(1) Refer to the diagram on the right. One may reflect $\triangle ABC$ about the line AC and see that $\triangle ABC$ is half of the equilateral triangle $\triangle BCB'$.

It is now clear that $AB = \dfrac{1}{2}BB' = \dfrac{1}{2}BC.$

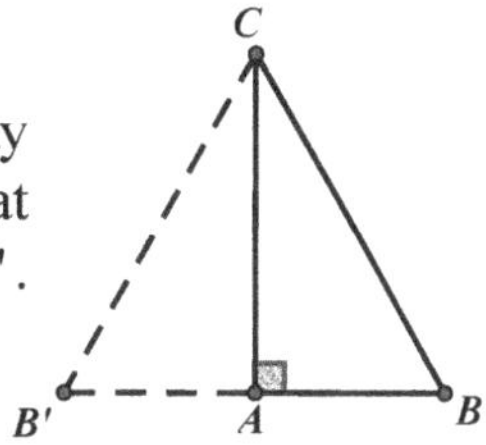

(2) Notice that the inverse also holds: given $\triangle ABC$ where $\angle A = 90°$, if $AB = \dfrac{1}{2}BC,$ then $\angle C = 30°.$ This is because $AM = BM = \dfrac{1}{2}BC$ $= AB$ by Theorem 1.4.6, where M is the midpoint of BC. Hence, $\triangle ABM$ is an equilateral triangle and $\angle B = 60°.$

Definition 1.4.9 A rhombus is a quadrilateral whose sides are of equal length.

We give the following equivalent ways to define a rectangle.

(1) A rhombus is a parallelogram with a pair of equal neighboring sides.
(2) A rhombus is a parallelogram whose diagonals are perpendicular to each other.

One may show that all these definitions are equivalent by the techniques of congruent triangles.

Example 1.4.10 Given a parallelogram $ABCD$ where $BC = 2AB,\ E,F$ are on the line AB such that $AE = AB = BF$. Connect CE, DF. Show that $CE \perp DF$.

Insight. Refer to the diagram on the right. We are given a parallelogram $ABCD$ and $AE = AB = BF$. Hence, we can see more parallelograms, for example $ACDE$ (because $AE = CD$ and $AE \,/\!/\, CD$).

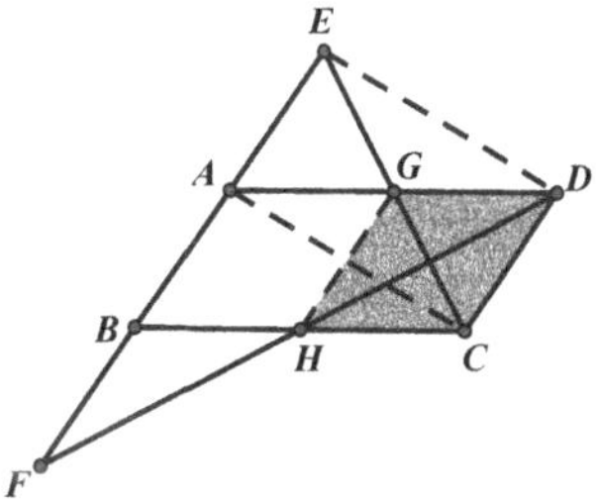

It follows that AD and CE bisect each other. Now we can see that the condition $BC = 2AB$ is useful. Can you see a rhombus in the diagram?

Proof.   Let AD, CE intersect at G and BC, DF intersect at H. Since $ABCD$ is a parallelogram, we have $AE // CD$ and $AE = AB = CD$, which imply $ACDE$ is also a parallelogram. Hence, AD, CE bisect each other. Since $AD = 2AB$, $DG = AB = CD$. Similarly, $CH = CD$. It follows that $CDGH$ is a rhombus and hence, $CE \perp DF$.   □

Note: One may find an alternative solution using the technique of angle bisectors, parallel lines and isosceles triangle (Example 1.1.10):
Since $AB = BF$, we have $AF = 2AB = AD$, i.e., $\triangle AFD$ is an isosceles triangle. Now $\angle CDF = \angle AFD = \angle ADF$, i.e., DF bisects $\angle ADC$. Similarly, CE bisects $\angle BCD$. One sees $CE \perp DF$ because

$$\angle DCE + \angle CDF = \frac{1}{2}\angle BCD + \frac{1}{2}ADC = \frac{1}{2} \cdot 180° = 90°.$$

Notice that the last step is closely related to Example 1.1.9 that the angle bisectors of neighboring supplementary angles are perpendicular to each other.

Definition 1.4.11   A square is a rectangle whose sides are of equal length.

A square is a parallelogram which is both a rectangle and a rhombus. Hence, a square has all the properties of rectangles and squares, including equal sides, equal angles and diagonals of equal length which perpendicularly bisect each other. Of course, one may write down a lot of statements which are equivalent to the definition of a square.

Example 1.4.12   Refer to the diagram on the right. $ABCD$ is a square. Two lines, ℓ_1 and ℓ_2, intersect $ABCD$ at E, F and G, H respectively. If $\ell_1 \perp \ell_2$, show that $EF = GH$.

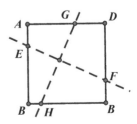

Insight. If ℓ_1, ℓ_2 are in the upright position, the conclusion is clear. Refer to the left diagram below.

Regrettably, we do not know the positions of ℓ_1 and ℓ_2 with respect to the square $ABCD$. Indeed, we are to show that for any $\ell_1 \perp \ell_2$, regardless of how they intersect $ABCD$, the conclusion holds.

Let us move ℓ_1, ℓ_2 around and observe. Refer to the middle diagram below. If we push EF upwards until E reaches A, we still have $EF = AF'$ because $AEFF'$ is a parallelogram. If we continue to push GH towards the right, we see that $GH = DH'$. Refer to the right diagram below. Hence, it suffices to show that $DH' = AF'$. This could be shown by congruent triangles.

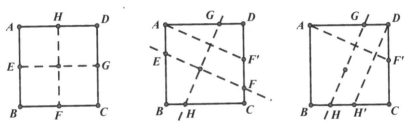

Proof. Draw $AF' /\!/ EF$, intersecting CD at F'. Draw $DH' /\!/ GH$, intersecting BC at H'. Since $AEFF'$ is a parallelogram, $EF = AF'$. Similarly, $GH = DH'$. It suffices to show that $AF' = DH'$.
Notice that $\angle DAF' = 90° - \angle ADH' = \angle CDH'$, $\angle ADC = 90° = \angle C$ and $AD = CD$. Hence, $\triangle ADF' \cong \triangle DCH'$ (A.A.S.) and $AF' = DH'$. □

Example 1.4.13 Refer to the diagram on the right. $ABCD$ is a square and $BDEF$ is a rhombus such that C, E, F are collinear. Find $\angle CBF$.

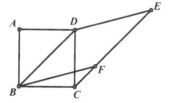

Insight. We have a square, a rhombus and the collinearity of C, E, F. One immediately sees that $\angle CBD = 45°$. Can we find $\angle DBF$? Notice that once $\angle DBF$ is known, the rhombus is uniquely determined. Which rhombus satisfies the conditions that C, E, F are collinear?

If we draw an arbitrary rhombus $BDE'F'$ based on BD, as shown in the diagram on the right, we will still have $BD \mathbin{/\mkern-5mu/} E'F'$, but C will not lie on the line $E'F'$, i.e., we **must** use the fact $CE \mathbin{/\mkern-5mu/} BC$ to show the conclusion.

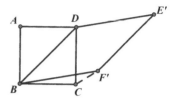

One may also observe that if F is chosen, i.e., $CF \mathbin{/\mkern-5mu/} BD$ and $BD = BF$, we do not need to draw E as it is not relevant to the problem anymore.

Hence, we may simplify the problem. Refer to the diagram on the right. Given $BD \mathbin{/\mkern-5mu/} CF$ and $BD = BF$, what can we deduce about $\angle DBF$? We know $\triangle BDF$ is an isosceles triangle, but calculating $\angle BDF$ or $\angle BFD$ is not easy. How can we use $BD = BF$ then? We know $AC = BD$. How is AC related to BF?

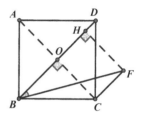

Given $BD \mathbin{/\mkern-5mu/} CF$, what is the distance between these two parallel lines? Can you see this distance is $\frac{1}{2}BD$? What if we introduce a perpendicular line to BD from F?

Ans. Let AC and BD intersect at O. Draw $FH \perp BD$ at H. Since $CF \mathbin{/\mkern-5mu/} BD$ and $AC \perp BD$, we have $FH = CO = \frac{1}{2}AC = \frac{1}{2}BD$. It follows that in the right angled triangle $\triangle BFH$, $\angle FBH = 30°$ (Example 1.4.8).

Now $\angle CBF = \angle CBD - \angle FBD = 45° - 30° = 15°$. $\qquad\qquad \square$

Note: Since the distance between BD and CF is $\frac{1}{2}BD$, it is natural to think of $\angle FBH = 30°$ in a right angled triangle. In fact, one may even draw the diagram accurately and see that $\angle FBH = 30°$. Even though such a drawing will **NOT** be accepted as part of the solution, it gives us a clue. Now constructing a right angled triangle with $\angle FBH = 30°$, i.e., where one leg is half of the hypotenuse, becomes a natural strategy.

Definition 1.4.14   A trapezium is a quadrilateral with **exactly** one pair of parallel sides.

By definition, a trapezium cannot be a parallelogram.

Example 1.4.15   In a trapezium $ABCD$ where $AD \,/\!/\, BC$, E is a point on AB. Show that $\angle ADE + \angle BCE = \angle CED$.

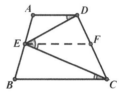

Proof.   Refer to the diagram on the right. Draw $EF \,/\!/\, AD$, intersecting CD at F. Notice that $\angle ADE = \angle DEF$ and $\angle BCE = \angle CEF$. Hence, $\angle ADE + \angle BCE = \angle DEF + \angle CEF$ $= \angle CDE$.   □

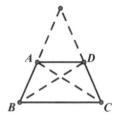

An isosceles trapezium is a trapezium whose unparalleled sides are of equal length. In fact, one obtains an isosceles triangle by extending the unparalleled sides. Refer to the diagram on the right where $ABCD$ is an isosceles trapezium with $AB = CD$. It can be shown easily that $\angle B = \angle C$ and $AC = BD$.

Example 1.4.16 $ABCD$ is an isosceles trapezium where $AD \,/\!/\, BC$ and $AB = CD$. Its diagonals AC, BD intersect at E and $\angle AED = 60°$. Let M, N be the midpoints of CE, AB respectively. Show that $MN = \dfrac{1}{2} AB$.

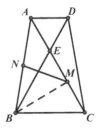

Proof.   Refer to the diagram on the right. Since $ABCD$ is an isosceles trapezium with $AB = CD$, we must have $\angle ABC = \angle BCD$. Hence, $\triangle ABC \cong \triangle DCB$ (S.A.S.), which implies $\angle BCE = \angle CBE$. Since $\angle BEC = \angle AED = 60°$, $\triangle BCE$ must be an equilateral triangle. Since M is the midpoint of CE, we must have $BM \perp CE$.

Since N is the midpoint of AB, MN is the median on the hypotenuse of $\triangle AMB$ and hence, $MN = \dfrac{1}{2}AB$ (Theorem 1.4.6). $\square$

1.5 Exercises

1. In a right angled triangle $\triangle ABC$ where $\angle A = 90°$, P is a point on BC. If $AP = BP$, show that $BP = CP$, i.e., P is the midpoint of BC.

2. Given $\triangle ABC$ where $\angle B = 2\angle C$, D is a point on BC such that AD bisects $\angle A$. Show that $AC = AB + BD$.

3. Refer to the left diagram below. Given $\triangle ABC$, draw squares $ABDE$ and $ACFG$ outwards from AB, AC respectively. Show that $BG = CE$ and $BG \perp CE$.

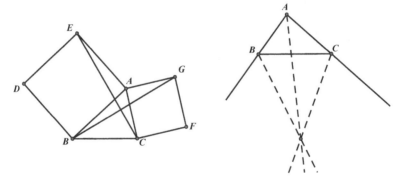

4. Refer to the right diagram above. Show that in $\triangle ABC$, the angle bisector of $\angle A$, the exterior angles bisectors of $\angle B$ and $\angle C$ are concurrent (i.e., they pass through the same point).

Note: This point is called the ex-center of $\triangle ABC$ opposite A. One may see that each triangle has three ex-centers.

5. Given $\triangle ABC$, J_1 and J_2 are the ex-centers (refer to Exercise 1.4) opposite B and C respectively. Let I be the incenter of $\triangle ABC$. Show that $J_1 J_2 \perp AI$.

6.   Let *ABCD* be a square. *E,F* are points on *BC,CD* respectively and ∠*EAF* = 45°. Show that *EF* = *BE* + *DF*.

7.   In the acute angled triangle △*ABC*, *BD* ⊥ *AC* at *D* and *CE* ⊥ *AB* at *E*. *BD* and *CE* intersect at *Q*. *P* is on *BD* extended such that *BP* = *AC*. If *CQ* = *AB*, find ∠*AQP*.

8.   Refer to the diagram on the right. △*ABC* is an equilateral triangle. *D* is a point inside △*ABC* such that *AD* = *BD*. Choose *E* such that *BE* = *AB* and *BD* bisects ∠*CBE*. Find ∠*BED*.

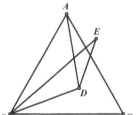

9.   Let *I* be the incenter of △*ABC*. *AI* extended intersects *BC* at *D*. Draw *IH* ⊥ *BC* at *H*. Show that ∠*BID* = ∠*CIH*.

10.   Given a quadrilateral *ABCD*, the diagonal *AC* bisects both ∠*A* and ∠*C*. If *AB* extended and *DC* extended intersect at *E*, and *AD* extended and *BC* extended intersect at *F*, show that for any point *P* on line *AC*, *PE* = *PF*.

11.   In △*ABC*, *AB* = *AC* and *D* is a point on *AB*. Let *O* be the circumcenter of △*BCD* and *I* be the incenter of △*ACD*. Show that *A*,*I*,*O* are collinear.

12.   Given a quadrilateral *ABCD* where *BD* bisects ∠*B*, *P* is a point on *BC* such that *PD* bisects ∠*APC*. Show that ∠*BDP* + ∠*PAD* = 90°.

13.   *ABCD* is a quadrilateral where *AD* // *BC*. Show that if *BC* − *AB* = *AD* − *CD*, then *ABCD* is a parallelogram.

14.   Given a square *ABCD*, ℓ_1 is a straight line intersecting *AB,AD* at *E,F* respectively and ℓ_2 is a straight line intersecting *BC,CD* at *G,H* respectively. *EH,FG* intersect at *I*. If ℓ_1 // ℓ_2 and the distance between ℓ_1,ℓ_2 is equal to *AB*, find ∠*GIH*.

Chapter 2

Similar Triangles

Similar triangles are the natural extension of the study on congruent triangles. While congruent triangles describe a pair of triangles with identical shape and size (area), similar triangles focus on the shape. The diagram below gives an illustration.

Indeed, similar triangles are even more powerful tools than congruent triangles. Many interesting properties and important theorems in geometry could be proved by similar triangles.

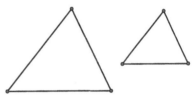

Two triangles of the same shape

One would see in this chapter that the Intercept Theorem plays a fundamental role in studying similar triangles, while the proof of this theorem is based on an even more fundamental concept: area.

2.1 Area of a Triangle

It is widely known that the area of $\triangle ABC$, denoted by $[\triangle ABC]$ or $S_{\triangle ABC}$, is given by $[\triangle ABC] = \dfrac{1}{2}BC \times h$, where h denotes the height on BC.

Of course, one may replace BC and h by any side of the triangle and the corresponding height on that side.

Notice that $[\triangle ABC] = \dfrac{1}{2}$ base $\times$ height implies that if two triangles have equal bases and heights, they must have the same area. Even though this is a simple conclusion, it has a number of (important) variations:

- In a trapezium $ABCD$ where $AD \, /\!/ \, BC$ and AC, BD intersect at E, we have $[\triangle ABC] = [\triangle DBC]$ because both triangles have a common base and equal heights.

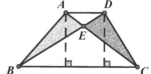

By substracting $[\triangle BCE]$ on both sides of the equation, we have $[\triangle ABE] = [\triangle CDE]$. Refer to the diagram above.

- In a triangle $\triangle ABC$ where M is the midpoint of BC, we must have $[\triangle ABM] = [\triangle ACM]$. Let D be any point on AM. We also have $[\triangle BDM] = [\triangle CDM]$. Refer to the diagram on the right.

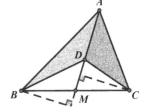

It follows that $[\triangle ABD] = [\triangle ACD]$. Since $\triangle ABD$ and $\triangle ACD$ have a common base AD, we conclude that the perpendicular distance from B, C respectively to the line AM is the same.

Notice that the conclusion above still holds even if D is a point on AM extended. Refer to the diagram on the right.

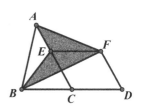

If M is the midpoint of BC, can you see $[\triangle ABD] = [\triangle ACD]$?

- Refer to the diagram on the right. Given a triangle $\triangle ABC$, extend BC to D such that $BC = CD$. E is a point on AC. Draw a parallelogram $CDFE$. Connect BE, BF and AF. One sees that the area of the shaded region is equal to the area of $\triangle ABC$.

This is because the shaded region consists of $\triangle AEF$ and $\triangle BEF$, which have the same base EF. Hence, the heights of the triangles, called h_1 and h_2, are the distances from A and B to the line EF respectively. Now $[AEBF] = \dfrac{1}{2} EF \cdot h_1 + \dfrac{1}{2} EF \cdot h_2 = \dfrac{1}{2} EF \cdot (h_1 + h_2)$.

Since $EF = CD = BC$ and $h_1 + h_2$ is equal to the distance from A to BC, we conclude that $[AEBF] = [\triangle ABC]$.

- Given a right angled triangle $\triangle ABC$ where $\angle C = 90°$, draw $CD \mathbin{//} AB$. Refer to the diagram on the right. Draw $AE \perp BD$ at E.

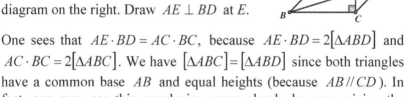

One sees that $AE \cdot BD = AC \cdot BC$, because $AE \cdot BD = 2[\triangle ABD]$ and $AC \cdot BC = 2[\triangle ABC]$. We have $[\triangle ABC] = [\triangle ABD]$ since both triangles have a common base AB and equal heights (because $AB \mathbin{//} CD$). In fact, one may see this conclusion more clearly by recognizing the trapezium $ABCD$. (You may rotate the page and hence, look at the trapezium from the "upright" position.)

Note that using areas of equal triangles is an important technique to show equal products (or ratios, $\dfrac{AE}{AC} = \dfrac{BC}{BD}$ in this example) of line segments.

Moreover, $[\triangle ABC] = \dfrac{1}{2} BC \times h$ implies that if two triangles, say $\triangle ABC$ and $\triangle A'B'C'$, have equal bases BC and $B'C'$, then $\dfrac{[\triangle ABC]}{[\triangle A'B'C']} = \dfrac{h}{h'}$, where h and h' are the respective heights. A similar conclusion could be drawn if two triangles have equal heights.

This is a very useful result because we may calculate the area of triangles indirectly by comparing its base and height with another triangle whose area is known.

Example 2.1.1   Given $\triangle ABC$, D is a point on BC such that $BC = 3BD$. E is a point on AD such that $AD = 4DE$. Show that:

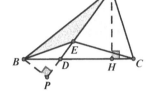

(1) $[\triangle ACE] = 2[\triangle ABE]$

(2) $[\triangle ABC] = 4[\triangle BCE]$

Proof.   Refer to the diagram on the right.

(1)   Notice that $\triangle ABD$ and $\triangle ACD$ has the same height AH.

> Hence, $\dfrac{[\triangle ABD]}{[\triangle ACD]} = \dfrac{BD}{CD} = \dfrac{1}{2}$, or $[\triangle ACD] = 2[\triangle ABD]$.

> Similarly, $[\triangle CDE] = 2[\triangle BDE]$. Hence, we have:

$$[\triangle ACE] = [\triangle ACD] - [\triangle CDE] = 2([\triangle ABD] - [\triangle BDE]) = 2[\triangle ABE].$$

(2)   Notice that $\triangle ABD$ and $\triangle BDE$ have the same height BP.

> Hence, $\dfrac{[\triangle ABD]}{[\triangle BDE]} = \dfrac{AD}{DE} = \dfrac{4}{1}$, or $[\triangle ABD] = 4[\triangle BDE]$.

> Similarly, $[\triangle ACD] = 4[\triangle CDE]$. Hence, we have:

$$[\triangle ABC] = [\triangle ABD] + [\triangle ACD] = 4([\triangle BDE] + [\triangle CDE]) = 4[\triangle BCE]. \quad \square$$

Note: One may see that similar arguments apply even if the ratios given (i.e., the positions of D and E) are different. Such an argument is commonly used in solving problems related to areas. In fact, experienced contestants in Mathematical Olympiads could see the conclusions almost instantaneously.

Example 2.1.2   Given $\triangle ABC$, D, E, F are points on BC, AC, AB respectively such that $BD = 2CD$, $AE = 3CE$ and $AF = 4BF$. If the area of $\triangle ABC$ is $240 \, \text{cm}^2$, find the area of $\triangle DEF$.

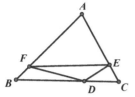 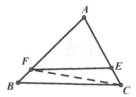

Insight. Refer to the left diagram above. Calculating $\triangle DEF$ directly will certainly be difficult because we do not know any of its bases or heights. We are given the area of $\triangle ABC$, but we do not know exactly how the areas of $\triangle DEF$ and $\triangle ABC$ are related. However, we could obtain the area of $\triangle DEF$ by subtracting the areas of $\triangle AEF, \triangle BDF$ and $\triangle CDE$ from $\triangle ABC$, where each of these triangles share a (part of) common side with $\triangle ABC$. Let us choose one of them, say $\triangle AEF$. Refer to the right diagram above. Connect CF.

Observe that $\dfrac{[\triangle AEF]}{[\triangle ACF]} = \dfrac{AE}{AC} = \dfrac{3}{4}$. We also have $\dfrac{[\triangle ACF]}{[\triangle ABC]} = \dfrac{AF}{AB} = \dfrac{4}{5}$. It follows that $\dfrac{[\triangle AEF]}{[\triangle ABC]} = \dfrac{3}{5}$, or $[\triangle AEF] = \dfrac{3}{5}[\triangle ABC] = \dfrac{3}{5} \times 240 = 144$.

Similarly, $[\triangle BDF] = \dfrac{2}{3} \times \dfrac{1}{5} \times 240 = 32$ and $[\triangle CDE] = \dfrac{1}{4} \times \dfrac{1}{3} \times 240 = 20$.

Now $[\triangle DEF] = [\triangle ABC] - [\triangle AEF] - [\triangle BDF] - [\triangle CDE]$

$= 240 - 144 - 32 - 20 = 44 \text{ cm}^2$.

Note:

(1) One sees that $[\triangle DEF] = \dfrac{11}{60}[\triangle ABC]$ always holds regardless of the area and the shape of $\triangle ABC$. This is solely determined by the relative positions of D, E, F on BC, AC, AB respectively.

(2) In general, given $\triangle ABC$ and D, E are on AB, AC respectively, we always have $\dfrac{[\triangle ADE]}{[\triangle ABC]} = \dfrac{AD}{AB} \cdot \dfrac{AE}{AC}$.

Refer to the diagram on the right.

One may see this conclusion by connecting CD and hence, relaying

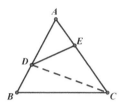

$$\frac{[\triangle ADE]}{[\triangle ACD]} = \frac{AE}{AC} \text{ and } \frac{[\triangle ACD]}{[\triangle ABC]} = \frac{AD}{AB} \text{ .}$$

Alternatively, one may apply Sine Rule, which we will discuss in Chapter 3.

(3) Using such a "relay" of area comparison is a useful technique because it links the unknown area to what is given. However, creating such a link literally requires a sequence of triangles, one after another which shares either a common side or a height. Of course, this may not be an easy task and one needs to draw one or more auxiliary lines wisely. Can you use this "relay" method to solve the following Example 2.1.3 and Example 2.1.4, without referring to the solution?

Example 2.1.3   Let $ABCD$ be a quadrilateral. E, F are on AB such that $AE = EF = BF = \dfrac{1}{3} AB$ and G, H are on CD such that $CG = GH = DH$ $= \dfrac{1}{3} CD$. Show that $[EFGH] = \dfrac{1}{3}[ABCD]$.

Proof.   Refer to the left diagram below. Since $AE = EF$, we must have $[\triangle EFH] = \dfrac{1}{2}[\triangle AFH]$. Similarly, $[\triangle FGH] = \dfrac{1}{2}[\triangle CFH]$.

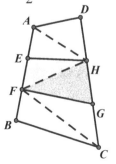
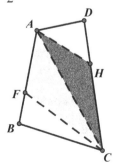

Hence, $[EFGH] = [\triangle EFH] + [\triangle FGH] = \dfrac{1}{2}[\triangle AFH] + \dfrac{1}{2}[\triangle CFH]$

$= \dfrac{1}{2}[AFCH].$

Now it suffices to show that $[AFCH] = \dfrac{2}{3}[ABCD]$. Refer to the previous right diagram. Since $AF = 2BF$ and $CH = 2DH$, we have

$[AFCH] = [\triangle ACF] + [\triangle ACH] = \dfrac{2}{3}[\triangle ABC] + \dfrac{2}{3}[\triangle ACD] = \dfrac{2}{3}[ABCD].$ □

Example 2.1.4 In $\triangle ABC$, D is a point on AB and $\dfrac{AD}{AC} = \dfrac{AC}{AB} = \dfrac{2}{3}$. M is the midpoint of CD while AM extended intersects BC at E. Find $\dfrac{CE}{BE}$.

Ans. Refer to the diagram on the right. Connect DE. Since $CM = DM$, one sees that

$[\triangle ACM] = [\triangle ADM]$ and $[\triangle CEM] = [\triangle DEM].$

It follows that $[\triangle ADE] = [\triangle ACE].$

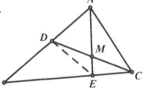

Since $\dfrac{[\triangle ADE]}{[\triangle ABE]} = \dfrac{AD}{AB} = \dfrac{AD}{AC} \cdot \dfrac{AC}{AB} = \dfrac{4}{9}$, we have $\dfrac{CE}{BE} = \dfrac{[\triangle ACE]}{[\triangle ABE]} = \dfrac{4}{9}.$ □

Note: We will see this example again in Section 3.2 and Section 3.4, where we will use two other methods (Intercept Theorem and Menelaus' Theorem) to solve it.

Example 2.1.5   Refer to the diagram on the right. In an acute angled triangle $\triangle ABC$ where $AB = AC$, M is the midpoint of BC. P is a point on AM and Q is a point on BP extended such that $QC \perp BC$ at C. Draw $QH \perp AB$ at H. Show that $\dfrac{AB}{AP} = \dfrac{BC}{HQ}$.

Insight.   We are to show $AB \cdot HQ = AP \cdot BC$. Since $AB \cdot HQ = 2[\triangle ABQ]$ and $AP \perp BC$, perhaps we can show the equality by area. Does $AP \cdot BC$ give the area of any triangle, or at least the area of a region in the diagram?

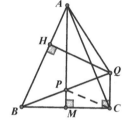

Proof.   Refer to the diagram on the right. Connect CP. Since M is the midpoint of BC and $AB = AC$, AM must be the perpendicular bisector of BC (Theorem 1.2.2). It follows that $BP = CP$ (Theorem 1.2.4).

Since $\angle BCQ = 90°$, we have $BP = PQ$ (Exercise 1.1), i.e., P is the midpoint of BQ.

Notice that $AB \cdot HQ = 2[\triangle ABQ] = 4[\triangle ABP]$ because $BQ = 2BP$.

We also have $AP \cdot BC = 2AP \cdot BM = 2 \times 2[\triangle ABP] = 4[\triangle ABP]$.

It follows that $AB \cdot HQ = AP \cdot BC$, or $\dfrac{AB}{AP} = \dfrac{BC}{HQ}$.   □

Note: Since $BM = CM$ and $MP /\!/ CQ$, one may obtain $BP = CQ$ easily by the Intercept Theorem. We will see this in the next section.

Pythagoras' Theorem

Pythagoras' Theorem is well known. Many of its popular proofs are based on the clever construction of a diagram. An example is given on the right. (We leave it to the reader to complete the proof based on this diagram.)

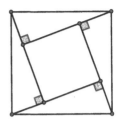

Proof of Pythagoras' Theorem

We shall introduce the classical proof of this theorem in Euclid's *Elements*. The proof is straightforward and is based on the area of triangles. It also illustrates a method applicable to many other problems related to areas of triangles.

Theorem 2.1.6 (Pythagoras' Theorem) *In $\triangle ABC$ where $\angle A = 90°$,*
$AB^2 + AC^2 = BC^2$.

Proof. Refer to the diagram below. We draw squares outwards from AB, AC, BC respectively. Since AB^2, AC^2, BC^2 represent the areas of squares, we are to show that the sum of the areas of the two small squares equals the area of the large square, i.e.,

$$[ABDE] + [ACFG] = [BCHI]. \quad (1)$$

Notice that $[ACFG] = 2[\triangle ACF]$ and $[\triangle ACF] = \dfrac{1}{2} CF \cdot AC = [\triangle BCF]$.

Since $\triangle BCF \cong \triangle HCA$ (Exercise 1.3), we must have $[\triangle ACH] = [\triangle BCF]$
$= [\triangle ACF] = \dfrac{1}{2}[ACFG]. \quad (2)$

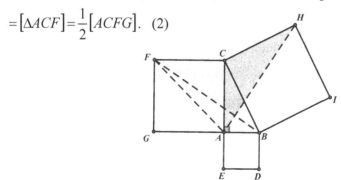

Similarly, $[\Delta ABI] = [\Delta BCD] = [\Delta ABD] = \dfrac{1}{2}[ABDE]$.   (3)

Refer to the left diagram below.

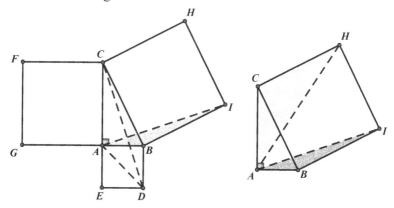

From (1), (2) and (3), it suffices to show $[\Delta ACH] + [\Delta ABI] = \dfrac{1}{2}[BCHI]$.

One sees that ΔACH and ΔABI have equal bases CH and BI with their respective heights added up to HI. Refer to the right diagram above. This completes the proof.   □

Example 2.1.7   *ABCD* is a trapezium such that $AD \,//\, BC$. If the two diagonals are perpendicular to each other, i.e., $AC \perp BD$, show that $AC^2 + BD^2 = (AD + BC)^2$.

Insight.   Refer to the left diagram below. Given $AC \perp BD$, we are asked about $AC^2 + BD^2$. Apparently, one should apply Pythagoras' Theorem. However, AC, BD are not intersecting at the endpoints. Can we bring them into a right angled triangle, say by moving the lines?

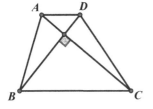
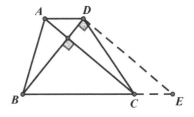

Proof. Draw $DE \parallel AC$, intersecting BC extended at E. Refer to the previous right diagram. Clearly, $ACED$ is a parallelogram and hence, $AC = DE$ and $AD = CE$. Now $AC^2 + BD^2 = DE^2 + BD^2 = BE^2$ because $DE \perp BD$. It is easy to see that $BE = AD + BC$ because $AD = CE$. This completes the proof. □

We know that in a right angled triangle $\triangle ABC$ where $\angle B = 90°$, if $\angle A = 30°$, then $AC = 2BC$ (Example 1.4.8). Hence, by Pythagoras' Theorem, $AB^2 = AC^2 - BC^2 = 3BC^2$, i.e., $AB = \sqrt{3}BC$.

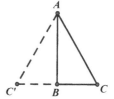

Refer to the diagram above where $\triangle ACC'$ is an equilateral triangle with a side of length a, i.e., $BC = \dfrac{1}{2}a$. We have $AB = \dfrac{\sqrt{3}}{2}a$ and hence, the area of the equilateral triangle is $\dfrac{\sqrt{3}}{4}a^2$.

Similarly, in a right angled triangle $\triangle ABC$ where $\angle B = 90°$, if $\angle A = 45°$, we must have $AB = BC$ and hence, $AC = \sqrt{2}AB$ by Pythagoras' Theorem.

Example 2.1.8 $ABCD$ is an isosceles trapezium where $AD \parallel BC$ and AC, BD intersect at P. If $BC = AC$ and $AC \perp BD$, show that $AD + BC = 2BP$.

Proof. Refer to the diagram on the right. Since $ABCD$ is an isosceles trapezium and $AC \perp BD$, both $\triangle PAD$ and $\triangle PBC$ are right angled isosceles triangles.

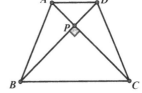

Let $AP = x$ and $CP = y$. We have $AD = \sqrt{2}x$ and $BC = \sqrt{2}y$.

Since $AC = BC$, we must have $x + y = \sqrt{2}y$. (1)

We are to show $AD + BC = 2BP$, i.e., $\sqrt{2}x + \sqrt{2}y = 2y$, but this can be obtained immediately from (1), by multiplying $\sqrt{2}$ on both sides. □

The inverse of Pythagoras' Theorem also holds, i.e., in $\triangle ABC$, if $AB^2 + AC^2 = BC^2$, then $\angle A = 90°$. This can be proved by contradiction.

The following result could be seen as an extension of the inverse of Pythagoras' Theorem.

Theorem 2.1.9 *Let A be a point outside the line BC and D is on the line BC. If $AB^2 - BD^2 = AC^2 - CD^2$, then $AD \perp BC$.*

Proof. Suppose otherwise. Refer to the diagram on the right. Draw $AP \perp BC$ at P. We may assume, without loss of generality, that $BD > BP$.

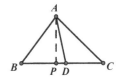

By Pythagoras' Theorem, $AP^2 = AB^2 - BP^2 = AC^2 - CP^2$.

Since $AB^2 - BD^2 = AC^2 - CD^2$, we have $BD^2 - CD^2 = AB^2 - AC^2 = BP^2 - CP^2$. This is impossible since $BD > BP$ and $CD < CP$, i.e., $BD^2 - CD^2 > BP^2 - CP^2$.

Note that the proof is not complete yet because one should also consider the cases where either D or P is outside the line segment BC. Refer to the following diagrams. Indeed, we have $BD^2 - CD^2 \neq BP^2 - CP^2$ in each case. We leave the details to the reader.

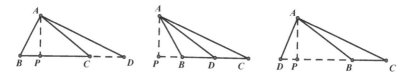

Indeed, Theorem 2.1.9 still holds even if D does not lie on the line BC. One may write down a similar proof by contradiction.

2.2 Intercept Theorem

Theorem 2.2.1 (Intercept Theorem) *Let ℓ_1, ℓ_2, ℓ_3 be a group of parallel lines which intersect two straight lines at A, B, C and D, E, F respectively. We have $\dfrac{AB}{BC} = \dfrac{DE}{EF}$.*

Proof. Refer to the diagram on the right.

Notice that $\dfrac{AB}{BC} = \dfrac{[\triangle ABE]}{[\triangle BCE]}$, since

$\triangle ABE$ and $\triangle BCE$ share the same height from E to the line AC.

Similarly, $\dfrac{DE}{EF} = \dfrac{[\triangle BDE]}{[\triangle BEF]}$.

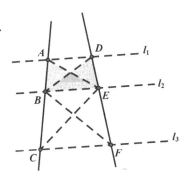

Notice that $[\triangle ABE] = [\triangle BDE] = \dfrac{1}{2} BE \times h$, where h is the height on BE.

The two triangles have the same height as $\ell_1 \, // \, \ell_2$.

Similarly, $[\triangle BCE] = [\triangle BEF]$.

It follows that $\dfrac{AB}{BC} = \dfrac{[\triangle ABE]}{[\triangle BCE]} = \dfrac{[\triangle BDE]}{[\triangle BEF]} = \dfrac{DE}{EF}$. □

Note:
(1) One may easily see that the Intercept Theorem applies when more than three parallel lines intercept two straight lines: the corresponding line segments will still be in ratio.
(2) There are a few cases where the Intercept Theorem applies for only two parallel lines. Refer to the following diagrams.

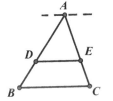

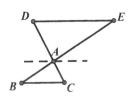

In these cases, we always have $\dfrac{AB}{AC} = \dfrac{AD}{AE}$.

Notice that one could always draw the third parallel line at A before applying the Intercept Theorem.

(3)   Notice that the inverse of the Intercept Theorem holds as follows: In $\triangle ABC$ where D, E are on AB, AC respectively, if $\dfrac{AD}{AB} = \dfrac{AE}{AC}$, we must have $DE \parallel BC$. This could be proved easily by contradiction:

Suppose otherwise. We draw $DE' \parallel BC$, intersecting AC at E'. Refer to the diagram on the right.

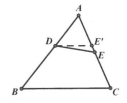

We have $\dfrac{AD}{AB} = \dfrac{AE'}{AC}$ by the Intercept Theorem.

Since $\dfrac{AD}{AB} = \dfrac{AE}{AC}$, we must have $AE = AE'$, i.e., E and E' coincide.

This completes the proof that $DE \parallel BC$.

Corollary 2.2.2 *In $\triangle ABC$, D, E are on AB, AC respectively such that $DE \parallel BC$. We have $\dfrac{AD}{AB} = \dfrac{AE}{AC} = \dfrac{DE}{BC}$.*

Proof.   Refer to the diagram on the right. Draw $AH \perp BC$ at H. Let AH intersect DE at G. Since $BC \parallel DE$, $AG \perp DE$.

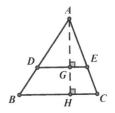

We have $\dfrac{AD}{AB} = \dfrac{AG}{AH}$ by the Intercept Theorem.

Let $\dfrac{AD}{AB} = \dfrac{AG}{AH} = k$, i.e., $AD = k \cdot AB$ and $AG = k \cdot AH$.

Pythagoras' Theorem gives $DG^2 = AD^2 - AG^2 = (k \cdot AB)^2 - (k \cdot AH)^2$
$= k^2(AB^2 - AH^2) = k^2 BH^2$, i.e., $\dfrac{DG}{BH} = k = \dfrac{AD}{AB}$.

Similarly, $\dfrac{EG}{CH} = k = \dfrac{AE}{AC} = \dfrac{AD}{AB}$.

Now $DE = DG + EG = k \cdot BH + k \cdot CH = k(BH + CH) = k \cdot BC$.

This implies $\dfrac{DE}{BC} = k = \dfrac{AD}{AB}$.   □

Note:

(1) The conclusion holds even if D, E lie on BA, CA extended respectively, i.e., when the lines BC, DE are on different sides of A. Refer to the diagrams in the remarks after Theorem 2.2.1.

(2) Refer to the diagram on the right where $BC /\!/ PQ$.

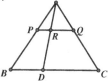

We have $\dfrac{PR}{QR} = \dfrac{BD}{CD}$, because $\dfrac{PR}{BD} = \dfrac{AP}{AB} = \dfrac{QR}{CD}$.

One familiar with similar triangles may see the conclusion almost immediately. We shall study similar triangles in the next section.

The Intercept Theorem and Corollary 2.2.2 are very useful in calculating the ratio of line segments.

Example 2.2.3   In $\triangle ABC$, D, E are on BC, AC respectively such that $BC = 3BD$ and $AC = 4AE$. If AD and BE intersect at F, find $\dfrac{BF}{EF}$.

Ans. Refer to the diagram on the right. Draw $EG \parallel BC$, intersecting AD at G.

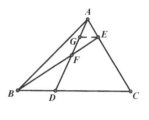

Since $\dfrac{EG}{CD} = \dfrac{AE}{AC} = \dfrac{1}{4}$ and $\dfrac{CD}{BD} = \dfrac{2}{1}$, we have

$\dfrac{EF}{BF} = \dfrac{EG}{BD} = \dfrac{EG}{CD} \cdot \dfrac{CD}{BD} = \dfrac{1}{2}$, i.e., $\dfrac{BF}{EF} = 2$. $\quad\square$

Note: This solution shows a standard method solving this type of questions. Once the positions of D and E are known, one could always use this method to find $\dfrac{BF}{EF}$. Can you use the same technique to show that $AF = DF$? (**Hint**: Draw $DP \parallel AC$, intersecting BE at P.)

Example 2.2.4 Given $\triangle ABC$, D, E, F are on AB, BC, CA respectively such that $AB = 3AF$, $BC = 3BD$ and $AC = 3CE$. Refer to the diagram on the right. Find $\dfrac{[\triangle PQR]}{[\triangle ABC]}$.

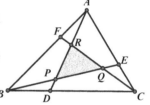

Insight. This is similar to Example 2.1.2. We can calculate $[\triangle PQR]$ by subtracting the unshaded areas from $[\triangle ABC]$. In order to calculate the area of the unshaded region, we may divide it into a few triangles, say $\triangle ABP$, $\triangle BCQ$ and $\triangle CAR$. How can we calculate $[\triangle ABP]$? We know $[\triangle ABD] = \dfrac{1}{3}[\triangle ABC]$ because $BD = \dfrac{1}{3}BC$. Notice that $\dfrac{[\triangle ABP]}{[\triangle ABD]} = \dfrac{AP}{AD}$.

We can use the method illustrated in Example 2.2.3 to find $\dfrac{AP}{AD}$.

Proof. Refer to the diagram on the right. Draw $DX \parallel AC$, intersecting BE at X.

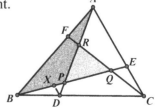

We have $\dfrac{DX}{CE} = \dfrac{BD}{BC} = \dfrac{1}{3}$ and $\dfrac{CE}{AE} = \dfrac{1}{2}$.

Hence, $\dfrac{DX}{AE}=\dfrac{1}{6}=\dfrac{PD}{AP}$, i.e., $\dfrac{AP}{AD}=\dfrac{6}{7}$.

Since $\dfrac{[\triangle ABP]}{[\triangle ABD]}=\dfrac{AP}{AD}=\dfrac{6}{7}$ and $[\triangle ABD]=\dfrac{1}{3}[\triangle ABC]$, we must have

$\dfrac{[\triangle ABP]}{[\triangle ABC]}=\dfrac{2}{7}$, or $[\triangle ABP]=\dfrac{2}{7}[\triangle ABC]$.

Similarly, one sees that $[\triangle BCQ]=[\triangle CAR]=\dfrac{2}{7}[\triangle ABC]$.

It follows that $[\triangle PQR]=\dfrac{1}{7}[\triangle ABC]$, or $\dfrac{[\triangle PQR]}{[\triangle ABC]}=\dfrac{1}{7}$. ☐

Recall Example 2.1.4.

In $\triangle ABC$, D is a point on AB and $\dfrac{AD}{AC}=\dfrac{AC}{AB}=\dfrac{2}{3}$. M is the midpoint of CD while AM extended intersects BC at E. Find $\dfrac{CE}{BE}$.

Can you solve it using the technique demonstrated above, drawing parallel lines and applying the Intercept Theorem?

Ans. Refer to the diagram below. Draw *DF // BC,* intersecting *AE* at *F.*

We have $\dfrac{DF}{CE}=\dfrac{DM}{CM}=1$ and $\dfrac{DF}{BE}=\dfrac{AD}{AB}=\dfrac{AD}{AC}\cdot\dfrac{AC}{AB}=\dfrac{4}{9}$.

It follows that $\dfrac{CE}{BE}=\dfrac{DF}{BE}=\dfrac{4}{9}$. ☐

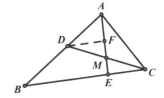

An important special case of Corollary 2.2.2 is the Midpoint Theorem.

Theorem 2.2.5 (Midpoint Theorem) *In $\triangle ABC$, D, E, F are midpoints of BC, AC, AB respectively. We have $EF \mathbin{/\mkern-5mu/} BC$, $EF = \dfrac{1}{2} BC$ and AD, BE, CF are concurrent.*

Proof. Since E, F are midpoints, $EF \mathbin{/\mkern-5mu/} BC$ by the Intercept Theorem. Now Corollary 2.2.2 implies $\dfrac{EF}{BC} = \dfrac{AF}{AB} = \dfrac{1}{2}$. Refer to the diagram on the right.

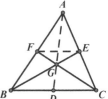

Suppose BE and CF intersect at G. We have $\dfrac{BG}{GE} = \dfrac{BC}{EF} = 2$, i.e., CF must intersect BE at the trisection point closer to E. Notice that this argument applies to AD as well, i.e., AD must also intersect BE at G where $\dfrac{BG}{BE} = \dfrac{2}{3}$. Indeed, AD, BE, CF are concurrent at G.   □

Note: One may derive the following important properties easily from the Midpoint Theorem.
(1) The medians of a triangle are concurrent (at the *centroid*) and the centroid is always at the lower one-third position of a median.
(2) A *midline*, i.e., a line segment connecting the midpoints of two legs, is always parallel to and has half of the length of the corresponding base of the triangle. Hence, drawing a midline is an important technique when solving problems related to midpoints as the line segments far apart could be brought together.

Example 2.2.6   Let $ABCD$ be a quadrilateral and E, F, G, H be the midpoints of AB, BC, CD, DA respectively. Show that $EFGH$ is a parallelogram.

Insight.   This is a simple application of the Midpoint Theorem. Refer to the diagram on the right. One easily sees that $EF \mathbin{/\mkern-5mu/} AC \mathbin{/\mkern-5mu/} GH$ and $EH \mathbin{/\mkern-5mu/} BD \mathbin{/\mkern-5mu/} FG$.

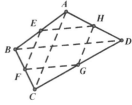

Example 2.2.7 Let D be a point inside $\triangle ABC$ such that AD bisects $\angle A$ and $AD \perp BD$. Let M be the midpoint of BC.
(1) If $AB = 11$ and $AC = 17$, find MD.
(2) Show that M cannot lie on AD extended.

Insight.

(1) We are to find MD where M is the midpoint of BC. If D is the midpoint of another line segment, perhaps we could apply the Midpoint Theorem. Is there a line segment whose midpoint is D?

Since AD is an angle bisector, it is a common technique to *reflect* $\triangle ABD$ about AD. This technique is even more useful here because $AD \perp BD$. Refer to the diagram on the right. Can you see $\triangle ABE$ is an isosceles triangle?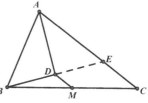

(2) Let E be the reflection of B about AD. If M lies on AD extended, can you see $BM = CM = EM$? What does it imply?

Ans.

(1) Let AD extended intersect AC at E. Since AD is the angle bisector and $AD \perp BE$, we have $\triangle ABD \cong \triangle AED$ (A.A.S.), which implies $BD = DE$ and $AE = AB$. Since $BM = CM$, we must have $MD = \frac{1}{2}CE$ by the Midpoint Theorem. It follows that

$$MD = \frac{1}{2}(AC - AE) = \frac{1}{2}(AC - AB) = 3.$$

(2) Suppose otherwise that M lies on AD extended. It is easy to see that $\triangle ABM \cong \triangle AEM$ (S.A.S.), which implies $BM = EM$.
Now $BM = CM = EM$ implies $\angle BEC = 90°$ (Example 1.1.8). This is absurd because $\triangle ABE$ is an isosceles triangle. □

Example 2.2.8 Given $\triangle ABC$, D is a point on AC such that $AB = CD$. Let M, N be the midpoints of AD, BC respectively. Show that MN is parallel to the angle bisector of $\angle BAC$.

Insight.   How can we apply $AB = CD$, where AB, CD are far apart? Since we are given the midpoints of AD, BC, if we connect BD and let P be the midpoint of BD, then $PM = \frac{1}{2}AB$ and $PN = \frac{1}{2}CD$.

Hence, $PM = PN$. Refer to the diagram on the right. Now $\triangle PMN$ is an isosceles triangle. Can we use the technique of the isosceles triangle and parallel line to obtain the angle bisector (Example 1.1.10)?

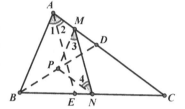

Proof.   Let P be the midpoint of BD. Notice that $PM = \frac{1}{2}AB = \frac{1}{2}CD$ $= PN$ by the Midpoint Theorem. Hence, $\angle 3 = \angle 4$.

Draw $AE \,/\!/\, MN$, intersecting BC at E. Since $AB \,/\!/\, PM$ and $AE \,/\!/\, MN$, one sees that $\angle 1 = \angle 3$ and similarly, $\angle 2 = \angle 4$. It follows that $\angle 1 = \angle 2$, i.e., AE bisects $\angle BAC$. This completes the proof.   □

Note:
(1) One could see $\angle 1 = \angle 3$ and $\angle 2 = \angle 4$ easily by recognizing corresponding angles, alternate angles and interior angles with respect to parallel lines.
(2) If one draws the angle bisector of $\angle BAC$ instead of $AE \,/\!/\, MN$, the proof is similar. One could show $\angle 3 + \angle 4 = \angle BAC$ (using parallel lines), which also leads to the conclusion.

Example 2.2.9   Refer to the diagram on the right. Given $\triangle ABC$, D is the midpoint of BC and AF bisects $\angle A$. Draw $BE \perp AF$ at E and $CF \perp AF$ at F. Show that $DE = DF$.

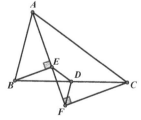

Insight.   Considering the midlines (and medians) could be a wise strategy because we are given not only midpoints, but also right angled triangles. For example, say P, Q are the midpoint of AB, AC

respectively, we have $QD = \frac{1}{2}AB$ by the Midpoint Theorem and

$PE = \frac{1}{2}AB$ because PE is the median on the hypotenuse of the right

angled triangle $\triangle ABE$. Hence, $QD = PE$. Can you see that $PD = QF$ as well?

Proof. Refer to the diagram on the right. Let P, Q be the midpoints of AB, AC

respectively. We have $PE = \frac{1}{2}AB = QD$

and $PD = \frac{1}{2}AC = QF.$

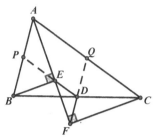

In the right angled triangle $\triangle ABE$, we have $AP = PE$ and hence, $\angle BPE = 2\angle BAE = \angle BAC$, which implies $PE /\!/ AC$. Since $PD /\!/ AC$, we must have P, D, E collinear. Similarly, D, F, Q are collinear. It follows that $DE = PD - PE = FQ - DQ = DF$.

Note: It seems from the diagram above that P, D, E are collinear, but one should **not** assume this without a proof. In fact, if an inaccurate diagram is casually drawn, one may even see $\triangle PDE \cong \triangle QFD$.

2.3 Similar Triangles

Congruent triangles are very useful in solving geometry problems, as a pair of congruent triangles are of not only the same size, but of identical *shape* as well. However, we may frequently encounter triangles which have identical shape, but differ in size.

For example, a height on the hypotenuse of a right angled triangle gives three triangles of the same shape. Refer to the diagram on the right.

Note that $\triangle ABD$, $\triangle CAD$ and $\triangle CBA$ show similarity in their shapes.

We say two triangles $\triangle ABC$ and $\triangle A'B'C'$ are similar if they have the same shape, or more precisely, if all the corresponding angles are the same and all the corresponding sides are of equal ratio, i.e., $\angle A = \angle A'$, $\angle B = \angle B'$, $\angle C = \angle C'$ and $\dfrac{AB}{A'B'} = \dfrac{AC}{A'C'} = \dfrac{BC}{B'C'}$. We denote this by $\triangle ABC \sim \triangle A'B'C'$.

One may verify similar triangles by definition. However, this is often unnecessary. It is taught in most secondary schools that one can verify similar triangles by the following criteria, the proof of which is based on the Intercept Theorem:

- If two pairs of corresponding angles are identical, then the two triangles are similar, i.e., if $\angle A = \angle A'$ and $\angle B = \angle B'$ (in which case one must have $\angle C = \angle C'$), then $\triangle ABC \sim \triangle A'B'C'$.

- If two pairs of corresponding sides are of equal ratio and the angles between them are identical, then the two triangles are similar, i.e., if $\dfrac{AB}{A'B'} = \dfrac{AC}{A'C'}$ and $\angle A = \angle A'$, then $\triangle ABC \sim \triangle A'B'C'$.

- If all the corresponding sides are of equal ratio, then the two triangles are congruent, i.e., if $\dfrac{AB}{A'B'} = \dfrac{AC}{A'C'} = \dfrac{BC}{B'C'}$, then $\triangle ABC \sim \triangle A'B'C'$.

One may also determine a pair of similar right angled triangles by legs and hypotenuses. This is similar to determining congruent triangles using H.L. and it can be justified easily by Pythagoras' Theorem.

Notice that if $\triangle ABC \sim \triangle A'B'C'$, **all** the corresponding angles are the same and the corresponding line segments are of the same ratio. Refer to the diagram on the right for an example.

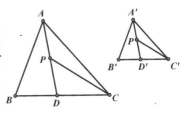

Given $\triangle ABC \sim \triangle A'B'C'$, let AD bisect $\angle A$ and $A'D'$ bisects $\angle A'$. If P, P' are the midpoints of $AD, A'D'$ respectively, we have $\dfrac{AB}{A'B'} = \dfrac{CP}{C'P'}$ and $\angle ACP = \angle A'C'P'$.

Now we can see that in a right angled triangle $\triangle ABC$ where $\angle A = 90°$ and AD is a height, $\triangle ABC \sim \triangle ABD \sim \triangle ACD$. Refer to the diagram on the right. In particular, the following result is useful.

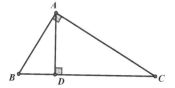

Example 2.3.1 $\triangle ABC$ is a right angled triangle where $\angle A = 90°$ and AD is a height. We have $AB^2 = BD \cdot BC$, $AC^2 = CD \cdot BC$ and $AD^2 = BD \cdot CD$.

Proof.   Since $\angle C = \angle BAD$, we have $\triangle ABC \sim \triangle DBA \sim \triangle DAC$, which gives $\dfrac{AB}{BD} = \dfrac{BC}{AB}$, $\dfrac{AC}{CD} = \dfrac{BC}{AC}$ and $\dfrac{AD}{BD} = \dfrac{CD}{AD}$.

It follows that $AB^2 = BD \cdot BC$, $AC^2 = CD \cdot BC$ and $AD^2 = BD \cdot CD$.   □

Note:

(1) Pythagoras' Theorem follows immediately from this example as
$$AB^2 + AC^2 = BD \cdot BC + CD \cdot BC = (BD + CD) \cdot BC = BC^2.$$

(2) One sees from this example that $\dfrac{AB^2}{AC^2} = \dfrac{BD}{CD}$. This is a very useful conclusion. You may compare it with the Angle Bisector Theorem (Theorem 2.3.7).

Recognizing similar triangles is a very important technique because a pair of similar triangles gives equal angles and ratios of line segments. One may seek similar triangles via the following clues:

- Parallel lines
- Angle bisectors
- Opposite angles
- Refer to the diagram on the right. If $\angle ACD = \angle B$, then $\triangle ACD \sim \triangle ABC$.

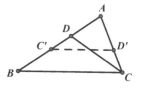

One may see this more clearly by reflecting $\triangle ACD$ about the angle bisector of $\angle A$, which gives $\triangle AC'D'$. It is easy to show $BC \, // \, C'D'$ and hence, $\triangle AC'D' \sim \triangle ABC$.

Notice that Example 2.3.1 could be seen as a special case of this result, where $\angle ACB = 90°$.

Example 2.3.2   Given $\triangle ABC$ where $\angle A = 120°$, D is a point of BC such that $BD = 15$, $CD = 5$ and $\angle ADB = 60°$. Find AC.

Ans. Refer to the diagram on the right.
Since $\angle ADB = 60°$, we have $\angle ADC = 120°$ $= \angle BAC$. It follows that $\triangle ABC \sim \triangle DAC$.

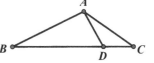

Now we have $\dfrac{CD}{AC} = \dfrac{AC}{BC}$, or $AC^2 = CD \cdot BC$. Since $CD = 5$ and $BC = BD + CD = 15 + 5 = 20$, we conclude that $AC = 10$.   □

Example 2.3.3   In $\triangle ABC$, $\angle A = 2\angle B$. Show that $BC^2 = AC \cdot (AB + AC)$.

Insight.   We are only given that $\angle A = 2\angle B$. Hence, it is natural to draw the angle bisector of $\angle A$ and we obtain equal angles $\angle B = \angle 1 = \angle 2$.

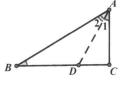

Refer to the diagram above. Perhaps we shall seek similar triangles and set up the ratio.

Since $\angle 1 = \angle B$, $\triangle CAD \sim \triangle CBA$. We have $\dfrac{AC}{BC} = \dfrac{AD}{AB} = \dfrac{CD}{AC}$. Hence, $AC \cdot AB = BC \cdot AD$ and $AC^2 = BC \cdot CD$. Since $AD = BD$, we have

$AC \cdot AB + AC^2 = BC \cdot BD + BC \cdot CD,$ simplifying which gives the conclusion.

Example 2.3.4   In $\triangle ABC$, $\angle A = 120°$ and $AB = AC$. Let D, E be trisection points of BC, i.e., $BD = DE = CE$. Show that $\triangle ADE$ is an equilateral triangle.

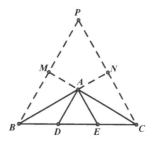

Proof.   Refer to the diagram on the right. Draw an equilateral triangle $\triangle PBC$ from BC such that A is inside $\triangle PBC$. It is easy to see that $\angle B = \angle C = 30°$, which implies that A is the incenter of $\triangle PBC$. Clearly, A is also the centroid of $\triangle PBC$.

Now $\dfrac{AC}{CM} = \dfrac{2}{3} = \dfrac{CD}{BC}$, which implies $AD \,/\!/\, PB$. Similarly, $AE \,/\!/\, PC$.

It follows that $\triangle ADE \sim \triangle PBC$ and hence the conclusion.   □

Note:
(1) $\triangle ABD$, $\triangle ACE$ and $\triangle BCA$ are similar.
(2) An isosceles triangle with $120°$ at the vertex is closely related to equilateral triangles. Besides the example above, one may also double a leg. Refer to the left diagram below. Extend BA to D such that $AB = AD$. Notice that $\triangle ACD$ is an equilateral triangle and $BC \perp CD$. Indeed, we are familiar with $\triangle BCD$, which is half of a larger equilateral triangle.

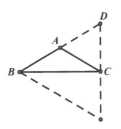

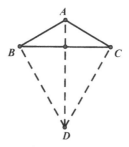

On the other hand, one may draw an equilateral triangle $\triangle BCD$ outwards. Refer to the right diagram above. Notice that both $\triangle ABD$ and $\triangle ACD$ are half of a larger equilateral triangle.

Example 2.3.5   In a right angled triangle $\triangle ABC$ where $\angle B = 90°$, D is a point on AC such that BD bisects $\angle B$. Draw $DE \perp AB$ at E and $DF \perp BC$ at F. Show that $BD^2 = 2AE \cdot CF$.

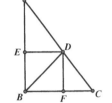

Insight.   Refer to the diagram on the right. It is given that BD bisects a right angle and DE, DF are perpendicular to AB, AC respectively. Can you see that $BEDF$ is a square!

How is BD related to AE and CF? We know $BD = \sqrt{2}BF$ and it is easy to relate AE, CF and BF (or DE) together by similar triangles.

Proof.   It is easy to see that $BFDE$ is a rectangle because $DE /\!/ BF$, $BE /\!/ DF$ and $\angle B = 90°$. We are given that $\angle ABD = \angle CBD = 45°$ and $\angle BED = 90°$. Hence, $BD = BE$, which implies that $BEDF$ is a square. It follows that $BD = \sqrt{2}BF$.

Clearly, $\triangle ADE \sim \triangle ACB$. Now $\dfrac{AE}{AB} = \dfrac{DE}{BC}$. Let $DE = BF = x$.

We have $\dfrac{AE}{AE + x} = \dfrac{x}{CF + x}$, simplifying which gives $x^2 = AE \cdot CF$.

It follows that $BD^2 = 2x^2 = 2AE \cdot CF$.   □

Example 2.3.6   Let P be a point inside the square $ABCD$. M, N are the feet of the perpendicular from P to BC, CD respectively. If $AP \perp MN$, show that either $AP = MN$, or $AP \perp BD$.

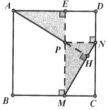

Insight.   Refer to the diagram on the right. Notice that there are a lot of right angles. Clearly, $CMPN$ is a rectangle and $MN = CP$. If $AP = MN$, we **should** have $AP = CP$, which implies P lies on BD. If $AP \perp BD$, then P lies on AC.

It seems from the diagram that $\triangle AEP \cong \triangle MPN$, which immediately gives $AP = MN$. However, this may **not** be true because it excludes the case for $AP \perp BD$. Nevertheless, we still have $\triangle AEP \sim \triangle MPN$ since $\angle PAE = \angle HPN = PMN$. Perhaps when $AP \neq MN$, we would have $AP \perp BD$. Notice that $AE + PN = PM + PE$ and $\dfrac{AE}{PE} = \dfrac{PM}{PN}$!

Proof. Let AP extended intersect MN at H and MP extended intersect AD at E. Since $PN \, /\!/ \, AD$, $\angle PAE = \angle HPN$. In the right angled triangle $\triangle PMN$, we must have $\angle HPN = PMN$. Hence, $\angle PAE = PMN$, which implies $\triangle AEP \sim \triangle MPN$.

Let $k = \dfrac{AE}{PE} = \dfrac{PM}{PN}$, i.e., $AE = k \cdot PE$ and $PM = k \cdot PN$. Since $AE + PN = PM + PE$, we have $k \cdot PE + PN = k \cdot PN + PE$, simplifying which gives $(k-1) \cdot PE = (k-1) \cdot PN$. Hence, either $k = 1$ or $PE = PN$.

If $k = 1$, we have $AE = PE$ and $PM = PN$. Now $AE = PE$ implies that $\angle PAE = 45°$, i.e., P lies on AC. $PM = PN$ implies $PMCN$ is a square and we must have $MN \, /\!/ \, BD$. Hence, $AP \perp BD$.
If $PE = PN$, we have $\triangle AEP \cong \triangle MPN$ and hence, $AP = MN$. □

Similar triangles are even more frequently seen when circle properties are introduced, which we will discuss in Chapter 4.

The following is an important property of angle bisectors.

Theorem 2.3.7 (Angle Bisector Theorem) *In $\triangle ABC$, the angle bisector of $\angle A$ intersects BC at D. We have* $\dfrac{AB}{AC} = \dfrac{BD}{CD}$.

Proof. Refer to the diagram on the right. Draw $DE \, /\!/ \, AB$, intersecting AC at E. We have $\angle BAD = \angle EDA$. Since AD bisects $\angle A$, $\angle EDA = \angle BAD = \angle EAD$. It follows that $AE = DE$.

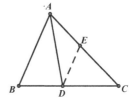

Since $DE /\!/ AB$, we have $\dfrac{BD}{CD} = \dfrac{AE}{CE} = \dfrac{DE}{CE}$. Notice that $\triangle ABC \sim \triangle EDC$.

Hence, $\dfrac{DE}{CE} = \dfrac{AB}{AC}$ and the proof is complete. □

Note:

(1) We are still using the strategy of constructing an isosceles triangle with the angle bisector and parallel lines.

(2) One may easily see that the inverse of the Angle Bisector Theorem holds: Given $\triangle ABC$ where D is a point of BC, if $\dfrac{AB}{AC} = \dfrac{BD}{CD}$, then AD bisects $\angle A$. Otherwise, let AD' be the angle bisector and we have $\dfrac{BD'}{CD'} = \dfrac{AB}{AC} = \dfrac{BD}{CD}$, which implies D and D' coincide.

(3) Notice that the conclusion $\dfrac{AB}{AC} = \dfrac{BD}{CD}$ still holds even if AD is an exterior angle bisector, i.e., when AD bisects the supplementary angle of $\angle A$. Refer to the diagram below.

The proof is similar. Draw $CE /\!/ AB$, intersecting AD at E.

One sees that $\triangle ACE$ is an isosceles triangle where $AC = CE$ (because $\angle 2 = \angle 1 = \angle CAE$).

Now $\dfrac{AB}{AC} = \dfrac{AB}{CE} = \dfrac{BD}{CD}$ by the Intercept Theorem.

Example 2.3.8   Let AD bisect $\angle A$ in $\triangle ABC$, intersecting BC at D. Show that $BD = \dfrac{ac}{b+c}$, where $BC = a$, $AC = b$ and $AB = c$.

Proof.   Refer to the diagram on the right. By the Angle Bisector Theorem, $\dfrac{BD}{CD} = \dfrac{AB}{AC} = \dfrac{c}{b}$.

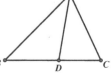

Since $a = BC = BD + CD = \left(1 + \dfrac{b}{c}\right)BD$, we must have $BD = \dfrac{ac}{b+c}$. $\quad\square$

Note: One may draw similar conclusions if AD, BE, CF are the angle bisectors of $\angle A, \angle B, \angle C$ respectively. This result is useful if angle bisectors are given and the ratios of sides are to be found.

2.4 Introduction to Trigonometry

Since any two right angled triangles are similar if they have an equal pair of acute angles, a right angled triangle with a given acute angle, say $\angle A$, must have constant ratios between the legs and the hypotenuse.

Refer to the diagram on the right.

We define $\sin \angle A = \dfrac{BC}{AB}$, $\cos \angle A = \dfrac{AC}{AB}$ and $\tan \angle A = \dfrac{BC}{AC}$.

Trigonometry is taught in most secondary schools. The most important and commonly used properties are as follows, which one may see easily from the definition.

- $\sin \angle A = \cos(90° - \angle A)$

- $\tan \angle A = \dfrac{\sin \angle A}{\cos \angle A}$

- $(\sin \angle A)^2 + (\cos \angle A)^2 = 1$ by Pythagoras' Theorem.

Trigonometric methods are widely applicable in geometric calculations, which we do not emphasize in this book. Nevertheless, we still encounter simple trigonometry occasionally in problem-solving and hence, one should be very familiar with the basic properties.

One important application is about the area of triangles. Refer to the diagram on the right.

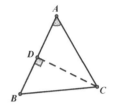

In $\triangle ABC$, given $CD \perp AB$ at D, we have $[\triangle ABC] = \dfrac{1}{2} AB \cdot CD$ and by definition,

$CD = AC \sin \angle A$. It follows that $[\triangle ABC] = \dfrac{1}{2} AB \cdot AC \sin \angle A$. Notice that heights are no longer involved in this formula.

If $\angle A > 90°$, we extend CA to D such that $AC = AD$. Refer to the diagram on the right.

Since $[\triangle ABC] = [\triangle ABD] = \dfrac{1}{2} AB \cdot AD \sin \angle BAD$ and $AC = AD$, if we define $\sin \angle A = \sin \angle BAD = \sin(180° - \angle A)$, we still have

$[\triangle ABC] = \dfrac{1}{2} AB \cdot AC \sin \angle A$. In particular, one sees that $\sin 90° = 1$.

Now $[\triangle ABC] = \dfrac{1}{2} AB \cdot AC \sin \angle A$ is consistent for any $\triangle ABC$.

Example 2.4.1 (HUN 10)   Let $ABCD$ be a quadrilateral whose area is S. Show that if $(AB + CD)(AD + BC) = 4S$, then $ABCD$ is a rectangle.

Insight. We have $(AB + CD)(AD + BC) = AB \cdot AD + AB \cdot BC + CD \cdot AD + CD \cdot BC$. How are these related to S?

Proof.   Notice that $S = [\triangle ABC] + [\triangle ACD]$ $= \dfrac{1}{2} AB \cdot BC \sin B + \dfrac{1}{2} AD \cdot CD \sin D$. Refer to the diagram on the right. Similarly, we have

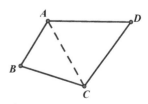

$$S = [\triangle ABD] + [\triangle BCD] = \frac{1}{2}AB \cdot AD \sin A + \frac{1}{2}BC \cdot CD \sin C, \text{ i.e.,}$$

$$4S = AB \cdot AD \sin A + AB \cdot BC \sin B + CD \cdot BC \sin C + CD \cdot AD \sin D.$$

It is given that $4S = (AB + CD)(AD + BC)$. One sees that this is only possible when $\sin A = \sin B \sin C = \sin D = 1$. We must have $\angle A = \angle B = \angle C = \angle D = 90°$ and hence, $ABCD$ is a rectangle. $\square$

Cosine Rule is one of the most elementary and commonly used results in trigonometry. One may see it as an extension of Pythagoras' Theorem.

Theorem 2.4.2 (Cosine Rule) *In $\triangle ABC$ where $BC = a$, $AC = b$ and $AB = c$, we have $a^2 = b^2 + c^2 - 2bc \cos A$.*

Proof. We use Pythagoras' Theorem to prove Cosine Rule. Refer to the right diagram below, where $\angle A$ is acute. Draw $CD \perp AB$ at D. Let $AD = x$. We have $BD = c - x$.

Pythagoras' Theorem gives $AC^2 - AD^2 = CD^2 = BC^2 - BD^2$, i.e., $b^2 - x^2 = a^2 - (c - x)^2$. Simplifying the equation, we obtain

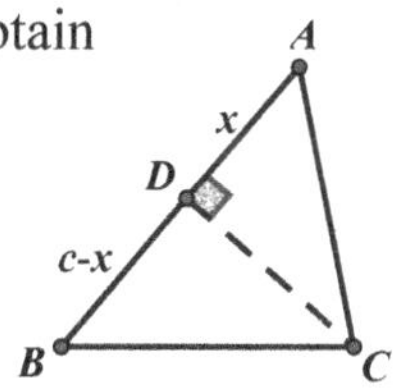

$$b^2 = a^2 - c^2 + 2cx, \text{ or } a^2 = b^2 + c^2 - 2cx.$$

The conclusion follows as $x = b \cos A$.

A similar argument applies if $\angle A$ is obtuse. Refer to the diagram on the right. We draw $CD \perp AB$, intersecting BA extended at D. Let $AD = x$. Pythagoras' Theorem gives

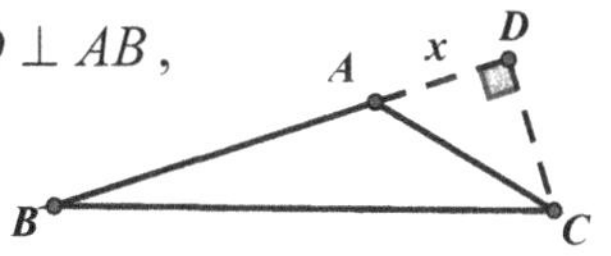

$$AC^2 - AD^2 = CD^2 = BC^2 - BD^2, \text{ i.e., } b^2 - x^2 = a^2 - (c + x)^2.$$

Simplifying the equation, we obtain $a^2 = b^2 + c^2 + 2cx$, where $x = b \cos \angle CAD = b \cos(180° - \angle A)$. One sees that the conclusion holds if we define $\cos \theta = -\cos(180° - \theta)$ for $\theta \geq 90°$ and in particular, $\cos 90° = 0$.

Now $a^2 = b^2 + c^2 - 2bc\cos A$ is consistent for any triangle $\triangle ABC$.   □

Note:

(1) If $\angle A = 90°$, $a^2 = b^2 + c^2$ is exactly Pythagoras' Theorem.

(2) One may perceive congruent triangles by Cosine Rule: Given a, b, c
     are the three sides of a triangle, we have $\cos A = \dfrac{b^2 + c^2 - a^2}{2bc}$.

     Hence, one may calculate $\angle A$, and similarly $\angle B$ and $\angle C$. Now
     $\triangle ABC$ is uniquely determined.
     On the other hand, if b, c and $\angle A$ are given, one may calculate a
     using Cosine Rule. Hence, $\triangle ABC$ is uniquely determined.
     Notice that these are consistent with the criteria determining
     congruent triangles, S.S.S. and S.A.S. respectively.

One may apply Cosine Rule to calculate the length of a median in a
given triangle.

Theorem 2.4.3 *In* $\triangle ABC$ *where* $BC = a$, $AC = b$, $AB = c$ *and M is*
the midpoint of BC, we have $AM^2 = \dfrac{1}{2}b^2 + \dfrac{1}{2}c^2 - \dfrac{1}{4}a^2$.

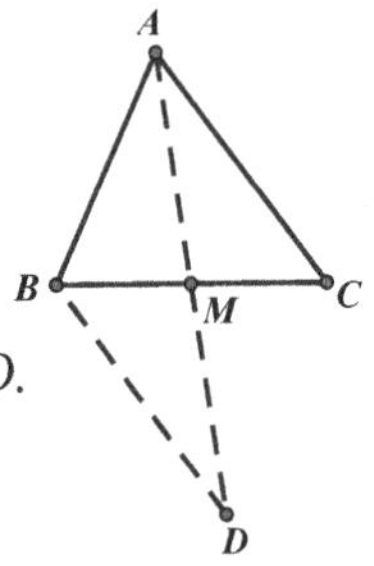

Proof. Refer to the diagram on the right.
Extend AM to D such that $AM = MD$.
By Cosine Rule, $AD^2 = AB^2 + BD^2 - 2AB \cdot BD\cos\angle ABD$.
Notice that AD and BC bisect each other, which implies
$ABDC$ is a parallelogram. Hence, $BD = AC = b$ and
$\angle ABD = 180° - \angle A$.

We have $AD^2 = b^2 + c^2 - 2bc\cos(180° - \angle A) = b^2 + c^2 + 2bc\cos A$.

Since $\cos A = \dfrac{b^2 + c^2 - a^2}{2bc}$, we have

$$AD^2 = b^2 + c^2 + 2bc \cdot \dfrac{b^2 + c^2 - a^2}{2bc} = 2b^2 + 2c^2 - a^2.$$

It follows that $AM^2 = \dfrac{1}{4}AD^2 = \dfrac{1}{2}b^2 + \dfrac{1}{2}c^2 - \dfrac{1}{4}a^2.$ □

Example 2.4.4 In $\triangle ABC$, $AB = 9$, $BC = 8$ and $AC = 7$. Let M be the midpoint of BC. Show that $AM = AC$.

Proof. By Theorem 2.4.3, $AM^2 = \dfrac{1}{2}\left(AB^2 + AC^2\right) - \dfrac{1}{4}BC^2$

$= \dfrac{1}{2} \cdot \left(9^2 + 7^2\right) - \dfrac{1}{4} \cdot 8^2 = 49$, i.e., $AM = 7 = AC$. □

2.5 Ceva's Theorem and Menelaus' Theorem

One important type of problems in geometry is on collinearity and concurrence. We know that any two points determine a unique straight line which passes through them. Hence, if we have three points say A, B, C, in general we can draw three lines AB, BC, CA, unless in the special case where A, B, C are collinear, i.e., they lie on the same line. Refer to the left diagram below.

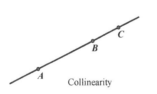
Collinearity

Concurrence

Similarly, we know that any two distinct and non-parallel lines intersect at exactly one point. If we have three such straight lines say ℓ_1, ℓ_2, ℓ_3, in general we should have three points of intersection, unless in the special case where ℓ_1, ℓ_2, ℓ_3 are concurrent, i.e., they pass through the same point. Refer to the right diagram above.

In many geometry questions, one may need to decide whether a given set of three points are collinear, or a given set of three lines are concurrent.

For example, one may recall that we show in any triangles, the perpendicular bisectors of the three sides are concurrent (at the circumcenter). We have also shown the existence of the incenter, the ex-centers and the centroid of a triangle. We shall introduce Ceva's Theorem and Menelaus' Theorem, which provide more general criteria to determine concurrency and collinearity.

Theorem 2.5.1 (Ceva's Theorem)   *In* $\triangle ABC$, *D, E, F are points on AB, AC, BC respectively such that AD, BE, CF are concurrent. We have* $\dfrac{AF}{BF} \cdot \dfrac{BD}{CD} \cdot \dfrac{CE}{AE} = 1.$

Note: The conclusion is not difficult to remember. First, write down the three sides of the triangle AB, BC, CA in this manner: $\dfrac{A*}{B*} \cdot \dfrac{B*}{C*} \cdot \dfrac{C*}{A*} = 1.$ Notice that each letter appears in the numerator and denominator exactly once. Next, replace $*$ by the point which divides the respective side: $\dfrac{AF}{BF}, \dfrac{BD}{CD}$ and $\dfrac{CE}{AE}.$ Notice that all the letters are "cancelled out"!

We use the area method to prove this theorem.

Proof.   Refer to the diagram on the right. Let AD, BE, CF intersect at P. Draw $BH_1 \perp AP$ at H_1 and $CH_2 \perp AP$ at H_2.

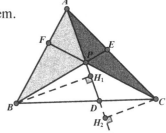

We have $\dfrac{[\triangle ABP]}{[\triangle ACP]} = \dfrac{\dfrac{1}{2} \cdot AP \cdot BH_1}{\dfrac{1}{2} \cdot AP \cdot CH_2} = \dfrac{BH_1}{CH_2} = \dfrac{BD}{CD}$, because $BH_1 /\!/ CH_2$.

Similarly, $\dfrac{[\triangle CBP]}{[\triangle ABP]} = \dfrac{CE}{AE}$ and $\dfrac{[\triangle ACP]}{[\triangle BCP]} = \dfrac{AF}{BF}.$

Now $\dfrac{AF}{BF}\cdot\dfrac{BD}{CD}\cdot\dfrac{CE}{AE}=\dfrac{[\triangle ACP]}{[\triangle BCP]}\cdot\dfrac{[\triangle ABP]}{[\triangle ACP]}\cdot\dfrac{[\triangle BCP]}{[\triangle ABP]}=1.$ □

Note:

(1) The inverse of Ceva's Theorem also holds: if D,E,F are points on BC,AC,AB respectively such that $\dfrac{AF}{BF}\cdot\dfrac{BD}{CD}\cdot\dfrac{CE}{AE}=1,$ then AD,BE,CF are concurrent.

This can be proved easily by contradiction: Suppose otherwise that AD,BE,CF are not concurrent. Refer to the diagram on the right. Let AD and BE intersect at P. Suppose CP extended intersects AB at F'.
Now AD,BE,CF' are concurrent.

By Ceva's Theorem, one must have $\dfrac{AF'}{BF'}\cdot\dfrac{BD}{CD}\cdot\dfrac{CE}{AE}=1.$

Since $\dfrac{AF}{BF}\cdot\dfrac{BD}{CD}\cdot\dfrac{CE}{AE}=1$, we must have $\dfrac{AF}{BF}=\dfrac{AF'}{BF'}$ which implies F and F' coincide.

(2) Ceva's Theorem also holds even if the points of division are on the extension of the sides of $\triangle ABC$. Refer to the diagrams below where AD,BE,CF are concurrent at P.

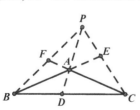
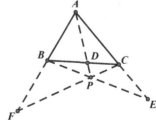

We still have $\dfrac{AF}{BF}\cdot\dfrac{BD}{CD}\cdot\dfrac{CE}{AE}=1$ in either case.

The proof is still by the area method. We leave the details to the reader. (**Hint:** Can you see that $\dfrac{AF}{BF} = \dfrac{[\triangle AFC]}{[\triangle BFC]} = \dfrac{[\triangle APC]}{[\triangle BPC]}$ in the diagram on the right? Notice that $\dfrac{[\triangle AFC]}{[\triangle APC]} = \dfrac{CF}{CP} = \dfrac{[\triangle BFC]}{[\triangle BPC]}$.)

Ceva's Theorem, especially its inverse, is very useful in showing concurrency. For example, the proof for the existence of the centroid of a triangle becomes trivial: if D, E, F are the midpoints of BC, AC, AB respectively, then $\dfrac{AF}{BF} \cdot \dfrac{BD}{CD} \cdot \dfrac{CE}{AE} = 1 \times 1 \times 1 = 1$. Hence, AD, BE, CF are concurrent.

One may also show the existence of the incenter using Ceva's Theorem (and the Angle Bisector Theorem). We leave it to the reader.

Example 2.5.2   In $\triangle ABC$, D is on BC. DE bisects $\angle ADC$, intersecting AC at E. Draw $DF \perp DE$, intersecting AB at F. Show that AD, BE, CF are concurrent.

Insight.   It suffices to show $\dfrac{AF}{BF} \cdot \dfrac{BD}{CD} \cdot \dfrac{CE}{AE} = 1$.
Since DE bisects $\angle ADC$ and $DF \perp DE$, DF bisects $\angle ADB$ (Example 1.1.9). Perhaps we should apply the Angle Bisector Theorem.

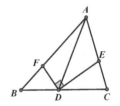

Proof.   Since DE bisects $\angle ADC$ and $DF \perp DE$, DF bisects $\angle ADB$ (Example 1.1.9). By the Angle Bisector Theorem, we have $\dfrac{AF}{BF} = \dfrac{AD}{BD}$ and $\dfrac{AE}{CE} = \dfrac{AD}{CD}$. Now $\dfrac{AF}{BF} \cdot \dfrac{BD}{CD} \cdot \dfrac{CE}{AE} = \dfrac{AD}{BD} \cdot \dfrac{BD}{CD} \cdot \dfrac{CD}{AD} = 1$. By Ceva's Theorem, AD, BE, CF are concurrent.   □

Example 2.5.3   Given a triangle $\triangle ABC$, draw equilateral triangles $\triangle ABF$, $\triangle BCD$, $\triangle ACE$ outwards based on AB, BC, AC respectively. Show that AD, BE, CF are concurrent.

Insight.   Refer to the diagram on the right. It seems an application of Ceva's Theorem, i.e., say AD intersects BC at P, BE intersects AC at Q and CF intersects AB at R, we are to show $\dfrac{BP}{CP} \cdot \dfrac{CQ}{AQ} \cdot \dfrac{AR}{BR} = 1$.

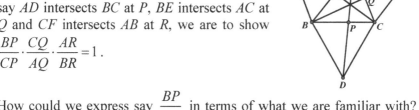

How could we express say $\dfrac{BP}{CP}$ in terms of what we are familiar with? We use areas of triangles when proving Ceva's Theorem, but we cannot use the same triangles once more because we do **not** know whether AD, BE, CF are collinear.

Notice that $\dfrac{BP}{CP} = \dfrac{\left[\triangle ABP\right]}{\left[\triangle ACP\right]} = \dfrac{\left[\triangle BDP\right]}{\left[\triangle CDP\right]}$. Hence, $\dfrac{BP}{CP} = \dfrac{\left[\triangle ABD\right]}{\left[\triangle ACD\right]}$

$$= \frac{\frac{1}{2} AB \cdot BD \sin \angle ABD}{\frac{1}{2} AC \cdot CD \sin \angle ACD} = \frac{AB \sin(\angle ABC + 60^\circ)}{AC \sin(\angle ACB + 60^\circ)} \text{ since } BD = CD.$$

Similarly, $\dfrac{CQ}{AQ} = \dfrac{BC \sin(\angle ACB + 60^\circ)}{AB \sin(\angle BAC + 60^\circ)}$ and $\dfrac{AR}{BR} = \dfrac{AC \sin(\angle BAC + 60^\circ)}{BC \sin(\angle ABC + 60^\circ)}$.

It follows that $\dfrac{BP}{CP} \cdot \dfrac{CQ}{AQ} \cdot \dfrac{AR}{BR} = 1$.

Example 2.5.4   In $\triangle ABC$, M is the midpoint of BC. AD bisects $\angle A$, intersecting BC at D. Draw $BE \perp AD$, intersecting AD extended at E. If AM extended intersect BE at P, show that $AB /\!/ DP$.

Insight. Refer to the diagram on the right.

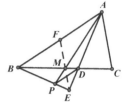

We **should** have $AB /\!/ DP$, i.e., $\dfrac{AD}{DE} = \dfrac{BP}{PE}$.

Hence, if EM extended intersects AB at F, we **should** have $\dfrac{AF}{BF} \cdot \dfrac{BP}{EP} \cdot \dfrac{ED}{AD} = 1$ by Ceva's Theorem, which implies $AF = BF$.

Given that M is the midpoint of BC, we **should** have $MF /\!/ AC$, or equivalently, $EM /\!/ AC$. How can we show it?
We have not used the condition $AE \perp BE$ and the angle bisector AE. It is a common technique to reflect $\triangle ABE$ about AE (Example 1.2.5) and obtain an isosceles triangle!

Proof. Refer to the diagram on the right, where BE extended intersect AC extended at X, and EM extended intersect AB at F. Since AE bisects $\angle BAX$ and $AE \perp BX$, $\triangle ABX$ must be an isosceles triangle where $AB = AX$. It follows that $BE = XE$.

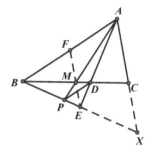

By the Midpoint Theorem, $ME /\!/ AX$, or equivalently, $FF /\!/ AX$.
It follows from the Intercept Theorem that F is the midpoint of AB.

By Ceva's Theorem, $\dfrac{AF}{BF} \cdot \dfrac{BP}{EP} \cdot \dfrac{ED}{AD} = 1$. Since $\dfrac{AF}{BF} = 1$, we must have $\dfrac{BP}{EP} = \dfrac{AD}{ED}$, which implies $PD /\!/ AB$ by the Intercept Theorem. $\square$

Note:
(1) One may easily show the following result by applying Ceva's Theorem. Refer to the diagram on the right. Given $\triangle ABC$ where D, E are on AC, AB respectively and

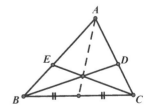

BD, CE intersect at P, we have $DE \, // \, BC$ if and only if AP extended passes through the midpoint of BC.

(2) Example 2.5.4 is not an easy problem. However, one may see the clues more clearly by dividing it into three sub-problems: reflecting $\triangle ABE$ about the angle bisector AE (Example 1.2.5), applying the Midpoint Theorem and the Intercept Theorem to the midline EM, and applying Ceva's Theorem with the median EF. Hence, one could understand how the auxiliary lines are constructed. (You may draw the diagrams separately for each sub-problem.)

Ceva's Theorem has a **trigonometric form**. Refer to the diagram below.

If AD, BE, CF *are concurrent, then* $\dfrac{\sin \angle 1}{\sin \angle 2} \cdot \dfrac{\sin \angle 3}{\sin \angle 4} \cdot \dfrac{\sin \angle 5}{\sin \angle 6} = 1$.

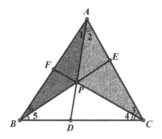

Proof. We still use the area method. Recall the area formula of a triangle: $[\triangle ABC] = \dfrac{1}{2} bc \sin A$.

We have $\dfrac{[\triangle ABP]}{[\triangle ACP]} = \dfrac{\dfrac{1}{2} \cdot AP \cdot AB \sin \angle 1}{\dfrac{1}{2} \cdot AP \cdot AC \sin \angle 2} = \dfrac{AB \sin \angle 1}{AC \sin \angle 2}$.

Similarly, $\dfrac{[\triangle ACP]}{[\triangle BCP]} = \dfrac{AC \sin \angle 3}{BC \sin \angle 4}$ and $\dfrac{[\triangle BCP]}{[\triangle BAP]} = \dfrac{BC \sin \angle 5}{AB \sin \angle 6}$.

Multiply the three equations and we obtain:

$$\frac{[\Delta ABP]}{[\Delta ACP]} \cdot \frac{[\Delta ACP]}{[\Delta BCP]} \cdot \frac{[\Delta BCP]}{[\Delta BAP]} = 1 = \frac{AB \cdot AC \cdot BC}{AC \cdot BC \cdot AB} \cdot \frac{\sin \angle 1}{\sin \angle 2} \cdot \frac{\sin \angle 3}{\sin \angle 4} \cdot \frac{\sin \angle 5}{\sin \angle 6},$$

which leads to the conclusion.   □

Applying the trigonometric form of Ceva's Theorem, it is easy to show that the three heights of a triangle are concurrent. Refer to the diagram on the right for the case of an acute angled triangle.

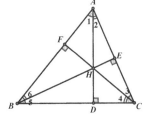

Notice that $\angle 1 = 90° - \angle AHF = \angle 4$. Similarly, $\angle 2 = \angle 5$ and $\angle 3 = \angle 6$.

It follows immediately that $\dfrac{\sin \angle 1}{\sin \angle 2} \cdot \dfrac{\sin \angle 3}{\sin \angle 4} \cdot \dfrac{\sin \angle 5}{\sin \angle 6} = 1$.

Hence, AD, BE, CF are concurrent, i.e., they pass through a common point H, which is called the **orthocenter** of ΔABC.

A similar argument applies for obtuse angled triangles. Refer to the obtuse angled triangle ΔHBC in the diagram on the right. Its orthocenter is A (while H is the orthocenter of ΔABC).

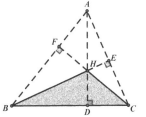

Example 2.5.5   Let H be the orthocenter of an acute angled triangle ΔABC. Show that $\angle BHC = 180° - \angle A$.

One easily sees the conclusion by considering the internal angles of the quadrilateral $AEHF$. Refer to the diagram on the right.

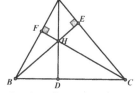

Note that there are a lot of pairs of equal angles in the diagram above. We will study more about the orthocenter of a triangle after we introduce the circle properties in Chapter 3.

Theorem 2.5.6 (Menelaus' Theorem) *Given $\triangle ABC$, a straight line intersects AB, AC and the extension of BC at D, E, F respectively. We have $\dfrac{AD}{BD} \cdot \dfrac{BF}{CF} \cdot \dfrac{CE}{AE} = 1$.*

Note: The conclusion of Menelaus' Theorem is similar to that of Ceva's Theorem: it is also of the form $\dfrac{A*}{B*} \cdot \dfrac{B*}{C*} \cdot \dfrac{C*}{A*} = 1$ where $*$ is to be replaced by the point which divides (internally or externally) the respective side of $\triangle ABC$. Notice that all the letters are "cancelled out"!

We also use the area method to prove Menelaus' Theorem.

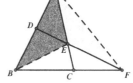

Proof. Connect AF and BE. We denote $S_1 = [\triangle ABE]$, $S_2 = [\triangle AEF]$ and $S_3 = [\triangle BEF]$. Refer to the diagram on the right.

Notice that $\dfrac{AD}{BD} = \dfrac{S_2}{S_3}$ because $\triangle AEF$ and $\triangle BEF$ share a common base EF and their heights on EF are of the ratio $\dfrac{AD}{BD}$. Similarly, $\dfrac{BC}{CF} = \dfrac{S_1}{S_2}$.

Hence, $\dfrac{BF}{CF} = \dfrac{BC + CF}{CF} = \dfrac{BC}{CF} + 1 = \dfrac{S_1}{S_2} + 1 = \dfrac{S_1 + S_2}{S_2}$.

We also have $\dfrac{CE}{AE} = \dfrac{[\triangle BCE]}{S_1} = \dfrac{[\triangle FCE]}{S_2} = \dfrac{[\triangle BCE] + [\triangle FCE]}{S_1 + S_2} = \dfrac{S_3}{S_1 + S_2}$.

Now $\dfrac{AD}{BD} \cdot \dfrac{BF}{CF} \cdot \dfrac{CE}{AE} = \dfrac{S_2}{S_3} \cdot \dfrac{S_1 + S_2}{S_2} \cdot \dfrac{S_3}{S_1 + S_2} = 1$. $\qquad\square$

Note:

(1) The inverse of Menelaus' Theorem also holds: if D, E, F are points on AB, AC and BC extended respectively and $\dfrac{AD}{BD} \cdot \dfrac{BF}{CF} \cdot \dfrac{CE}{AE} = 1$, then D, E, F are collinear.

This can be proved easily by contradiction: Suppose otherwise, say DE extended intersects BC extended at F'.

By Menelaus' Theorem, we have $\dfrac{AD}{BD} \cdot \dfrac{BF'}{CF'} \cdot \dfrac{CE}{AE} = 1$.

Hence, $\dfrac{BF'}{CF'} = \dfrac{BD}{AD} \cdot \dfrac{AE}{CE} = \dfrac{BF}{CF}$ by the condition given. We conclude that F and F' coincide.

Applying Menelaus' Theorem, especially its inverse, is an important method when showing collinearity.

(2) Menelaus' Theorem applies regardless of the relative positions of the division points, i.e., the division points can be on the extension of the sides of a triangle. Refer to the right diagram where the line DE does **not** intersect $\triangle ABC$.

We still have $\dfrac{AD}{BD} \cdot \dfrac{BF}{CF} \cdot \dfrac{CE}{AE} = 1$.

One may prove it by the similar area method. We leave the details to the reader.

(3) Although the conclusions of Ceva's Theorem and Menelaus' Theorem are highly similar, one may see their different geometric meanings easily from the diagrams.

One may apply Menelaus' Theorem and calculate the ratio of line segments very efficiently. Recall Example 2.1.4.

In $\triangle ABC$, D is a point on AB and $\dfrac{AD}{AC} = \dfrac{AC}{AB} = \dfrac{2}{3}$. M is the midpoint of CD while AM extended intersects BC at E. Find $\dfrac{CE}{BE}$.

Ans.   Refer to the diagram on the right. Apply Menelaus' Theorem when the line AE intersects $\triangle BCD$. We have $\dfrac{CE}{BE} \cdot \dfrac{BA}{DA} \cdot \dfrac{DM}{CM} = 1$.

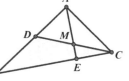

Since $\dfrac{BA}{DA} = \dfrac{BA}{AC} \cdot \dfrac{AC}{DA} = \left(\dfrac{3}{2}\right)^2 = \dfrac{9}{4}$ and $\dfrac{DM}{CM} = 1$, we have $\dfrac{CE}{BE} = \dfrac{4}{9}$. $\square$

Note: Choosing an appropriate triangle and a line intersecting it is very important when applying Menelaus' Theorem. For example, if we choose the line CD intersecting $\triangle ABE$ in this example, we will not be able to obtain $\dfrac{CE}{BE}$.

Example 2.5.7 Given $\triangle ABC$, D is a point on BC such that AD bisects $\angle A$. E, F are on AB, AC respectively such that DE, DF bisect $\angle ADB$ and $\angle ADC$ respectively. If EF extended intersects the line BC at P, show that $AP \perp AD$.

Insight. Refer to the diagram on the right. It seems we should consider the line EF intersecting $\triangle ABC$ and apply Menelaus' Theorem.

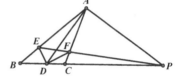

Can we apply the Angle Bisector Theorem for $\dfrac{AE}{BE}$ and $\dfrac{CF}{AF}$?

On the other hand, since we are to show $AP \perp AD$, AP **should** be the exterior angle bisector of $\angle BAC$ (Example 1.1.9). Hence, we **should** have $\dfrac{BP}{CP} = \dfrac{AB}{AC}$ by the Angle Bisector Theorem.

Proof. By Menelaus' Theorem, $\dfrac{AE}{BE} \cdot \dfrac{BP}{CP} \cdot \dfrac{CF}{AF} = 1$. (*)

Since DE, DF are angle bisectors, we must have $\dfrac{AE}{BE} = \dfrac{AD}{BD}$ and $\dfrac{CF}{AF} = \dfrac{CD}{AD}$ by the Angle Bisector Theorem. Now $\dfrac{AE}{BE} \cdot \dfrac{CF}{AF} = \dfrac{CD}{BD} = \dfrac{AC}{AB}$ because AD bisects $\angle BAC$. It follows from (*) that $\dfrac{BP}{CP} = \dfrac{AB}{AC}$.

Hence, AP is the exterior angle bisector of $\angle BAC$. We conclude that $AP \perp AD$ (Example 1.1.9). □

Example 2.5.8   In $\triangle ABC$, M, N are points on AB, AC respectively such that the centroid G of $\triangle ABC$ lies on MN. Show that $AM \cdot CN + AN \cdot BM = AM \cdot AN$.

Insight.   Let AG intersect BC at D. Notice that $\dfrac{AG}{AD} = \dfrac{2}{3}$. Since G lies on MN, if $MN // BC$, $\dfrac{AM}{AB} = \dfrac{AN}{AC} = \dfrac{2}{3}$ and $\dfrac{CN}{AC} = \dfrac{BM}{AB} = \dfrac{1}{3}$. Refer to the diagram on the right.

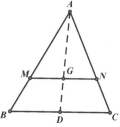

It follows that $AM \cdot CN + AN \cdot BM = \dfrac{2}{3} AB \cdot \dfrac{1}{3} AC + \dfrac{2}{3} AC \cdot \dfrac{1}{3} AB$

$$= \frac{4}{9} AB \cdot AC = \frac{2}{3} AB \cdot \frac{2}{3} AC = AM \cdot AN.$$

Otherwise, say MN extended intersects BC extended at P. Refer to the diagram on the right. We see that the line MP intersects several triangles. Moreover, we know $BD = CD$ and $AG = 2DG$. Hence, applying Menelaus' Theorem would probably help us to find the relationship among those line segments.

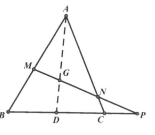

It is also noteworthy that the common factors AM and AN appear on both sides of the equation in the conclusion. Hence, we may consider dividing both sides by $AM \cdot AN$.

Proof.   It is easy to show the conclusion when $MN // BC$. Otherwise, say MN extended intersects BC extended at P. By dividing $AM \cdot AN$ on both sides of the equation, it suffices to show that $\dfrac{CN}{AN} + \dfrac{BM}{AM} = 1$.

Apply Menelaus' Theorem when the line MN intersects $\triangle ACD$:

$$\frac{AG}{DG} \cdot \frac{DP}{CP} \cdot \frac{CN}{AN} = 1 \text{, i.e., } \frac{CN}{AN} = \frac{CP}{2DP} \text{ since } \frac{AG}{DG} = 2 \text{.}$$

Apply Menelaus' Theorem when the line MN intersects $\triangle ABD$:

$$\frac{AM}{BM} \cdot \frac{BP}{DP} \cdot \frac{DG}{AG} = 1 \text{, i.e., } \frac{BM}{AM} = \frac{BP}{2DP} \text{.}$$

Hence, we are to show $\dfrac{CP}{2DP} + \dfrac{BP}{2DP} = 1$, or $CP + BP = 2DP$.

This is clear because $CP + BP = CP + CP + BC = 2CP + 2DC = 2DP$. $\square$

Example 2.5.9 (USA 11)   In a non-isosceles acute angled triangle $\triangle ABC$ where AD, BE, CF are heights, H is the orthocenter. AD and EF intersect at S. Draw $AP \perp EF$ at P and $HQ \perp EF$ at Q. If the lines DP and QH intersect at R, show that $HQ = HR$.

Insight.   Refer to the diagram on the right. Besides the feet of perpendicular D, E, F and the orthocenter H, the diagram is constructed by drawing perpendicular lines and we also have $AP // QR$. In particular, for any given $\triangle ABC$, Q and R are uniquely determined.

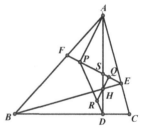

How could we show $\dfrac{HQ}{HR} = 1$? Menelaus' Theorem could be very useful in such a diagram which is purely constructed by the intersection of straight lines.

Since $AP // QR$, we have $\dfrac{HQ}{AP} = \dfrac{HS}{AS}$ and $\dfrac{HR}{AP} = \dfrac{HD}{AD}$. It suffices to show

that $\dfrac{HS}{AS} = \dfrac{HD}{AD}$. Which triangle (and the line intersecting it) should we

apply Menelaus' Theorem to?

Proof.　Refer to the diagram on the right.
Apply Menelaus' Theorem to $\triangle AHC$ and EF.

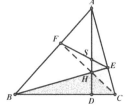

We have $\dfrac{AS}{HS} \cdot \dfrac{HF}{CF} \cdot \dfrac{CE}{AE} = 1$.　(1)

Since $AP \,/\!/\, QR$, we have $\dfrac{HQ}{AP} = \dfrac{HS}{AS}$ and $\dfrac{HR}{AP} = \dfrac{HD}{AD}$. We claim that

$\dfrac{HS}{AS} = \dfrac{HD}{AD}$, which implies $\dfrac{HQ}{AP} = \dfrac{HR}{AP}$ and hence, $HQ = HR$.

By (1), it suffices to show $\dfrac{AD}{HD} \cdot \dfrac{HF}{CF} \cdot \dfrac{CE}{AE} = 1$.　(2)

Let $S_1 = [\triangle ABH]$, $S_2 = [\triangle BCH]$ and $S_3 = [\triangle ACH]$.

We have $\dfrac{AD}{HD} = \dfrac{S_1 + S_2 + S_3}{S_2}$, $\dfrac{HF}{CF} = \dfrac{S_1}{S_1 + S_2 + S_3}$ and $\dfrac{CE}{AE} = \dfrac{S_2}{S_1}$.

Now it is easy to see that (2) holds. This completes the proof.　□

Note:　One may perceive (2) as Ceva's Theorem applied to $\triangle AHC$ where lines AF, HE, CD are concurrent at B. Of course, beginners may find difficulties in recognizing Ceva's Theorem when the point of concurrency is outside the triangle. In such cases, one may always use the area method. We can see from the proof above that this is not difficult.

As an application of Menelaus' Theorem, we will show Desargues' Theorem, which is also an important result in showing collinearity and concurrency.

Theorem 2.5.10 (Desargues' Theorem)　*Given $\triangle ABC$ and $\triangle A'B'C'$ such that the lines $AB, A'B'$ intersect at P, the lines $BC, B'C'$ intersect at Q and the lines $AC, A'C'$ intersect at R, if the lines AA', BB', CC' are concurrent, then P, Q, R are collinear.*

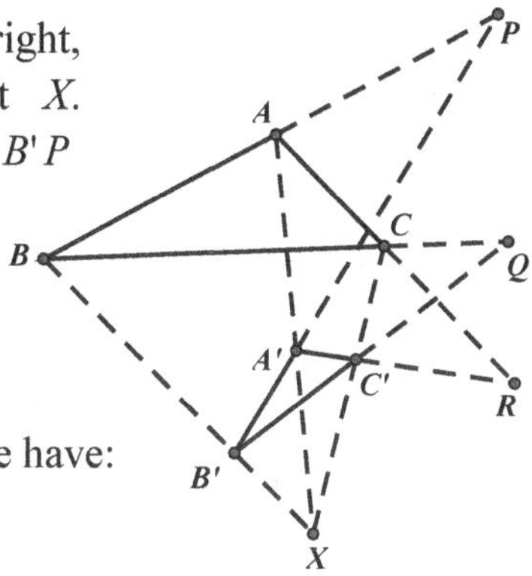

Proof.   Refer to the diagram on the right, where AA', BB', CC' are concurrent at X. Apply Menelaus' Theorem when $B'P$ intersects $\triangle XAB$ and we obtain:

$$\frac{XB'}{BB'} \cdot \frac{BP}{AP} \cdot \frac{AA'}{XA'} = 1. \quad (1)$$

Similarly, when $B'Q$ intersects $\triangle XBC$, we have:

$$\frac{BB'}{XB'} \cdot \frac{CQ}{BQ} \cdot \frac{XC'}{CC'} = 1. \quad (2)$$

When $A'R$ intersects $\triangle XAC$, we have $\dfrac{CC'}{XC'} \cdot \dfrac{XA'}{AA'} \cdot \dfrac{AR}{CR} = 1. \quad (3)$

Multiplying (1), (2) and (3) gives $\dfrac{CQ}{BQ} \cdot \dfrac{BP}{AP} \cdot \dfrac{AR}{CR} = 1$, which implies

P, Q, R are collinear by Menelaus' Theorem.   □

Note:
(1) We apply Menelaus' Theorem extensively in this proof, which does not depend on the relative positions of $\triangle ABC$ and $\triangle A'B'C'$.
(2) The inverse of Desargues' Theorem also holds, i.e., if P, Q, R are collinear, then lines AA', BB', CC' are concurrent (or parallel to each other). One may follow a similar argument as above: given P, Q, R collinear and the lines AA', BB' intersect at X, show that C, C', X are collinear by Menelaus' Theorem.

Applying Desargues' Theorem changes the conclusion of concurrency to an equivalent one of collinearity, or vice versa. This may be a wise strategy when solving difficult problems, say if the conclusion to be shown seems unrelated to the conditions given. We will see examples in Chapter 6.

Ceva's Theorem and Menelaus' Theorem are very useful in showing concurrency and collinearity. However, we shall point out there are many other ways to show concurrency and collinearity.

- Collinearity: Showing equal or supplementary angles is the most fundamental and straightforward method. Refer to the diagrams below.

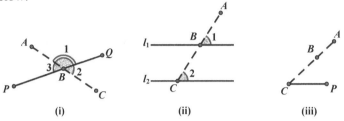

(i)	(ii)	(iii)

(i) PQ is a straight line where B lies. We have A,B,C collinear if $\angle 1 + \angle 2 = 180°$ or $\angle 2 = \angle 3$.

(ii) B,C are on ℓ_1, ℓ_2 respectively and $\ell_1 /\!/ \ell_2$. We have A,B,C collinear if $\angle 1 = \angle 2$.

(iii) We have A,B,C collinear if $\angle ACP = \angle BCP$.

Another commonly used method is via the properties of similar triangles. Refer to the diagram on the right for an example, where C is a point on BD and $AB /\!/ DE$. Now A,C,E are collinear if $\dfrac{AC}{CE} = \dfrac{BC}{CD}$.

- Concurrency: One may suppose two lines meet at a point and show that the third line also passes through that point. We used this method to show the existence of the incenter, circumcenter, centroid and ex-centers (Exercise 1.4) of a triangle. Another commonly used method is via the properties of similar triangles, an example of which is given below.

Theorem 2.5.11   *Given* $\triangle ABC$ *and* $\triangle DEF$ *such that* $AB /\!/ DE$, $BC /\!/ EF$ *and* $AC /\!/ DF$, *then* AD, BE, CF *are either parallel or concurrent.*

Proof. Notice that there are two possible cases regarding the relative positions of $\triangle ABC$ and $\triangle DEF$. Refer to the diagrams below.

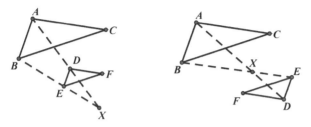

It is easy to see that $\triangle ABC \sim \triangle DEF$ because all the corresponding angles are equal. Suppose AD and BE intersect at X. It suffices to show that CF passes through X as well, i.e., C, F, X are collinear.

Connect CX, FX. Since $AB /\!/ DE$, we must have $\dfrac{AX}{DX} = \dfrac{AB}{DE} = \dfrac{AC}{DF}$.

Clearly, $\angle CAX = \angle FDX$. We conclude that $\triangle ACX \sim \triangle DFX$ and hence, $\angle DXF = \angle AXC$. Now C, X, F are collinear.

Notice that the proof does not depend on the diagram. $\qquad\square$

2.6 Exercises

1. Refer to the diagram on the right. Given $\triangle ABC$, extend AB to D such that $AB = BD$, extend BC to E such that $BC = 2CE$ and extend CA to F such that $AF = 2AC$. Draw parallelograms $BCXD$, $ACEY$ and $ABZF$. If the total area of these three parallelograms is $175\,\mathrm{cm}^2$, find the area of $\triangle ABC$ in cm^2.

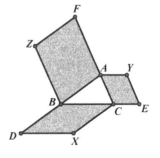

2. Given $\triangle ABC$, draw squares $ABDE$ and $ACFG$ outwards from AB, AC respectively. Let O_1, O_2 denote the centers of squares $ABDE$ and $ACFG$ respectively. If M, N are the midpoints of BC, EG respectively, show that MO_1NO_2 is a square.

3. In a quadrilateral $ABCD$, $AB \perp AD$ and $BC \perp CD$. F is a point on CD such that AF bisects $\angle BAD$. If BD and AF intersect at E and $AF \mathbin{/\mkern-5mu/} BC$, show that $AE < \dfrac{1}{2}CD$.

4. In a right angled triangle $\triangle ABC$, $\angle A = 90°$ and D, E are on AB, AC respectively. If M, N, P, Q are the midpoints of DE, BC, BE, CD respectively, show that $MN = PQ$.

5. Let $ABCD$ be a quadrilateral and E, F, G, H be the midpoints of AB, BC, CD, DA respectively. Let M be the midpoint of GH and P be a point on EM such that $FG = PG$. Show that $PF \perp EM$.

6. Given a square $ABCD$, E, F are the midpoints of AB, BC respectively. Let CE, DF intersect at P. Connect AP. Show that $AP = AB$.

7. Let G be the centroid of $\triangle ABC$. Show that if $BG \perp CG$, then $AB^2 + AC^2 = 5BC^2$.

8. Given a triangle $\triangle ABC$, a line $\ell_1 \mathbin{/\mkern-5mu/} BC$ intersects AB, AC at D, D' respectively, a line $\ell_2 \mathbin{/\mkern-5mu/} AC$ intersects BC, AB at E, E' respectively and a line $\ell_3 \mathbin{/\mkern-5mu/} AB$ intersects AC, BC at F, F' respectively. Show that $[\triangle DEF] = [\triangle D'E'F']$.

9. Let $\triangle ABC$ be an equilateral triangle and D is a point on BC. The perpendicular bisector of AD intersects AB, AC at E, F respectively. Show that $BD \cdot CD = BE \cdot CF$.

10. Given an acute angled triangle $\triangle ABC$ where H is the orthocenter, show that $\dfrac{BC}{AH} = \tan \angle A$.

11. In a right angled triangle $\triangle ABC$ where $\angle A = 90°$, D, E are on BC such that $BD = DE = CE$. Show that $AD^2 + AE^2 = \dfrac{5}{9} BC^2$.

12. In $\triangle ABC$, M is the midpoint of AB and D is a point on AC. Draw $CE \parallel AB$, intersecting BD extended at E. Show that lines AE, BC, MD are concurrent.

13. Given $\triangle ABC$, draw squares $ABDE$, $BCFG$ and $CAHI$ outwards based on AB, BC, AC respectively. Let P, Q, R be the midpoints of DE, FG, HI respectively. Show that AQ, BR, CP are concurrent.

14. Refer to the diagram on the right. $\triangle ABC$ is a non-isosceles triangle. AD, BE, CF are the exterior angle bisectors of $\angle A, \angle B, \angle C$ respectively, intersecting the lines BC, AC, AB at D, E, F respectively. Show that D, E, F are collinear.

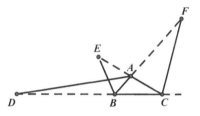

15. Given an isosceles triangle $\triangle ABC$ where $AB = AC$, M is the midpoint of BC. A line ℓ passing through M intersects AB at D and intersects AC extended at E. Show that $\dfrac{1}{AD} + \dfrac{1}{AE} = \dfrac{2}{AB}$.

Chapter 3

Circles and Angles

A circle is uniquely determined by its center and radius, i.e., if two circles have the same center and radius, they must coincide. We use $\odot O$ to denote a circle centered at O.

It is widely known that given a circle with radius r, its perimeter equals $2\pi r$ and the area of the disc is πr^2. Indeed, there are many more interesting properties about circles. In this chapter, we will focus on the properties of angles related to circles.

3.1 Angles inside a Circle

Theorem 3.1.1 *An angle at the center of a circle is twice of the angle at the circumference.*

Proof. Refer to the diagram on the right. We are to show $\angle BOC = 2\angle BAC$. Extend AO to D. Since O is the center of the circle, we have $AO = BO$. Now $\angle B = \angle OAB$ in $\triangle AOB$, and the exterior angle $\angle BOD = \angle B + \angle OAB = 2\angle OAB$.

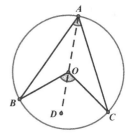

Similarly, $\angle COD = \angle C + \angle OAC = 2\angle OAC$.
Now $\angle BOC = \angle BOD + \angle COD = 2\angle OAB + 2\angle OAC = 2\angle BAC$.

Notice that the proof is **not** completed yet: there is another possible situation as illustrated in the diagram on the right. Notice that the proof above does not apply in this situation, but an amended version following the same idea (using subtraction instead of addition) leads to the conclusion. We leave it to the reader.

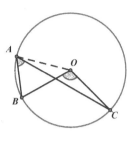

Example 3.1.2 Let O be the circumcenter of $\triangle ABC$. We have:
(1) $\angle BOC = 2\angle A$
(2) $\angle OBC = 90° - \angle A$

Proof. (1) follows directly from Theorem 3.1.1.

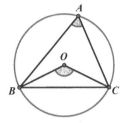

(2) is because $\angle OBC = \dfrac{1}{2}\left(180° - \angle BOC\right)$

$$= \dfrac{1}{2}\left(180° - 2\angle A\right) = 90° - \angle A .$$ □

Theorem 3.1.1 has a few immediate corollaries which are very important in circle geometry.

Corollary 3.1.3 *Angles in the same arc are the same.*

Refer to the left diagram below. $\angle 1 = \angle 2$ because they are both equal to half of the angle at the center of the circle.

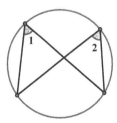

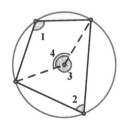

Angles in the same arc   Opposite angles in a cyclic quadrilateral

We call a quadrilateral **cyclic** if it is inscribed inside a circle.

Corollary 3.1.4   *Opposite angles of a cyclic quadrilateral are supplementary, i.e., their sum is* $180°$.

Refer to the previous right diagram. We have $\angle 1 + \angle 2$
$$= \frac{1}{2}\angle 3 + \frac{1}{2}\angle 4 = \frac{1}{2} \cdot 360° = 180°.$$

Notice that $\angle 3$ in the diagram is greater than 180°, but one can easily show that Theorem 3.1.1 still applies.

Corollary 3.1.5   *An exterior angle of a cyclic quadrilateral is equal to the corresponding opposite angle.*

Refer to the diagram on the right where $\angle 1 = \angle 2$. This is immediately from Corollary 3.1.4.

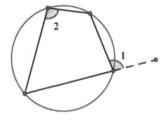

Exterior angle of a cyclic quadrilateral

In Section 2.5, we studied the relationship between points and lines, i.e., collinearity and concurrence. Similarly, we will study the relationship between points and circles in this chapter. First, one sees that any three non-collinear points uniquely determine a circle: for points A, B, C not collinear, there exists a unique circle passing through A, B, C. This is simply the circumcircle of $\triangle ABC$.

In general, four points do not lie on the same circle. Hence, it is noteworthy if the contrary happens, in which case we say the four points are concyclic. Refer to the diagram on the right for an example.

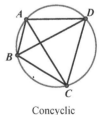

Concyclic

Showing concyclicity seems harder than collinearity or concurrence. For example, one may prove collinearity by showing the neighboring angles are supplementary, or prove concurrence by showing the intersection of two lines lies on the third. Are there any similar and *straightforward* techniques applicable to show concyclicity?

We have to accept that circles are not as *straight* as lines. Nevertheless, circle geometry has a rich structure which provides us abundant methods in showing concyclicity. For example, one sees that the inverse statements of Corollaries 3.1.3 to 3.1.5 also hold, which can be shown easily by contradiction. Now we have simple and effective criteria to determine concyclicity. Refer to the diagrams below. In any of these cases, A, B, C, D are concyclic.

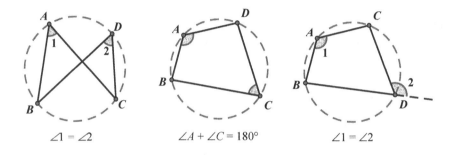

$$\angle 1 = \angle 2 \qquad\qquad \angle A + \angle C = 180° \qquad\qquad \angle 1 = \angle 2$$

Example 3.1.6   In an acute triangle $\triangle ABC$, AD, BE, CF are heights. Show that the line AD is the angle bisector of $\angle EDF$.

Proof.   Refer to the diagram on the right. Since $\angle BFH = \angle BDH = 90°$, B, D, H, F are concyclic by the inverse of Corollary 3.1.4. Hence, $\angle 1 = \angle 3$. Similarly, C, D, H, E are concyclic and we have $\angle 2 = \angle 4$.

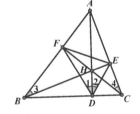

Since $\angle BFC = \angle BEC = 90°$, B, C, E, F are concyclic by the inverse of Corollary 3.1.3. It follows that $\angle 1 = \angle 3 = \angle 4 = \angle 2$.   □

Note:
(1) Since $AD \perp BC$, $\angle 1 = \angle 2$ also implies $\angle BDF = \angle CDE$. Since A, C, D, F are concyclic, we also have $\angle BDF = \angle CDE = \angle BAC$ (Corollary 3.1.5).

(2) One sees that a lot of concyclicity appear in this diagram. In fact, experienced contestants know this diagram very well and are able to recall those basic facts almost instantaneously.

(3) The conclusion implies that *H*, the orthocenter of $\triangle ABC$, is the incenter of $\triangle DEF$.

Example 3.1.7 Let *ABCD* be a cyclic quadrilateral. A line ℓ parallel to *BC* intersects *AB, CD* at *E, F* respectively. Show that *A, D, F, E* are concyclic.

Proof. Refer to the diagram on the right. Since *EF // BC*, $\angle 1 = \angle C$. Notice that $\angle A + \angle C = 180°$ by Corollary 3.1.4. Hence, $\angle A + \angle 1 = 180°$, which implies *A, D, F, E* are concyclic. $\square$

Example 3.1.8 $\odot O_1$ and $\odot O_2$ intersect at *P* and *Q*. If O_1P extended intersects $\odot O_2$ at *B* and O_2P extended intersects $\odot O_1$ at *A*, show that O_1, O_2, A, B, Q are concyclic.

Insight. We are to show five points are concyclic. So many of them! Perhaps we can show four points are concyclic first, say O_1, O_2, A, B. Refer to the diagram on the right.

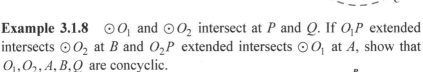

The simplest method is to show that $\angle 1 = \angle 2$. Are there any equal angles in the diagram? Yes, say $\angle 1 = \angle 3$ (because $O_1A = O_1P$) and similarly $\angle 2 = \angle 4$. We also have opposing angles $\angle 3 = \angle 4$. Job done!

Next, we may show that O_1, O_2, A, Q are concyclic. Let us draw the quadrilateral. Refer to the diagram on the right. Can we show $\angle 1 + \angle O_1 Q O_2 = 180°$? This seems not difficult.

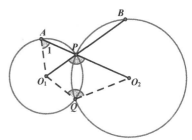

Observe that $\angle O_1QO_2 = \angle O_1PO_2$ ($\triangle O_1PO_2 \cong \triangle O_1QO_2$), $\angle 1 = \angle APO_1$ and $\angle APO_1 + \angle O_1PO_2 = 180°$. Job done!

In conclusion, both O_1, O_2, A, B and O_1, O_2, A, Q are concyclic, which means that B and Q lie on the circumcircle of $\triangle O_1AO_2$. Indeed, O_1, O_2, A, B, Q are concyclic.

Note: One may show that O_1, O_2, A, Q are concyclic and hence, O_1, O_2, B, Q are concyclic by similar reasoning. This would also complete the proof.

Example 3.1.9   Refer to the diagram on the right. A, B, C are points on the circle. PC extended intersects the circle at D. Q is a point on CD such that $\angle DAQ = \angle PBC$. Show that $\angle DBQ = \angle PAC$.

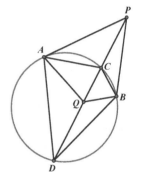

Insight.   We are given a circle and a pair of equal angles. Could we find more pairs of equal angles? How are they related to our conclusion $\angle DBQ = \angle PAC$?
One may see the difficulty as $\angle PAC$ (and $\angle PBC$) are **not** extended by an arc. Perhaps we should relate $\angle PBC$ to another angle on the circumference besides $\angle DAQ$ and seek clues. How about $\angle PBC = \angle BCD - \angle BPD$? We may connect AB. Now $\angle BCD = \angle BAD$ is also related to $\angle DAQ$!

Proof.   Refer to the diagram on the right. We have $\angle PBC = \angle BCD - \angle BPC$. Connect AB. Notice that $\angle BCD = \angle BAD$ (angles in the same arc). It is given that $\angle DAQ = \angle PBC$. Hence, $\angle DAQ = \angle BAD - \angle BPC$, or $\angle BPC = \angle BAD - \angle DAQ = \angle BAQ$. This implies P, A, Q, B are concyclic. Now $\angle DBQ = \angle PQB - \angle CDB = \angle PAB - \angle CAB = \angle PAC$.   □

Example 3.1.10   Given an equilateral $\triangle ABC$ and its circumcircle, M is a point on the minor arc $\overset{\frown}{BC}$. Show that $MA = MB + MC$.

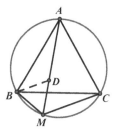

Insight.   We are to show $MA = MB + MC$. Hence, it is a common technique to "cut" MB from MA and see whether the remaining portion equals to MC, i.e., we choose D on MA such that $MB = MD$ and attempt to show $MC = AD$. Refer to the diagram on the right. Notice that there are many equal sides and angles due to the equilateral triangle and the circle. Can you find congruent triangles?

Proof.   Choose D on MA such that $MB = MD$. It suffices to show that $AD = MC$. Notice that $\angle AMB = \angle ACB = 60°$ (angles in the same arc). Hence, $\triangle MBD$ is an isosceles triangle with the vertex angle $60°$, i.e., an equilateral triangle. Now $BD = BM$ and $\angle DBM = 60°$, which implies $\angle CBM = 60° - \angle CBD = \angle DBA$. It follows that $\triangle CBM \cong \triangle ABD$ (S.A.S.). Hence, $AD = MC$ and the conclusion follows.   □

Example 3.1.11   In a quadrilateral $ABCD$, $AB = AD$ and $BC \neq CD$. If CA bisects $\angle BCD$, then A, B, C, D are concyclic.

Insight.   Refer to the diagram on the right. If A, B, C, D are concyclic, we have $\angle 1 = \angle 4$ $= \angle 3 = \angle 2$. It seems exactly right! Perhaps we can show the conclusion by contradiction: what if A, B, C, D are not concyclic?

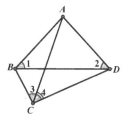

Proof.   Suppose otherwise that A, B, C, D are not concyclic. Let the circumcircle of $\triangle ABD$ intersect the line AC at P. Refer to the diagram on the right. Notice that $\angle 1 = \angle APD$ and $\angle 2 = \angle APB$ (angles in the same arc).

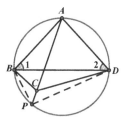

Since $AB = AD$, $\angle 1 = \angle 2$ and hence, $\angle APB = \angle APD$. We are also given that $\angle ACB = \angle ACD$. Hence, AP is the perpendicular bisector of BD (Example 1.2.10). This is impossible because $BC \neq CD$. □

Note:
(1) This proof does not depend on the diagram, i.e., it still holds if C is outside the circle.
(2) One may also show $BC = CD$ by $\triangle PBC \cong \triangle PDC$ (A.A.S.).

Example 3.1.12   Let $ABCD$ be a cyclic quadrilateral where the angle bisectors of $\angle A$ and $\angle B$ intersect at E. Draw a line passing through E parallel to CD, intersecting AD, BC at P, Q respectively. Show that $PQ = PA + QB$.

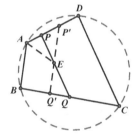

Insight.   Given angle bisectors and parallel lines, can we have isosceles triangles? Not exactly in this case because $PQ // CD$: if $PQ // AB$, we will obtain isosceles triangles. Hence, we may draw $P'Q' // AB$, intersecting AD, BC at P', Q' respectively. Refer to the diagram on the right.
Since AE bisects $\angle A$, we have $\angle P'AE = \angle BAE = \angle P'EA$, which implies $P'A = P'E$. Similarly, $Q'B = Q'E$. We have $P'Q' = P'A + Q'B$. How are PQ and $P'Q'$ related? If we randomly draw a line PQ passing through E, we shall **not** have $PQ = PA + QB$. Notice that we have not used the conditions $PQ // CD$ and A, B, C, D concyclic!

Proof.   Draw $P'Q' // AB$, intersecting AD, BC at P', Q' respectively. Since $\angle P'AE = \angle BAE = \angle P'EA$, we have $P'A = P'E$ and similarly, $Q'B = Q'E$. Hence, $P'Q' = P'A + Q'B$. (1)
Since $P'Q' // AB$, $PQ // CD$ and A, B, C, D are concyclic, we have $\angle PP'Q = 180° - \angle A = \angle C = \angle PQQ'$. Similarly, $\angle P'PQ = \angle P'Q'Q$.

Let the lines AD and BC intersect at X. Refer to the diagram on the right. Observe that E is the ex-center of $\triangle XAB$ opposite X.

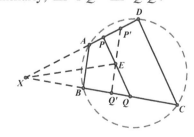

Hence, XE bisects $\angle AXB$. One easily sees that $\triangle XP'E \cong \triangle XQE$ (A.A.S.) and $\triangle XPE \cong \triangle XQ'E$ (A.A.S.). It follows that $P'E = QE$, $PE = Q'E$ and $PP' = QQ'$. Now $PQ = PE + QE = P'E + Q'E = P'Q'$ (2) and $P'A + Q'B = PA + Q'B + PP' = PA + Q'B + QQ' = PA + QB$.   (3) (1), (2) and (3) imply that $PQ = PA + QB$.

Note that the proof still holds if the lines AD and BC intersect at the other side of PQ, in which case E is the incenter of $\triangle XAB$ instead of the ex-center, and we still have XE bisects $\angle X$.   □

Note:

(1) Once it is shown that the corresponding angles in $\triangle PP'E$ and $\triangle Q'QE$ are the same, we should probably have $\triangle PP'E \cong \triangle Q'QE$ (which leads to the conclusion immediately). Hence, it is natural to consider the intersection of the lines AD and BC, which gives congruent triangles with common sides.

(2) Another strategy to solve the problem is via "cut and paste": since we are to show $PQ = PA + QB$, we choose F on PQ such that $BQ = FQ$ and we attempt to show $AP = FP$. Refer to the diagram on the right. Since $PQ // CD$, we have A, B, Q, P concyclic (Example 3.1.7).

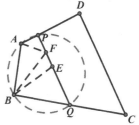

Since $BQ = FQ$, we have $\angle BFE = \angle FBQ = \dfrac{1}{2}\angle CQP = \dfrac{1}{2}\angle BAD$ $= \angle BAE$, i.e., A, B, E, F are concyclic. We are to show that $\angle PAF = \angle PFA = \dfrac{1}{2}\angle DPQ$, while $\dfrac{1}{2}\angle DPQ = \dfrac{1}{2}\angle ABQ = \angle ABE$. Since A, B, E, F are concyclic, we must have $\angle PFA = \angle ABE$ (Corollary 3.1.5). This completes the proof.

One should also take note of another immediate corollary from Theorem 3.1.1 that the diameter of the circle always extends a right angle on the circumference. This is a common method in identifying right angles.

Corollary 3.1.13   *If AB is the diameter of ⊙O and P is a point on the circle, then ∠APB = 90°.*

Proof.   Refer to the diagram on the right. Notice that $\angle AOB = 180°$.

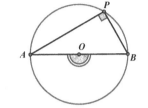

By Theorem 3.1.1, $\angle APB = \dfrac{1}{2} \angle AOB = 90°$ □

Note: The inverse of this corollary also holds, i.e., if a chord *AB* extends an angle of 90° on the circumference, then *AB* is the diameter (which passes through the center of the circle).

Example 3.1.14   Refer to the diagram on the right. Given a circle where *AB* is a diameter, *C, D, E* are on the circle such that *C, E* are on the same side of *AB* while *D* is on the other side. Show that $\angle C + \angle E = 90°$.

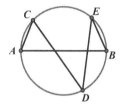

Proof.   Refer to the diagram on the right. Connect *AE*. Since *AB* is a diameter, we have $\angle E = 90° - \angle AED$. Notice that $\angle AED = \angle C$ (angles in the same arc) and the conclusion follows.

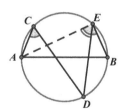

Example 3.1.15   Given an acute angled $\triangle ABC$ where $AD \perp BC$ at *D*, *M, N* are the midpoints of *AB, AC* respectively. Let ℓ be a line passing through *A*. Draw $BE \perp \ell$ at *E* and $CF \perp \ell$ at *F*. If the lines *EM, FN* intersect at *P*, show that *D, E, F, P* are concyclic.

Insight.   Refer to the diagram on the right. We could probably show the concyclicity by equal angles. Can you see *A, D, B, E* (and similarly *A, D, C, F*) are concyclic?

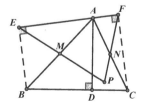

What do we know about *P*? *P* is obtained by intersecting *EM* and *FN*. Notice that *EM*, *FN* are medians on the hypotenuses of right angled triangles. This gives us more equal angles!

Proof. Since $\angle AEB = \angle ADB = 90°$, A, D, B, E are concyclic and in particular, *M* is the center of the circle. Clearly, $\angle AEM = \angle EAM$. Similarly, $\angle AFN = \angle FAN$. Now $\angle P = 180° - (\angle AEM + \angle AFN)$
$= 180° - (\angle EAM + \angle FAN) = \angle BAC$.
On the other hand, we have $\angle 1 = \angle 2$ and
$\angle 3 = \angle 4$ (angles in the same arc). Refer to
the diagram on the right.

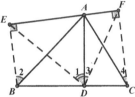

It follows that $\angle EDF = \angle 1 + \angle 3 = \angle 2 + \angle 4$
$= \angle BAC$, since *BCFE* is a trapezium (Example 1.4.15).
Now $\angle P = \angle EDF$, which implies D, E, F, P are concyclic. □

Example 3.1.16 Let *ABCD* be a square. *E*, *F* are points on *BC*, *CD* respectively and $\angle EAF = 45°$. Draw $EP \perp AC$ at *P* and $FQ \perp AC$ at *Q* (*P*, *Q* do not coincide). Show that the circumcenter of $\triangle BPQ$ lies on *BC*.

Insight. How shall we use the condition $\angle EAF = 45°$? One may recall Exercise 1.6. However, rotating $\triangle ABE$ seems not useful this time.

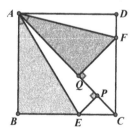 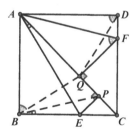

Notice that $\angle BAE = 45° - \angle CAE = \angle CAF$. Refer to the left diagram above. It follows that $\triangle ABE \sim \triangle AQF$ and $\angle AEB = \angle AFQ$. In fact, one may find other pairs of equal angles due to symmetry. Refer to the right diagram above. We have $\angle ABQ = \angle ADQ = \angle AFQ$ (since A, D, F, Q are concyclic where $\angle ADF = \angle AQF = 90°$). Similarly, $\angle PAE = \angle PBE$ because A, B, P, E are concyclic.

Now we have $\angle ABQ = \angle AFQ = \angle AEB = 90° - \angle BAE$, which implies $\angle ABQ + \angle BAE = 90°$, i.e., $BQ \perp AE$.

We are to show the circumcenter of $\triangle BPQ$ lies on BC. Let us draw the circumcircle. Refer to the diagram below.

Let the circumcircle of $\triangle BPQ$ intersect BC at R. Now it suffices to show that BR is a diameter, i.e., $\angle BQR = 90°$.

Note that this is equivalent to showing $QR \parallel AE$. We have already shown $\angle CAE = \angle CBP$. Since $\angle CBP = \angle PQR$ (angles in the same arc), we have $\angle CAE = \angle PQR$ and $AE \parallel QR$. This completes the proof.

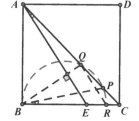

Occasionally, one may need to apply circle properties to solve a problem, even though no circle is given explicitly.

Example 3.1.17   Let P be a point inside $\triangle ABC$ such that $\angle BPC = 90°$ and $\angle BAP = \angle BCP$.   Let M, N be the midpoints of AC, BC respectively.   Show that if $BP = 2PM$, then A, P, N are collinear.

Insight.   We are given a few conditions about the point P. However, neither $\angle BAP = \angle BCP$ nor $BP = 2PM$ seems helpful in determining the position of P. On the other hand, M, N are midpoints. If we can find a triangle where PM is a midline, the Midpoint Theorem will give a line segment equal to $2PM$!

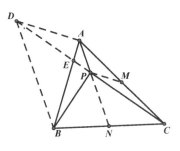

Refer to the diagram above. If we extend CP to D such that $CP = DP$, then $AD = 2PM = BP$. Since A, P, N **should** be collinear, $ADBP$ **should** be an isosceles trapezium, i.e., A, D, B, P **should** be concyclic and we **should** have $\angle BAP = \angle BDP$. Now the condition $\angle BAP = \angle BCP$ seems useful and we may complete the proof by showing that $\triangle BCD$ is isosceles.

Proof. Extend CP to D such that $CP = DP$. Let CD intersect AB at E. Since M, N are the midpoints of AC, BC respectively, by the Midpoint Theorem, we have $AD = 2PM = BP$ and $PN \parallel BD$. (*)
Since $\angle BPC = 90°$, we have $\triangle BCP \cong \triangle BDP$ (S.A.S.). It follows that $\angle BDP = \angle BCP = \angle BAP$ and hence, A, D, B, P are concyclic. Since $AD = BP$, one sees that $\triangle ADE \cong \triangle PBE$ (A.A.S.) and hence, $ADBP$ is an isosceles trapezium where $BD \parallel AP$. By (*), A, P, N are collinear. $\square$

Note: An experienced contestant may write down an elegant proof starting with "Let the circumcircle of $\triangle ABP$ intersect CP extended at D. ..." Of course, beginners may feel puzzled because the motivation of constructing the circumcircle of $\triangle ABP$ is not clear. Nevertheless, by showing $BC = BD$ and $ADBP$ is an isosceles trapezium, one sees that this is equivalent to the given proof.

As shown in the examples above, Corollary 3.1.3 to Corollary 3.1.5, including their inverse, are useful in showing equal angles and concyclicity. One may also use these simple results to show the following theorem.

Theorem 3.1.18 (Simson's Line) *Let P be a point on the circumcircle of $\triangle ABC$. Let D, E, F be the feet of the perpendiculars from P to the lines BC, AC, AB respectively. We have D, E, F collinear, called the Simson's line of $\triangle ABC$ with respect to P.*

Proof. Refer to the diagram on the right. Notice that P, D, C, E are concyclic because $\angle PDC = \angle PEC = 90°$. Hence, we have $\angle 1 = \angle 2$ (Corollary 3.1.3). Notice that $\angle 2 = \angle 3$ (Corollary 3.1.5). Now $\angle 1 = \angle 3 = 180° - \angle PDF$ (Corollary 3.1.4.).
This implies $\angle 1 + \angle PDF = 180°$, or D, E, F are collinear. $\square$

Note:
(1) The inverse of this theorem also holds, i.e., if P is a point such that the feet of its perpendicular to the sides of $\triangle ABC$ are collinear, then

P lies on the circumcircle of $\triangle ABC$. This can be shown by reversing the reasoning: if D, E, F are collinear, we have $\angle 1 + \angle PDF = 180°$. Hence, $\angle 3 = 180° - \angle PDF = \angle 1 = \angle 2$, which implies A, B, C, P are concyclic.

(2) Naturally, beginners may find it difficult to recognize pairs of equal angles, especially when the diagram is complicated. Such angle-chasing skills can only be enhanced via practice. For example, can you see $\angle 1 = \angle 2 = \angle 3$ from the diagram without referring to the proof? (**Hint**: One may occasionally *erase* extra lines and simplify the diagram.)

Example 3.1.19   A quadrilateral $ABCD$ is inscribed inside a circle and $AD \perp CD$. Draw $BE \perp AC$ at E and $BF \perp AD$ at F. Show that the line EF passes through the midpoint of the line segment BD.

Insight.   From the first glance, it is not clear how EF is related to the midpoint of BD. Refer to the diagram on the right. What do we know about the midpoint of BD? One may easily see that BD is the hypotenuse of the right angled triangle $\triangle BDF$. In fact, the only clues we have are the given right angles!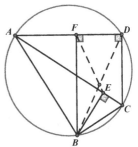
Can we show $\angle EFD = \angle BDF$? This may not be easy because $\angle EFD$ is neither an angle on the circumference nor closely related to other angles.

Perhaps the other right angles can help us. Since $\angle BFD = \angle CDF = 90°$, we see that BD is *almost* the diagonal of a rectangle, except that $BCDF$ is not a rectangle yet while one of the corners is cut. What if we fix it?

Refer to the diagram on the right. We draw $BP \perp CD$ at P. If EF indeed passes through the midpoint of BD, EF should be part of the other diagonal of the rectangle $BPDF$. Indeed, that diagonal is PF and what we need to show is that P, E, F are collinear. Do you recognize a Simson's line?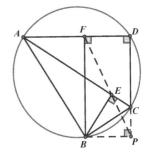

Proof. Draw $BP \perp CD$ at P. Since $AD \perp PD$ and $BF \perp AD$, we have $AD \parallel BP$ and $BF \parallel PD$, i.e., $BPDF$ is a parallelogram (and a rectangle). Since P, E, F are the feet of the perpendiculars from B to the sides of $\triangle ACD$ respectively, we must have P, E, F collinear (Simson's Line). Now the conclusion follows as the diagonals of a parallelogram bisect each other, i.e., EF passes through the midpoint of BD. $\qquad\square$

We mention the following elementary but very useful theorem as the end of this section. It is widely applicable when solving problems related to a few circles intersecting each other.

Theorem 3.1.20 *If $\odot O_1$ and $\odot O_2$ intersect at A, B, then O_1O_2 is the perpendicular bisector of AB.*

Proof. Refer to the diagram on the right. Notice that $\triangle O_1AO_2 \cong \triangle O_1BO_2$ (S.S.S.). $\quad\square$

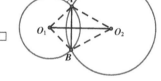

3.2 Tangent of a Circle

Definition 3.2.1 A line AB is tangent to (or touches) a circle $\odot O$ at A if $\angle OAB = 90°$. In this case, A is called the point of tangency.

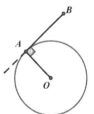

It is easy to see that a tangent line cannot intersect the circle more than once. Otherwise, we will have a triangle with two right angles!

Notice that $\odot O_1$ and $\odot O_2$ are tangent to each other (i.e., touch exactly once) at P if and only if P introduces a common tangent to both circles. Refer to the diagrams below.

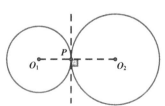
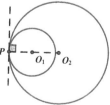

Notice that O_1O_2 is perpendicular to the common tangent in either case. One may consider this as an extreme case of Theorem 3.1.20.

Example 3.2.2   Refer to the left diagram below. The area of the ring between two concentric circles is 16π cm^2. AB is a chord of the larger circle and is tangent to the smaller circle. Find AB.

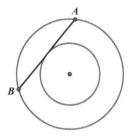 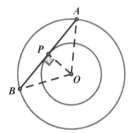

Ans.   Refer to the right diagram above. Let the center of the circles be O and the point of tangency be P. Since $OA = OB$ and $OP \perp AB$, one sees that $\triangle OAP \cong \triangle OBP$ (H.L.). Hence, $AB = 2AP$.

The area of the ring is the difference between the areas of two discs, i.e., $\pi \cdot OA^2 - \pi \cdot OP^2 = 16\pi$. Hence, $16 = OA^2 - OP^2 = AP^2$ by Pythagoras' Theorem. It follows that $AP = 4$ cm and $AB = 8$ cm.   □

Note: If AB is a chord in $\odot O$ and M is the midpoint of AB, we always have $OM \perp AB$ because $\triangle OAB$ is an isosceles triangle.

Theorem 3.2.3   *Let P be a point outside a circle and PA, PB are tangent to the circle at A, B respectively. We have $PA = PB$ (called equal tangent segments).*

Proof.   Refer to the diagram on the right. Connect OA, OB, OP. Since $OA = OB$, one observes that $\triangle PAO \cong \triangle PBO$ (H.L.). The conclusion follows.   □

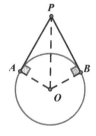

Note: An immediate corollary from the proof above is that $OP \perp AB$. In fact, OP is the perpendicular bisector of AB (Theorem 1.2.4).

We say a circle is *inscribed* inside a polygon if it touches (i.e., is tangent to) every side of the polygon. For example, every triangle has an inscribed circle, called the *incircle* of the triangle, centered at the incenter of the triangle (where angle bisectors meet). Refer to the proof of Theorem 1.3.2.

Example 3.2.4 *ABCD* is a quadrilateral with an inscribed circle. Show that $AB + CD = AD + BC$.

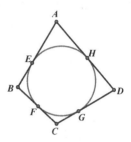

Proof. Refer to the diagram on the right. Let E, F, G, H be the points of tangency. Note that $AE = AH$ (equal tangent segments). Similarly, $BE = BF$, $CF = CG$, $DG = DH$. Now $AB + CD = AE + BE + CG + DG$ $= AH + BF + CF + DH = BC + AD$. □

Note: This is called Pitot's Theorem. However, as the result is simple and well-known, the name of the theorem is seldom mentioned.

Example 3.2.5 *ABCD* is a trapezium with $AD \,/\!/\, BC$ and $\odot O$ is inscribed inside *ABCD*. Show that $AO \perp BO$.

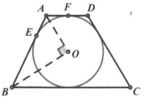

Proof. Refer to the diagram on the right. Let $\odot O$ touch AB, AD at E, F respectively. It is easy to see that $\triangle AOE \cong \triangle AOF$ (H.L.) and hence, AO bisects $\angle BAD$. Similarly, BO bisects $\angle ABC$.

Since $AD \,/\!/\, BC$, $\angle BAD + \angle ABC = 180°$. It follows that

$$\angle BAO + \angle ABO = \frac{1}{2}\angle BAD + \frac{1}{2}\angle ABC = 90°,$$ i.e., $AO \perp BO$. □

Example 3.2.6 A circle is inscribed inside $\triangle ABC$ and it touches the three sides BC, AC, AB at D, E, F respectively. Show that the lines AD, BE, CF are concurrent.

Insight. By Ceva's Theorem, we only need to show $\dfrac{AF}{BF} \cdot \dfrac{BD}{CD} \cdot \dfrac{CE}{AE} = 1$.

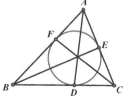

This is true because $AF = AE$, $CE = CD$ and $BD = BF$ (equal tangent segments).

Example 3.2.7 (IWYMIC 10)   A straight line divides a square into two polygons, each of which has an inscribed circle. One of the circles has a radius of 6 cm while the other has an even longer radius. If the line intersects the square at A and B, find the difference, in cm, between the side length of the square and twice the length of the line segment AB.

Ans.   There are a few cases when a line intersects a square.
Case I: Both A, B are vertices of the square.
One obtains two equal triangles and the radii of the inscribed circles must be the same. This contradicts the conditions given.

Case I   Case II   Case III

Case II: Only A is a vertex of the square.
One obtains a triangle and a quadrilateral. Notice that the quadrilateral cannot have an inscribed circle as the two pairs of opposite sides do not have equal sums (Example 3.2.4).

Case III: A, B lie on opposite sides of the square.
Similarly, the quadrilaterals obtained cannot have inscribed circles.

Case IV: A, B lie on neighboring sides of the square.
One obtains a triangle and a pentagon. Notice that the circle inscribed inside the pentagon is exactly the incircle of the square.

Refer to the right diagram below. We focus on the bottom right quarter of the square.

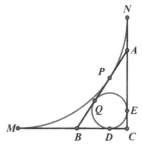

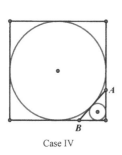

Case IV

The square has a side length $2CM = CM + CN$.

Now $CM + CN - 2AB = CM + CN - (AP + BP) - AB$

$= CM + CN - (AN + BM) - AB = (CM - BM) + (CN - AN) - AB$

$= BC + AC - AB = BC + AC - (AQ + BQ)$

$= BC + AC - (AE + BD) = CD + CE = 12$.

Note that we applied equal tangent segments repeatedly. □

Example 3.2.8 (CGMO 13) In a trapezium $ABCD$, $AD \, // \, BC$. Γ_1 is a circle inside the trapezium and is tangent to AB, AD, CD, touching AD at E. Γ_2 is a circle inside the trapezium and is tangent to AB, BC, CD, touching BC at F. Show that the lines AC, BD, EF are concurrent.

Insight. Refer to the diagram on the right. We know that Ceva's Theorem is useful in showing concurrency, but those three lines given are not inside a triangle. Perhaps we should use another method.

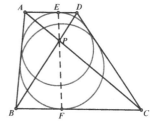

Notice that $ABCD$ is an ordinary trapezium with no special properties. Hence, we shall show that E, F, P are collinear. Can we show that $\dfrac{AE}{CF} = \dfrac{DE}{BF}$? Notice that AE, DE, BF, CF are tangent segments of the circles and they could be expressed by the radii of the circles and the related angles.

Proof. Refer to the diagram on the right. Let Γ_1 be centered at O_1 with the radius $O_1E = R_1$. Let $\angle BAD = 2\alpha$. We have $AE = R_1 \tan \angle O_1AD = R_1 \tan \alpha$. Let $\angle CDA = 2\beta$. $DE = R_1 \tan \beta$.

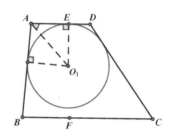

Similarly, if Γ_2 has a radius R_2, we have

$$BF = R_2 \tan\left(\frac{1}{2}\angle ABC\right), \text{ where } \frac{1}{2}\angle ABC = \frac{1}{2}(180° - 2\alpha) = 90° - \alpha.$$

Hence, $BF = R_2 \tan(90° - \alpha)$. Similarly, $CF = R_2 \tan(90° - \beta)$.

Notice that $\tan \alpha \tan(90° - \alpha) = 1$ by definition. Hence, we have

$$AE \cdot BF = R_1R_2 \tan \alpha \tan(90° - \alpha) = R_1R_2. \text{ Similarly, } DE \cdot CF = R_1R_2.$$

It follows that $\dfrac{AE}{CF} = \dfrac{DE}{BF}$, which implies AC, BD, EF are concurrent. □

The following theorem describes the properties of the points of tangency and the radius of the incircle of a triangle.

Theorem 3.2.9 *Let I be the incenter of $\triangle ABC$ where $AB = c$, $AC = b$ and $BC = a$. Let the incircle of $\triangle ABC$ touch BC, AC, AB at D, E, F respectively. We have:*

(1) $BD = \dfrac{1}{2}(a - b + c)$

(2) $DI = \dfrac{2S}{a+b+c}$, *where* $S = [\triangle ABC]$.

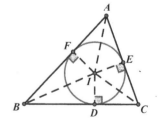

Proof. Refer to the diagram on the right.
(1) By equal tangent segments, $AE = AF = x$ say.
    Similarly, let $BD = BF = y$ and $CD = CE = z$.
    Notice that $a + b + c = 2(x + y + z)$ and $AE + CE = x + z = b$.

Hence, $y = \frac{1}{2}(a+b+c) - b = \frac{1}{2}(a-b+c)$.

(2) Let $DI = EI = FI = r$. Notice that DI, EI, FI are heights of $\triangle BCI$, $\triangle ACI$ and $\triangle ABI$ respectively.

Hence, $S = [\triangle BCI] + [\triangle ACI] + [\triangle ABI] = \frac{1}{2}r \cdot a + \frac{1}{2}r \cdot b + \frac{1}{2}r \cdot c$

$= \frac{1}{2}r \cdot (a+b+c)$. It follows that $r = \dfrac{2S}{a+b+c}$. □

The following is another important circle property. It says the angle between the tangent and chord equals the angle in the alternate segment.

Theorem 3.2.10 *Let AP touch ⊙O at A. B is a point on the circle such that B, P are on the same side of the line OA. Then* $\angle BAP = \frac{1}{2}\angle AOB$.

Proof. Refer to the diagram on the right. Since AP is tangent to ⊙O, we have $OA \perp AP$. Now $\angle BAP = 90° - \angle OAB$. Since $OA = OB$, $\angle AOB = 180° - 2\angle OAB$. It follows that $\angle BAP = \frac{1}{2}\angle AOB$. □

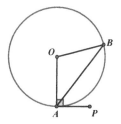

Note: By Theorem 3.1.1, we must have $\angle BAP = \angle ACB$ for any point C on the major arc $\overset{\frown}{AB}$. Refer to the diagram on the right. This is another commonly used result to show equal angles besides Corollaries 3.1.3 to 3.1.5.

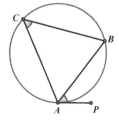

It is easy to see that the inverse of this statement is also true, i.e., if $\angle BAP = \angle ACB$, then AP is tangent to the circle.

Example 3.2.11   Let AB be a diameter of $\odot O$. P is a point outside $\odot O$ such that PB, PC touch $\odot O$ at B and C respectively. Show that $AC \parallel OP$.

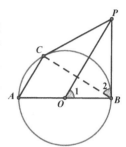

Proof.   Refer to the diagram on the right. It suffices to show $\angle A = \angle 1$. Connect BC. Since PB is tangent to $\odot O$, we have $\angle A = \angle 2$ (Theorem 3.2.10). Since $AB \perp PB$ and $OP \perp BC$ (Theorem 3.2.3), we have $\angle 1 = 90° - \angle OPB = \angle 2$. It follows that $\angle A = \angle 1$.   □

Note: It is a common technique to connect AB if PA, PB are tangent to $\odot O$. Refer to the diagram on the right. By connecting OA, OB, one obtains right angled triangles with the heights on the hypotenuses. Moreover, we also see angles at the center of the circle, tangent lines and equal tangent segments, which, together with other conditions, may help us in finding equal angles.

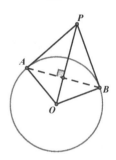

Example 3.2.12   Refer to the diagram on the right. AB is a diameter of $\odot O$ and C, D are two points on the circle. P is a point outside the circle such that PC, PD touch $\odot O$ at C, D respectively.
Show that $\angle CPD = 180° - 2\angle CAD$.

Proof.   Since the sum of the interior angles of the quadrilateral $CODP$ is $360°$ and $\angle OCP = \angle ODP = 90°$, we have $\angle CPD = 180° - \angle COD$. The conclusion follows as $\angle COD = 2\angle CAD$ (Theorem 3.1.1).   □

Note: One sees that the diameter AB is not useful. In particular, the point B complicates the diagram unnecessarily and should be deleted. One may also connect CD and see that $\angle PCD = \angle PDC = \angle CAD$ (Theorem 3.2.10), which also leads to the conclusion.

Example 3.2.13   Given $\odot O$ with radius R, A, B are two points on $\odot O$ and AB is NOT the diameter. C is a point on $\odot O$ distinct from A and B. $\odot O_1$ passes through A and is tangent to the line BC at C. $\odot O_2$ passes through B and is tangent to the line AC at C. If $\odot O_1$ and $\odot O_2$ intersect at C and D, show that $CD \le R$.

Insight.   Refer to the left diagram below. It may not be easy to see the relationship between CD and R immediately. Notice that $OO_1 \perp AC$ and $OO_2 \perp BC$ (Theorem 3.1.20). Given that BC, AC are tangent to $\odot O_1$, $\odot O_2$ respectively, it is easy to see that OO_1CO_2 is a parallelogram!

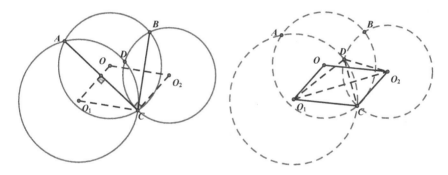

Let us focus on this parallelogram. Refer to the right diagram above. We are to show $CD \le R = CO$. Can you see that CD is *vertical* and CO is *oblique* with respect to O_1O_2? Can you see that $\angle ODC = 90°$?

Proof.   One sees that $OO_1 \perp AC$ and $O_2C \perp AC$. Hence, we have $OO_1 \mathbin{/\!/} O_2C$. Similarly, we have $OO_2 \mathbin{/\!/} O_1C$, which implies that OO_1CO_2 is a parallelogram.

It is easy to see that $\triangle O_1DO_2 \cong \triangle O_1CO_2 \cong \triangle O_1OO_2$, which implies OO_1O_2D is an isosceles trapezium. Hence, we have $OD \mathbin{/\!/} O_1O_2$, which implies $OD \perp CD$. It follows that $CD \le CO = R$.   □

3.3 Sine Rule

Theorem 3.3.1 (Sine Rule) *In* $\triangle ABC$, *we have*

$$\frac{AB}{\sin \angle C} = \frac{BC}{\sin \angle A} = \frac{AC}{\sin \angle B} = 2R, \text{ where } R \text{ is the circumradius of } \triangle ABC.$$

Proof. First, we show that $\dfrac{AB}{\sin \angle C} = 2R$.

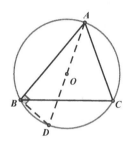

Let O be the circumcenter of $\triangle ABC$. Refer to the diagram on the right. Let AD be a diameter of the circumcircle of $\triangle ABC$. Connect BD.

Clearly, $AD = 2R$ and we have $\angle ABD = 90°$.

By definition, $\dfrac{AB}{2R} = \sin \angle D$. Since $\angle C = \angle D$ (angles in the same arc),

we have $\dfrac{AB}{\sin \angle C} = 2R$. Similarly, $\dfrac{BC}{\sin \angle A} = 2R$ and $\dfrac{AC}{\sin \angle B} = 2R$. $\square$

Note: Sine Rule is taught in most secondary schools. However, the last equality, which links it to the circumradius (i.e., the radius of the circumcircle) of the triangle, is usually not included.

Corollary 3.3.2   *Let* AB, CD *be two chords in a circle. If* AB, CD *extend the same angle at the circumference, then* $AB = CD$.

Proof. Let the radius of the circle be R. Refer to the diagram on the right.

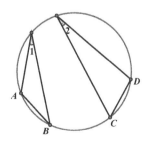

By Sine Rule, we have $\dfrac{AB}{\sin \angle 1} = 2R$, i.e., $AB = 2R \sin \angle 1$. Similarly, $CD = 2R \sin \angle 2$.

The conclusion follows as $\angle 1 = \angle 2$. $\square$

Note:

(1) One sees that the corollary still holds if two chords extend the same angle at the center: Apply Theorem 3.1.1, or simply show that $\triangle AOB \cong \triangle COD$.

(2) The corollary still holds if we are given equal minor arcs $\overarc{AB} = \overarc{CD}$. This is because the arc length is proportional to the angle extended at the center (or on the circumference).

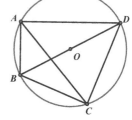

Refer to the diagram on the right, which illustrates a variation of Corollary 3.3.2. $ABCD$ is a quadrilateral inscribed in $\odot O$ where BD is a diameter. We have $AC = 2R\sin \angle D$. Notice that $2R = BD$. Hence, $AC = BD\sin \angle D = BD\sin \angle B$.

This is a useful fact. One shall see this conclusion even if $\odot O$ is not shown explicitly, say if we are only given $AB \perp AD$ and $BC \perp CD$.

Corollary 3.3.3 *Given $\triangle ABC$ and its circumcircle, show that the angle bisector of $\angle A$ passes through the midpoint of the minor arc $\overarc{BC}$.*

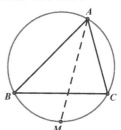

Proof. This follows immediately from the remarks above. Refer to the diagram on the right where AM bisects $\angle A$. One sees that $\overarc{BM} = \overarc{CM}$ because they extend equal angles on the circumference, i.e., $\angle BAM = \angle CAM$.

Example 3.3.4 Refer to the diagram on the right. Two circles intersect at A and B. A common tangent line touches the two circles at M, N respectively. Show that $\triangle MAN$ and $\triangle MBN$ have the same circumradius.

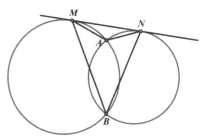

Insight. The two triangles have a common side MN. Sine Rule states that $\dfrac{MN}{\sin \angle MAN} = 2R$ where R is the circumradius of $\triangle AMN$. Can you see that it suffices to show $\sin \angle MAN = \sin \angle MBN$?

Clearly, $\angle MAN \neq \angle MBN$ as one is acute and the other obtuse. How about $\angle MAN + \angle MBN = 180°$? Perhaps the tangent line would give us equal angles.

Proof. Refer to the diagram on the right. We have $\angle 1 = \angle 2$ and $\angle 3 = \angle 4$ (Theorem 3.2.10). Since $\angle 1 + \angle 3 + \angle MAN = 180°$, we must have $\angle 2 + \angle 4 + \angle MAN = 180°$, i.e., $\angle MBN + \angle MAN = 180°$.

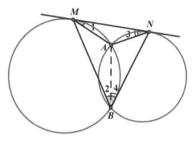

Hence, $\sin \angle MAN = \sin \angle MBN$. Let R_1, R_2 denote the circumradii of the two triangles. By Sine Rule, $\dfrac{MN}{\sin \angle MAN} = 2R_1$ and $\dfrac{MN}{\sin \angle MBN} = 2R_2$.

It follows that $R_1 = R_2$.   □

Example 3.3.5 Given an acute angled triangle $\triangle ABC$ where $\angle A = 60°$, O and H are the circumcenter and orthocenter of $\triangle ABC$ respectively. Show that $AO = AH$.

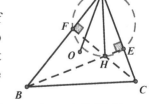

Insight. Refer to the diagram on the right. Of course, the most straightforward method is to show that $\angle AOH = \angle AHO$, but this is not easy because we do not know much about the line OH.

Let $BE \perp AC$ at E and $CF \perp AB$ at F. We know that A, E, H, F are concyclic. In particular, AH is the diameter of this circle (because $\angle AEH = 90°$, Corollary 3.1.13). Now it suffices to show that the radius of the circumcircle of $\triangle ABC$ is twice of the radius of the circumcircle of

$\triangle AEH$. We may show this by Sine Rule. Notice that the right angled triangle with an internal angle of $60°$ gives sides of ratio $1:2$.

Proof. Refer to the right diagram below where BE, CF are the heights in $\triangle ABC$. Since $\angle AEH = \angle AFH = \angle 90°$, A, E, H, F are concyclic. We denote R and r as the radii of the circumcircles of $\triangle ABC$ and $\triangle AEH$ respectively.

By Sine Rule, $\dfrac{AB}{\sin \angle ACB} = 2R$ and $\dfrac{AE}{\sin \angle AFE} = 2r.$

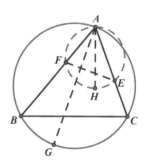

Hence, $\dfrac{R}{r} = \dfrac{AB}{AE} \cdot \dfrac{\sin \angle AFE}{\sin \angle ACB}.$

Notice that B, C, E, F are concyclic, which implies that $\angle ACB = \angle AFE$ (Corollary 3.1.5). We also have $AB = 2AE$ in the right angled triangle $\triangle ABE$ since $\angle A = 60°$ (Example 1.4.8).

Hence, $\dfrac{R}{r} = 2$, or $R = 2r$, which implies the radius of the circumcircle of $\triangle ABC$ equals the diameter of the circumcircle of $\triangle AEH$. Since AH is the diameter of the circumcircle of $\triangle AEH$, we have $AO = AH$. □

Note: Refer to the diagram on the right. Let BE, CF be the heights of $\triangle ABC$. One sees that $\triangle ABC \sim \triangle AEF$. If BE, CF intersect at H and AG is a diameter of the circumcircle of $\triangle ABC$, we must have $\dfrac{AB}{AE} = \dfrac{AG}{AH}$ because these are corresponding line segments with respect to the similar triangles.

In Chapter 1, we learnt the criteria determining congruent triangles, among which S.A.S. requires two pairs of equal sides and one pair of equal angles **between** the sides. Otherwise, we cannot apply S.A.S. Nevertheless, given $\triangle ABC$ and $\triangle A'B'C'$, if $AB = A'B'$, $AC = A'C'$ and $\angle B = \angle B'$, we have either $\angle C = \angle C'$ (which implies $\triangle ABC \cong \triangle A'B'C'$)

or $\angle C = 180° - \angle C'$. This is because Sine Rule gives $\dfrac{AB}{\sin\angle C} = \dfrac{AC}{\sin\angle B}$ $= \dfrac{A'C'}{\sin\angle B'} = \dfrac{A'B'}{\sin\angle C}$ and hence, $\sin\angle C = \sin\angle C'$, which implies either $\angle C = \angle C'$ or $\angle C = 180° - \angle C'$.

Example 3.3.6 (CGMO 03)   In a non-isosceles triangle $\triangle ABC$, AD, BE, CF are angle bisectors of $\angle A, \angle B, \angle C$ respectively, intersecting BC, AC, AB at D, E, F respectively. Show that if $DE = DF$, then $\dfrac{a}{b+c} = \dfrac{b}{a+c} + \dfrac{c}{a+b}$, where $BC = a$, $AC = b$ and $AB = c$.

Insight.   Refer to the diagram on the right. It is given that AD bisects $\angle A$ and $DE = DF$. Consider $\triangle ADE$ and $\triangle ADF$. We have either $\angle AED = \angle AFD$ or $\angle AED = 180° - \angle AFD$.

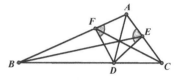

If $\angle AED = \angle AFD$, we have $\triangle ADE \cong \triangle ADF$ and it seems the diagram is symmetric about AD, probably contradicting the fact that $\triangle ABC$ is non-isosceles. (Show it!) Perhaps we should work with the condition that $\angle AED = 180° - \angle AFD$.

Notice that the conclusion is about the ratio $\dfrac{a}{b+c}$, $\dfrac{b}{a+c}$ and $\dfrac{c}{a+b}$. Is it reminiscent of the Angle Bisector Theorem? For example, $CE = \dfrac{ab}{a+c}$ and $BF = \dfrac{ac}{a+b}$ (Example 2.3.8). In fact, CE and BF are the only choices related to $\dfrac{b}{a+c}$ and $\dfrac{c}{a+b}$. Perhaps we can show that $CE + BF$ equals to a length of $\dfrac{a^2}{b+c}$.

However, CE and BF are far apart. Can we put them together? Since $DE = DF$ and $\angle AED = 180° - \angle AFD$, we may rotate $\triangle BDF$ so that BF and CE are on the same line.

Proof. By Sine Rule, $\dfrac{\sin \angle AFD}{\sin \angle DAF} = \dfrac{AD}{DF} = \dfrac{AD}{DE} = \dfrac{\sin \angle AED}{\sin \angle DAE}$ because $DE = DF$. Since $\angle DAE = \angle DAF$, we have $\sin \angle AFD = \sin \angle AED$, i.e., either $\angle AFD = \angle AED$ or $\angle AFD + \angle AED = 180°$.

If $\angle AFD = \angle AED$, we immediately have $\triangle ADF \cong \triangle ADE$ (A.A.S.) and hence, $AE = AF$. Notice that $AE = \dfrac{bc}{a+c}$ and $AF = \dfrac{bc}{a+b}$. It follows that $b = c$, or $AC = AB$, contradicting the fact that $\triangle ABC$ is non-isosceles. Hence, $\angle AFD \neq \angle AED$. We have $\angle AFD + \angle AED = 180°$.

Now $\angle CED = \angle AFD$ and hence, we may choose P on CE extended such that $\angle CPD = \angle ABC$. Refer to the diagram on the right. It is easy to see that $\triangle DEP \cong \triangle DFB$ (A.A.S.).

Hence, $PE = BF = \dfrac{ac}{a+b}$.

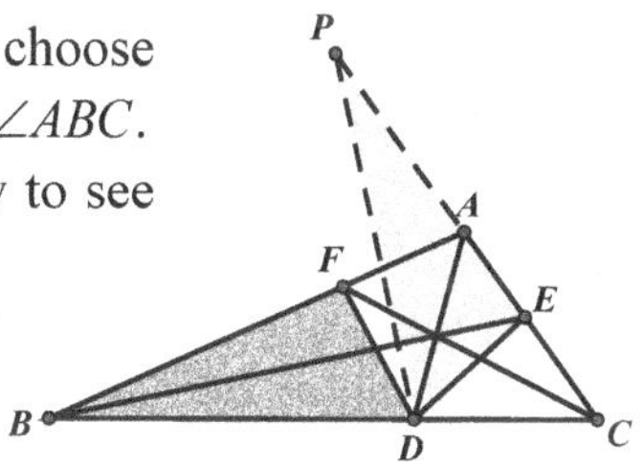

Since $CE = \dfrac{ab}{a+c}$, it suffices to show that $PE + CE = PC = \dfrac{a^2}{b+c}$.

Since $\angle CPD = \angle ABC$, we have $\triangle PCD \sim \triangle BCA$.

Hence, $\dfrac{PC}{BC} = \dfrac{CD}{AC}$, which gives $PC = BC \cdot \dfrac{CD}{AC} = a \cdot \dfrac{\frac{ab}{b+c}}{b} = \dfrac{a^2}{b+c}$.

This completes the proof. □

Example 3.3.7 In a non-isosceles acute angled triangle $\triangle ABC$, BE, CF are heights on AC, AB respectively. Let D be the midpoint of BC. The angle bisectors of $\angle BAC$ and $\angle EDF$ intersect at P. Show that the circumcircles of $\triangle BFP$ and $\triangle CEP$ has an intersection on BC.

Insight. Refer to the (simplified) diagram on the right. How can we show the concurrency of two circles and a line? Perhaps we can show that X, the intersection of the two circles, lie on BC, i.e., B, C, X are collinear. Thus, it suffices to show $\angle BXP + \angle CXP = 180°$.

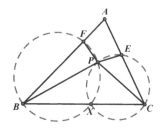

We do not know many properties of X, but given the circles, we know $\angle BXP = \angle AFP$ and $\angle CXP = \angle AEP$, where $\angle AEP$ and $\angle AFP$ are inside the quadrilateral $AEDF$ and the angle bisectors may give useful properties of those angles. Now we are to show $\angle AFP + \angle AEP = 180°$. One may attempt to show A, E, P, F are concyclic, but it could be difficult (*) because we do not know much about the angles except for AP bisecting $\angle EAF$. How about considering $\triangle AEP$ and $\triangle AFP$? The angle bisector AP could be useful if we apply Sine Rule, which gives

$$\frac{AP}{\sin\angle AFP} = \frac{PF}{\sin\angle FAP} \text{ and } \frac{AP}{\sin\angle AEP} = \frac{PE}{\sin\angle EAP}.$$

Since $\angle EAP = \angle FAP$ and we **should** have $\sin\angle AFP = \sin\angle AEP$, it seems we are to show $PE = PF$. Notice that $DE = DF$ (Example 1.4.7) and P is on the angle bisector of $\angle EDF$.

(*) One familiar with commonly used facts in circle geometry could see that if we are to show A, E, P, F are concyclic, it suffices to show $PE = PF$ (Example 3.1.11).

Proof. In the right angled triangle $\triangle BCE$,

$$DE = \frac{1}{2}BC. \text{ Similarly, } DF = \frac{1}{2}BC = DE.$$

Refer to the diagram on the right. Since DP bisects $\angle EDF$, we have $\triangle DPE \cong \triangle DPF$ (S.A.S.) and hence, $PE = PF$. Apply Sine Rule to $\triangle AFP$ and $\triangle AEP$:

$$\frac{AP}{\sin\angle AFP} = \frac{PF}{\sin\angle FAP} \text{ and } \frac{AP}{\sin\angle AEP} = \frac{PE}{\sin\angle EAP}. \text{ Since } AP \text{ is the}$$

angle bisector of $\angle EAF$, we must have $\sin\angle AFP = \sin\angle AEP$.

Case I: $\angle AFP = \angle AEP$

We have $\triangle AFP \cong \triangle AEP$ (A.A.S.) and hence, $AE = AF$. This implies $\triangle ABE \cong \triangle ACF$ (A.A.S.) and hence, $AB = AC$. This contradicts the fact that $\triangle ABC$ is non-isosceles.

Case II: $\angle AFP = 180° - \angle AEP$

Let the circumcircle of $\triangle BFP$ intersect BC at X. We must have $\angle BXP = \angle AFP = 180° - \angle AEP = \angle CEP$. Hence, C, E, P, X are concyclic, i.e., X lies on the circumcircle of $\triangle CEP$. This completes the proof. □

Example 3.3.8 Refer to the diagram on the right. $\odot O_1$ touches $\odot O_2$ at Q. BC is tangent to $\odot O_1$ at P. Show that if $\angle BAO_1 = \angle CAO_1$, then $\angle PAO_1 = \angle QAO_1$.

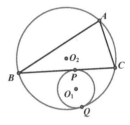

Insight. Refer to the left diagram below. Let AO_1 extended intersect $\odot O_2$ at M. Since $\angle BAO_1 = \angle CAO_1$, M is the midpoint of $\overarc{BC}$. Hence, $O_2M \perp BC$. We are to show $\angle PAO_1 = \angle QAO_1$. Notice that $O_1P = O_1Q$. We **should** have A, P, O_1, Q concyclic (Example 3.1.11). How can we show this? Notice that $O_1P \perp BC$, i.e., $O_1P // O_2M$. Perhaps the concyclicity and the parallel lines could give us equal angles.

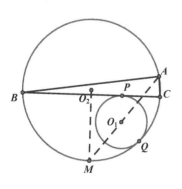

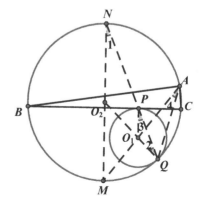

Proof. Refer to the previous right diagram. Let AO_1 extended intersect $\odot O_2$ at M. Since $\angle BAO_1 = \angle CAO_1$, M is the midpoint of $\overset{\frown}{BC}$ and hence, $O_2M \perp BC$. (*)

Let MN be a diameter of $\odot O_2$. We are given that $\odot O_1$ touches $\odot O_2$ at Q. Hence, O_2 lies on QO_1 extended. Since $O_1P \perp BC$, we must have $O_1P \,/\!/ \, MN$. Now the isosceles triangles $\triangle O_1PQ$ and $\triangle O_2NQ$ are similar and we must have $\angle 1 = \angle 2 = \angle 3$ where P, Q, N are collinear. Notice that $\angle 1 = \angle 4$ (angles in the same arc). We have $\angle 3 = \angle 4$ and hence, A, P, O_1, Q are concyclic. The conclusion follows as $O_1P = O_1Q$ (Corollary 3.3.2). □

Note: (*) Since $O_2B = O_2C$ and $BM = CM$, O_2M is the perpendicular bisector of BC (Theorem 1.2.4). It is a simple but useful technique to introduce a perpendicular bisector of a chord, which passes through the center of the circle. We will illustrate this technique more in Chapter 5.

3.4 Circumcenter, Incenter and Orthocenter

We have learned the basic properties of the circumcenter, incenter and orthocenter of a triangle. In this section, we will study a few results related to these special points of a triangle using circle geometry techniques. These results are important and frequently referred to as lemmas in various competitions.

Example 3.4.1   Let O be the circumcenter of an acute angled triangle $\triangle ABC$. If $AD \perp BC$ at D, show that $\angle CAD = \angle BAO$.

Proof.   Refer to the diagram on the right. Recall that $\angle BAO = 90° - \angle C$ (Example 3.1.2). Clearly $\angle CAD = 90° - \angle C$. The conclusion follows.   □

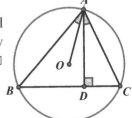

Note: We also have $\angle CAO = \angle BAD$.

Example 3.4.2 Let I be the incenter of $\triangle ABC$. If BI extended intersects the circumcircle of $\triangle ABC$ at P, show that $AP = CP = PI$.

Insight. One sees that $AP = CP$ follows directly from Corollary 3.3.3. To show $AP = PI$, we may consider showing $\angle AIP = \angle PAI$, as there are many equal angles in the diagram due to the incenter (i.e., angle bisectors) and the circumcircle.

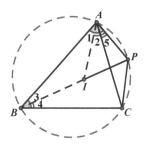

Proof. Refer to the diagram above. Since BP bisects $\angle B$, we have $AP = CP$ by Corollary 3.3.3. In $\triangle ABI$, we have the exterior angle $\angle AIP = \angle 1 + \angle 3$. On the other hand, $\angle PAI = \angle 2 + \angle 5$ where $\angle 1 = \angle 2$ and $\angle 3 = \angle 4 = \angle 5$ (angles in the same arc). Now $\angle AIP = \angle 1 + \angle 3 = \angle 2 + \angle 5 = \angle PAI$. Hence, $AP = PI$. This completes the proof. □

Example 3.4.3 Let H be the orthocenter of an acute angled triangle $\triangle ABC$. Let D be the foot of the perpendicular from A to BC. If AD extended intersects the circumcircle of $\triangle ABC$ at E, show that $DH = DE$.

Insight. Refer to the diagram on the right. Given that $BD \perp AE$, since we are to show $DH = DE$, we **should** have $\triangle BEH$ isosceles, i.e., $BE = BH$. Both the circumcircle and the orthocenter give equal angles. Hence, one may show that $\angle CBH = \angle CBE$.

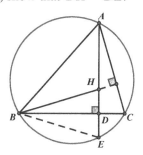

Proof. Notice that $\angle CBH = 90° - \angle BHD = \angle CAE$. Since $\angle CAE = \angle CBE$ (angles in the same arc), $\angle CBH = \angle CBE$. The conclusion follows as $\triangle DBH \cong \triangle DBE$ (A.A.S.). □

Example 3.4.4 Let H be the orthocenter of an acute angled triangle $\triangle ABC$. Let M be the midpoint of BC. If HM extended intersects the circumcircle of $\triangle ABC$ at A', show that:
(1) $HBA'C$ is a parallelogram
(2) AA' is a diameter of the circumcircle of $\triangle ABC$.

Insight. (1) follows from Example 2.5.5 and Example 1.4.3.
(2): It suffices to show that either $\angle ABA'$ or $\angle ACA'$ is $90°$.

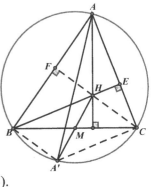

Proof. Refer to the diagram on the right.
(1) Since H is the orthocenter of $\triangle ABC$, we must have $\angle BHC = 180° - \angle BAC$ (Example 2.5.5).
Hence, $\angle BHC = \angle BA'C$ (Corollary 3.1.4).
Consider the quadrilateral $A'BHC$. Since $BM = CM$, we conclude that $A'BHC$ is a parallelogram (Example 1.4.3).
(2) Since $CH \perp AB$ and by (1), $CH \,/\!/\, A'B$, we must have $A'B \perp AB$, i.e., $\angle ABA' = 90°$. The conclusion follows.   □

Example 3.4.5 In an acute angled triangle $\triangle ABC$, BD, CE are heights. If the line DE intersects the circumcircle of $\triangle ABC$ at P, Q respectively, show that $AP = AQ$.

Insight. One may show $\angle APQ = \angle AQP$ since there are many equal angles due to the circles. Notice that B, C, D, E are concyclic.

Proof. Refer to the diagram on the right. Since $\angle BDC = \angle BEC = 90°$, we must have B, C, D, E concyclic. Hence, $\angle 1 = \angle C$.
Now $\angle APQ = \angle 1 - \angle 2 = \angle C - \angle 3$ because $\angle 2 = \angle 3$ (angles in the same arc). On the other hand, $\angle AQP = \angle AQB - \angle 3$. Since $\angle C = \angle AQB$, we conclude that $\angle APQ = \angle AQP$ and hence, $AP = AQ$.   □

Notice that the argument still applies whenever B, C, D, E are concyclic: it is not necessary that BD, CE are heights of $\triangle ABC$.

Example 3.4.1 to Example 3.4.5 are very useful results. One familiar with these results may find it much easier to see the insight when solving geometry problems related to the circumcenter, incenter and orthocenter of a triangle.

Example 3.4.6   Let O and H be the circumcenter and orthocenter of an acute angled triangle $\triangle ABC$ respectively. Let M be the midpoint of BC. Show that $AH = 2OM$.

Insight.   Refer to the diagram on the right. We do not know much about the properties of AH or how it is related to OM. For example, it is not easy to find a line segment with length $\dfrac{1}{2}AH$.

However, one sees that OM is related to $\dfrac{1}{2}$: M is the midpoint of BC and O is the midpoint of a diameter of the circle. If we draw the diameter BD, we immediately have $OM = \dfrac{1}{2}CD$. Now it suffices to show that $AH = CD$.

Recall that Example 3.4.4 states that $ADCH$ is a parallelogram, which completes the proof. (Beginners may spend a while to see how Example 3.4.4 is applied in the diagram above.) We leave the details to the reader.

Note: Since M is the midpoint of BC, AM is a median of $\triangle ABC$ and G, the centroid of $\triangle ABC$, lies on AM.

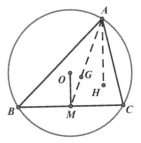

We have $\dfrac{AH}{OM} = \dfrac{AG}{GM} = \dfrac{2}{1}$ (Midpoint Theorem).

It follows that O, G, H are collinear (because $\triangle AGH \sim \triangle MGO$), which is called the Euler Line of $\triangle ABC$.

Example 3.4.7   Given $\triangle ABC$ and it circumcircle, P, Q, R are midpoints of minor arcs $\overarc{BC}$, $\overarc{AC}$ and $\overarc{AB}$ respectively. If PR intersects AB at D and PQ intersects AC at E, show that $DE \parallel BC$.

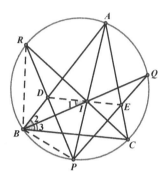

Insight.   It is easy to see that AP, BQ, CR are angle bisectors of $\triangle ABC$. Recall Example 3.4.2 which is about angle bisectors intersecting the circumcircle. Can you see $BP = PI$? (AI extended intersects the circumcircle at P.) Similarly, $BR = RI$. Hence, PR must be the perpendicular bisector of BI. Refer to the diagram on the right.

This implies $\angle 1 = \angle 2$. It is given that $\angle 2 = \angle 3$. Hence, $\angle 1 = \angle 3$, i.e., $DI \parallel BC$. A similar argument gives $EI \parallel BC$, which implies D, I, E collinear and $DE \parallel BC$.

Alternatively, one may solve the problem without applying Example 3.4.2. Notice that angle bisectors in a circle give a lot of equal angles. Refer to the left diagram below. One sees that $\angle 1 = \angle 2 = \angle 3$ (angles in the same arc). This implies A, I, D, R are concyclic. Refer to the right diagram below. Now $\angle 4 = \angle 5 = \angle 6$ (angles in the same arc), which implies $DI \parallel BC$. A similar argument gives $EI \parallel BC$. The conclusion follows.

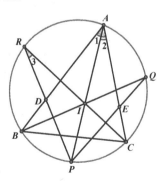
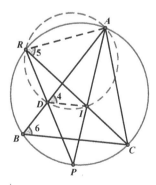

Note:
(1) The first method is also an illustration of the relationship among the angle bisector, parallel lines and the isosceles triangle.
(2) It is important to draw the diagram properly. One may see the incenter appears between D and E, giving an inspiration that D, I, E might be collinear.

Example 3.4.8 (TUR 09)   In an acute angled triangle $\triangle ABC$, D, E, F are the midpoints of BC, CA, AB respectively. Let H and O be the orthocenter and the circumcenter of $\triangle ABC$ respectively. Extend HD, HE, HF to intersect the circumcircle of $\triangle ABC$ at A', B', C' respectively. Let H' be the orthocenter of $\triangle A'B'C'$. Show that O, H and H' are collinear.

Insight.   A well-constructed diagram is important. Refer to the diagram on the right. One may see that $\triangle ABC$ and $\triangle A'B'C'$ are highly symmetric by a rotation of $180°$. If we can show this is true, it is not far away from the conclusion.

On the other hand, the orthocenter and the midpoints remind us of Example 3.4.4, which states that AA' is a diameter of the circumcircle. Similarly, BB' and CC' are also diameters. Now it is not difficult to show that $\triangle ABC$ and $\triangle A'B'C'$ are symmetric about O, the center of the circumcircle.

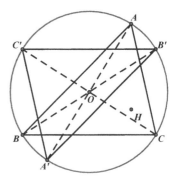

Proof.   By Example 3.4.4, we conclude that AA', BB', CC' are the diameters of $\odot O$, the circumcircle of $\triangle ABC$. Refer to the diagram on the right. Since AA' and BB' bisect each other, we conclude that $ABA'B'$ is a parallelogram (and in fact, a rectangle).   Hence, $AB = A'B'$ and $AB \,/\!/\, A'B'$.

Similarly, we have $BC = B'C'$, $BC // B'C'$ and $AC = A'C'$. It follows that $\triangle ABC \cong \triangle A'B'C'$ (S.S.S.).

Refer to the diagram on the right. We claim that $AHA'H'$ is a parallelogram. Since $\triangle ABC \cong \triangle A'B'C'$, we must have $AH = A'H'$ because H and H' are corresponding points in $\triangle ABC$ and $\triangle A'B'C'$ respectively. Since AH and $A'H'$ are heights and $BC // B'C'$, we have $AH // A'H'$. Hence, $AHA'H'$ is a parallelogram and HH' must pass through O, the midpoint of AA'.   □

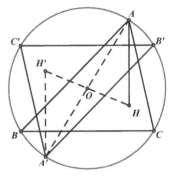

Example 3.4.9 (IMO 10)   In an acute angled triangle $\triangle ABC$, AD, BE, CF are heights. EF extended intersects the circumcircle of $\triangle ABC$ at P. BP extended and DF extended intersect at Q. Show that $AP = AQ$.

Proof.   Let the line EF intersect the circumcircle of $\triangle ABC$ at P, P'. Refer to the diagram on the right. By Example 3.4.5, $AP = AP'$. It suffices to show that $AP' = AQ$. Notice that $\angle ABP = \angle AP'P$ $= \angle APP' = \angle ABP'$, i.e., BA is the angle bisector of $\angle P'BQ$. We also have $\angle BFP = \angle AFE = \angle BFD$ (Example 3.1.6).

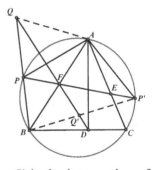

It follows that $\triangle FPB \cong \triangle FQ'B$ (A.A.S.), where Q' is the intersection of BP' and DF. We conclude that $AP' = AQ$ (Exercise 1.10, or simply by congruent triangles). This completes the proof.   □

Note:

(1) Notice that $\angle ABP = \angle ABP'$ and $\angle BFP = \angle BFD$ imply P and Q' are symmetric about the line AB, and so are P' and Q. Hence, $AP' = AQ$. Such an argument based on symmetry is acceptable in competitions. However, beginners are recommended to write down a complete argument via congruent triangles.

(2) Notice that A, F, P, Q are concyclic since $\angle APQ = \angle ACB$ $= \angle AFP' = \angle AFQ$. Hence, one may show the conclusion by applying Sine Rule to $\triangle AFQ$ and $\triangle AFP$. (Can you show it?)

3.5 Nine-point Circle

First, we shall attempt the following examples.

Example 3.5.1   Let AB be the diameter of the semicircle centered at O. P is a point outside the semicircle and PC, PD are tangent to the semicircle at C, D respectively. If the chords AC, BD intersect at E, show that $PE \perp AB$.

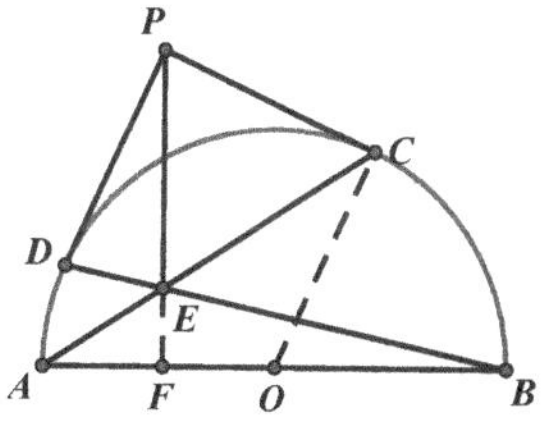

Insight.   Of course, the most straightforward method is to show that $\angle A + \angle AEF = 90°$. Refer to the diagram on the right, where PE extended intersects AB at F.

Since $OA = OC$ and $OC \perp PC$, we have $\angle A = \angle OCA = 90° - \angle PCE$. On the other hand, $\angle AEF = \angle PEC$. Hence, we **should** have $PE = PC$. Similarly, we **should** have $PD = PE$, i.e., $PC = PD = PE$. This implies that P **should** be the circumcenter of $\triangle CDE$. Can we show it?

If P is the circumcenter of $\triangle CDE$, Theorem 3.1.1 and Corollary 3.1.4 imply that $\angle P = 2 \cdot (180° - \angle CED)$. Can we show this, or equivalently, $\angle CED + \dfrac{1}{2} \angle P = 180°$? Notice that in the isosceles triangle $\triangle PCD$,

$$180° - \frac{1}{2} \angle P = 90° + \angle PCD.$$

Proof.   We claim that P is the circumcenter of $\triangle CDE$. Notice that $\angle CED = \angle BCE + \angle CBE$, where $\angle BCE = 90°$ (AB is the diameter) and $\angle CBE = \angle PCD$ (Theorem 3.2.10). Hence, $\angle CED = 90° + \angle PCD$.   (1)

In the isosceles triangle $\triangle PCD$, $\angle PCD = 90° - \dfrac{1}{2} \angle P$.   (2)

(1) and (2) give $\angle CED + \dfrac{1}{2}\angle P = 180°$, or $\angle P = 2 \cdot (180° - \angle CED)$.

Since $\angle P$ is twice the supplementary angle of $\angle CDE$ and $PC = PD$, we claim that P is the circumcenter of $\triangle CDE$. Otherwise, say O is the circumcenter of $\triangle CDE$, we must have $\angle O = 2 \cdot (180° - \angle CED) = \angle P$. Notice that O and P both lie on the perpendicular bisector of CD, and they are on the same side of CD because $\angle CED$ is obtuse. This is impossible.

In conclusion, P is the circumcenter of $\triangle CDE$ and hence, $PC = PD = PE$. It follows that $\angle A + \angle AEF = \angle ACO + \angle PEC = \angle ACO + \angle PCE = 90°$, i.e., $PE \perp AB$. $\qquad\square$

Example 3.5.2 (CWMO 10) Let AB be the diameter of the semicircle centered at O. P is a point outside the semicircle and PC, PD are tangent to the semicircle at C, D respectively. If the chords AC, BD intersect at E, and PE extended intersects AB at F, show that P, C, F, D are concyclic.

Proof. Refer to the diagram on the right. Clearly, P, C, O, D are concyclic because $OC \perp PC$ and $OD \perp PD$. We also have P, D, F, O concyclic since $PF \perp AB$ (Example 3.5.1). Now P, D, F, O, C are concyclic. $\qquad\square$

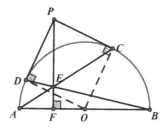

We shall review the diagrams in Example 3.5.1 and Example 3.5.2. Suppose AD extended and BC extended intersect at X. Since $BD \perp AX$ and $AC \perp BX$, E is indeed the orthocenter of $\triangle ABX$, i.e., $XE \perp AB$. Since $PE \perp AB$, X, P, E, F are collinear. Refer to the diagram on the left.

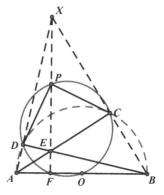

Example 3.5.2 states that P, D, F, O, C are concyclic. In fact, we may remove the semicircle centered at O and focus on $\triangle ABX$. Refer to the diagram on the right. C, D, F are the feet of the altitudes in $\triangle ABX$ and the circumcircle of $\triangle CDF$ passes through O, the midpoint of AB. Similarly, this circle should pass through the midpoints of AX, BX as well.

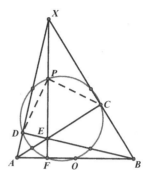

On the other hand, since $PC = PD = PE$, one can show that P is the midpoint of XE. (**Hint:** Consider the right angled triangle $\triangle XDE$. Apply Exercise 1.1.) By similar arguments, we see that the circumcircle of $\triangle ABC$ must pass through the midpoints of AE, BE as well. This circle is called the nine-point circle of $\triangle ABC$.

Theorem 3.5.3 (Nine-point Circle) *In any triangle, the following nine points are concyclic: the midpoints of the three sides, the feet of the three altitudes and the midpoints of the line segments connecting each vertex to the orthocenter of the triangle.*

As shown above, one may derive this result from Example 3.5.2. The following is an alternative proof.

Proof. Refer to the diagram on the right. Let D, E, F be the feet of the altitudes on BC, AC, AB respectively, L, M, N be the midpoints of BC, AC, AB respectively and P, Q, R be the midpoints of AH, BH, CH respectively, where H is the orthocenter of $\triangle ABC$.

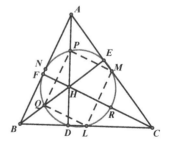

Notice that PM is a midline in $\triangle AHC$, i.e., $PM = \dfrac{1}{2}CH$ and $PM \,/\!/\, CH$.

Similarly, QL is a midline in $\triangle BCH$: $QL = \dfrac{1}{2}CH$ and $QL \,/\!/\, CH$.

Hence, $PMLQ$ is a parallelogram.

We also notice that PQ is a midline in $\triangle ABH$ and $PQ /\!/ AB$. Since $CH \perp AB$ and $CH /\!/ PM$, we have $PM \perp PQ$. This implies that $PMLQ$ is a rectangle and hence, P, M, L, Q are concyclic.

Similarly, $PRLN$ is a rectangle and we have $\angle PNL = 90° = \angle PQL$. Hence, N lies on the circumcircle of $\triangle PQL$. By similar arguments, we conclude that P, M, R, L, Q, N are concyclic. Refer to the left diagram below.

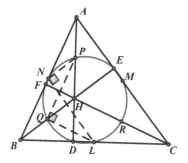 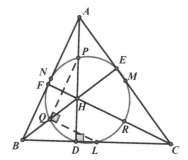

On the other hand, $\angle PDL = 90° = \angle PQL$, which implies D lies on the circumcircle of $\triangle PQL$. Similarly, E, F also lie on the circumcircle of $\triangle PQL$. Refer to the right diagram above.

In conclusion, $P, M, R, L, Q, N, D, E, F$ are concyclic.   □

Note: Since $\angle PDL = 90°$, PL is a diameter of the nine-point circle. Hence, the midpoint of PL is the center of the nine-point circle. In particular, the lines PL, QM, RN are concurrent (since they all pass through the center of the nine-point circle).

Notice that the nine-point circle of a triangle could be determined by any three of the nine points, among which the most commonly seen ones are midpoints and feet of altitudes. Recall Example 3.1.15. Can you see that P lies on the nine-point circle of $\triangle ABC$? (**Hint**: Show that $\angle P = \angle BAC = \angle MDN$. Now P lies on the circumcircle of $\triangle DMN$, which is indeed the nine-point circle of $\triangle ABC$.)

3.6 Exercises

1. (a) Given a parallelogram $ABCD$, show that $ABCD$ is cyclic if and only if it is a rectangle.
   (b) Given a trapezium $ABCD$, show that $ABCD$ is cyclic if and only if it is an isosceles trapezium.

2. Let $ABCD$ be a trapezium with $AD \mathbin{/\mkern-5mu/} BC$. Let E, F be on AB, CD respectively such that $\angle BAF = \angle CDE$. Show that $\angle BFA = \angle CED$.

3. In $\triangle ABC$, I is the incenter and J is the ex-center opposite B. Show that A, I, C, J are concyclic.

4. Let AB be the diameter of a semicircle. Let the chords AC, BD intersect at P. Draw $PE \perp AB$ at E. Show that P is the incenter of $\triangle CDE$.

5. Let P be a point outside $\odot O$ and PA, PB are tangent to $\odot O$ at A, B respectively. Show that the incenter of $\triangle PAB$ is the midpoint of $\overarc{AB}$.

6. Let $\triangle ABC$ be an acute angled triangle, where O, H are the circumcenter and the orthocenter respectively.
   (a) If B, C, O, H are concyclic, find $\angle A$.
   (b) Show that the circumcircles of $\triangle ABC$ and $\triangle BCH$ have the same radius.

7. Given $\triangle ABC$ and its circumcircle $\odot O$, D is the midpoint of BC and DO extended intersects AB at M. P is a point outside $\odot O$ such that PA, PB are tangent to $\odot O$ at A, B respectively. Show that $PM \mathbin{/\mkern-5mu/} BC$.

8. **(CGMO 07)** Let D be a point inside $\triangle ABC$ such that $\angle DAC = \angle DCA = 30°$ and $\angle DBA = 60°$. Let E be the midpoint of BC and F be a point on AC such that $AF = 2FC$, show that $DE \perp EF$.

9. Given $\triangle ABC$ where $\angle A > 90°$, its circumcenter and orthocenter are O and H respectively. Draw $\odot O_1$ where CH is a diameter. $\odot O_1$ and $\odot O$ intersect at C and D. If HD extended intersects AB at M, show that $AM = BM$.

10. Refer to the diagram on the right. Let AB be the diameter of a semicircle and C be a point on AB. Draw two semicircles with diameters AC, BC respectively.

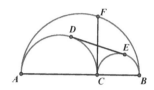

Let D, E be points on these two semicircles respectively such that DE is a common tangent. Draw $CF \perp AB$, intersecting the large semicircle at F. Show that $CDFE$ is a rectangle.

11. Given $\triangle ABC$ where $\angle B = 2\angle C$, D is a point on BC such that AD bisects $\angle A$. Let I be the incenter of $\triangle ABC$, show that the circumcenter of $\triangle CDI$ lies on AC.

12. (CZE-SVK 10) In a right angled triangle $\triangle ABC$ where $\angle A = 90°$, P, Q, R are on the side BC such that $BP = PQ = QR = RC = \dfrac{1}{4}BC$. The circumcircles of $\triangle ABP$ and $\triangle ACR$ intersect at A and M. Show that A, M, Q are collinear.

13. In an acute angled triangle $\triangle ABC$, AD, BE are the heights. Let A' be the reflection of A about the perpendicular bisector of BC and B' be the reflection of B about the perpendicular bisector of AC. Show that $A'B' \parallel DE$.

14. Let I be the incenter of $\triangle ABC$. Show that the circumcenter of $\triangle BIC$ lies on the circumcircle of $\triangle ABC$.

15. Given $\triangle ABC$, its incenter I and ex-centers J_1, J_2, J_3, show that the midpoints of the line segments $IJ_1, IJ_2, IJ_3, JJ_1, JJ_2, JJ_3$ all lie on the circumcircle of $\triangle ABC$.

16. Let $AXYZB$ be a convex pentagon inscribed in a semicircle centered at O with the diameter AB. Let P, Q, R and S denote the feet of the perpendiculars from point Y to the lines AX, BX, AZ and BZ respectively. Let PQ and RS intersect at C. Show that $\angle PCS = \dfrac{1}{2} \angle XOZ$.

17. (CHN 06) Let $ABCD$ be a trapezium such that $AD /\!/ BC$. Γ_1 is a circle tangent to the lines AB, CD, AD and Γ_2 is a circle tangent to the lines AB, BC, CD. Let ℓ_1 be the tangent line from A to Γ_2 (different from AB) and ℓ_2 be the tangent line from C to Γ_1 (different from CD). Show that $\ell_1 /\!/ \ell_2$.

Chapter 4

Circles and Lines

In Chapter 3, we learnt various properties about angles in circles. Indeed, one may also find important properties about line segments when straight lines intersect (or touch) a circle, or when triangles and quadrilaterals are inscribed in circles. We will study these properties in this chapter.

4.1 Circles and Similar Triangles

We have seen in Chapter 3 that straight lines intersecting a circle give equal angles. Hence, similar triangles could be constructed via circles. We will see a number of examples of circles and similar triangles in this section. Notice that one needs to be familiar with both circle and similar triangle properties in order to solve such problems.

Example 4.1.1 Refer to the diagram on the right. Γ_1 and Γ_2 are two circles touching each other at A. AB is a chord in Γ_1, intersecting Γ_2 at D. BC is a chord in Γ_1 which is tangent to Γ_2 at E. AE extended intersects Γ_1 at F.

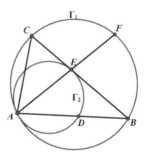

Show that $AB \cdot AC = AE \cdot AF$.

Insight. Given two circles and two tangent lines (including a common tangent of the two circles), one should be able to see many pairs of equal angles. Since the conclusion is equivalent to $\dfrac{AB}{AF} = \dfrac{AE}{AC}$, we may show it by similar triangles, for example, $\triangle ABE \sim \triangle AFC$.

It is easy to see that $\angle ABE = \angle AFC$. Hence, we **should** have $\triangle ABE \sim \triangle AFC$. Can we show it by finding another pair of equal angles?

Proof. Refer to the diagram on the right.

Let AC intersect Γ_2 at P. Connect PE, CF and draw a common tangent of Γ_1 and Γ_2 at A. Since BC is tangent to Γ_2 at E, we have $\angle AEB = \angle APE = \angle 1 = \angle ACF$ by applying Theorem 3.2.10 repeatedly.

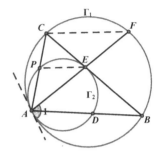

Since $\angle B = \angle F$ (angles in the same arc), we have $\triangle ABE \sim \triangle AFC$.

It follows that $\dfrac{AB}{AF} = \dfrac{AE}{AC}$ and hence the conclusion.   □

Note: One may also see $\angle AEB = \angle 1$ by equal tangent segments. Notice that the tangent line at A and the line BC are symmetric about the perpendicular bisector of AE.

Example 4.1.2   Let O be the circumcenter of an acute angled triangle $\triangle ABC$ and AO extended intersects BC at D. BE, CF are heights of $\triangle ABC$. Let O_1 be the circumcenter of $\triangle AEF$ and AO_1 intersects EF at P. Show that $AP \cdot BC = AD \cdot EF$.

Insight. Refer to the diagram on the right. Since $\angle BEC = \angle BFC = 90°$, B, C, E, F are concyclic and hence, $\angle ABC = \angle AEF$ (Corollary 3.1.5). We have $\triangle ABC \sim \triangle AEF$. This is a standard result which an experienced contestant would recall instantaneously.

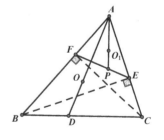

We are to show $AP \cdot BC = AD \cdot EF$. Since $AD \perp EF$ and $AP \perp BC$, one may think of using the area method. However, $AP \cdot BC$ seems not the area of any existing triangle. Notice that AP, BC, AD, EF are in the similar triangles $\triangle ABC$ and $\triangle AEF$. Can we show $\dfrac{AP}{AD} = \dfrac{EF}{BC}$ by the properties of similar triangles?

Proof. It is easy to see that B, C, E, F are concyclic, which implies $\angle ABC = \angle AEF$ and hence, $\triangle ABC \sim \triangle AEF$. Notice that AP and AD are corresponding line segments in $\triangle AEF$ and $\triangle ABC$. It follows that $\dfrac{AP}{AD} = \dfrac{EF}{BC}$ and hence the conclusion. $\qquad\qquad\square$

Note:

(1) Using the fact that the corresponding line segments are also in ratio as the corresponding sides in similar triangles is an effective technique. Beginners who are not familiar with this technique may also show $\dfrac{AP}{AD} = \dfrac{EF}{BC}$ as follows: First, we have $\triangle AOB \sim \triangle AO_1E$ because both are isosceles triangles and $\angle OAB = 2\angle ACB = 2\angle AFE = \angle O_1AE$. Now $\angle OBD = \angle O_1EP$ and $\angle BOD = \angle EO_1P$ imply that $\triangle OBD \sim \triangle O_1EP$. It follows that $\dfrac{EF}{BC} = \dfrac{AE}{AB} = \dfrac{AO_1}{AO} = \dfrac{EO_1}{BO} = \dfrac{O_1P}{OD}$. Hence, $\dfrac{EF}{BC} = \dfrac{AO_1 + O_1P}{AO + OD} = \dfrac{AP}{AD}$.

(2) One may see from the diagram that the lines AP, BE, CF are concurrent, i.e., AP passes through H, the orthocenter of $\triangle ABC$. This is because $\angle CAP = \angle BAO = \angle CAH$ (Example 3.4.1).

Example 4.1.3 Given a circle and a point P outside the circle, draw tangents PA, PB touching the circle at A, B respectively. C is a point on the minor arc $\overset{\frown}{AB}$ and PC extended intersects the circle at D. Show that $\dfrac{BC}{AC} = \dfrac{BD}{AD}$.

Proof. Refer to the following diagram. Since $\angle PAC = \angle PDA$, we have $\triangle PAC \sim \triangle PDA$.

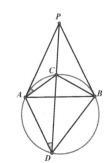

Hence, $\dfrac{PA}{PD} = \dfrac{AC}{AD}$. Similarly, $\dfrac{PB}{PD} = \dfrac{BC}{BD}$.

Since $PA = PB$, we must have $\dfrac{AC}{AD} = \dfrac{BC}{BD}$

and the conclusion follows.   □

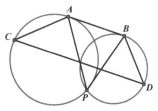

Example 4.1.4   Refer to the diagram on the right. AB is a common tangent of the two circles where A, B are the points of tangency. Given $CD /\!/ AB$, show that $\dfrac{AC}{BD} = \dfrac{AP}{BP}$.

Insight.   Given the tangent line and parallel lines, it is natural to search for equal angles and similar triangles since we are to show $\dfrac{AC}{BD} = \dfrac{AP}{BP}$. It would be great if we can show $\triangle ACP \sim \triangle BDP$. However, this is not true ($\angle ACP = \angle BAP$ and $\angle BDP = \angle ABP$, but $\angle BAP$ and $\angle ABP$ are not necessarily the same). Can you see any pair of similar triangles which put AC, BD, AP and BP together?

It seems not easy. Apparently, the tangent line and the parallel lines do not give equal angles which leads to the similar triangle we need. Notice that we have not used the condition that AB is a **common** tangent. This implies AB is perpendicular to the diameters of both circles.

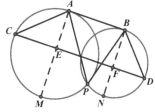

Refer to the diagram on the right. Let AM, BN be the diameters of the two circles. Notice that the diameter AM gives a right angled triangle $\triangle ACM$ where CE is the height on the hypotenuse.

Hence, $AC^2 = AE \cdot AM$ (Example 2.3.1).

Similarly, $BD^2 = BF \cdot BN$ by considering $\triangle BDN$.

Recognize that $AEFB$ is a rectangle, which implies $AE = BF$ and hence, $\left(\dfrac{AC}{BD}\right)^2 = \dfrac{AM}{BN}$. Perhaps we can show $\left(\dfrac{AP}{BP}\right)^2 = \dfrac{AM}{BN}$ as well. This should not be difficult since AP, BP are also related to AM, AN by right angled triangles.

Proof. Let AM, BN be the diameters of the two circles respectively, Clearly, $AM \perp AB$ and $BN \perp AB$. Let AM, BN intersect CD at E, F respectively. Since $CD \mathbin{/\!/} AB$, we have $AEFB$ a rectangle and $AE = BF$.

Since AM is a diameter, we have $\angle ACM = 90°$. Since $CE \perp AM$, we have $AC^2 = AE \cdot AM$ (Example 2.3.1). Similarly, $BD^2 = BF \cdot BN$. Now $AE = BF$ gives $\left(\dfrac{AC}{BD}\right)^2 = \dfrac{AM}{BN}$. (1)

On the other hand, $AP = AM \cos \angle MAP = AM \sin \angle BAP$. Similarly, $BP = BN \sin \angle ABP$. It follows that $\dfrac{AP}{BP} = \dfrac{AM \sin \angle BAP}{BN \sin \angle ABP}$. Notice that $\dfrac{\sin \angle BAP}{\sin \angle ABP} = \dfrac{BP}{AP}$ by Sine Rule. Hence, $\left(\dfrac{AP}{BP}\right)^2 = \dfrac{AM}{BN}$. (2)

The conclusion follows from (1) and (2). $\qquad\square$

Note:

(1) We intended to search for similar triangles but failed, and we completed the proof based on right angled triangles. This is because the tangent line and parallel lines did not give us equal angles directly, but a rectangle. Nevertheless, we managed to find the clues by carefully examining the conditions and setting up intermediate steps which lead to the conclusion. Without such repeated (and mostly failed) attempts, the insight will not appear spontaneously!

(2) One may also show $\left(\dfrac{AP}{BP}\right)^2 = \dfrac{AM}{BN}$ by drawing a line ℓ passing through P and parallel to AB. Applying Example 2.3.1 to the right angled triangles $\triangle APM$ and $\triangle BPN$ leads to the conclusion.

(3) If the two circles intersect at P and Q, one may show $\dfrac{AP}{BP} = \dfrac{AQ}{BQ}$.

Indeed, a similar argument applies when showing $\left(\dfrac{AQ}{BQ}\right)^2 = \dfrac{AM}{BN}$.

Example 4.1.5 (CHN 10)   Let AB be the diameter of a semicircle. C, D are points on the semicircle such that the chords AD, BC intersect at E. Let F, G be points on AC extended and BD extended respectively such that $AF \cdot BG = AE \cdot BE$. Let H_1, H_2 be the orthocenters of $\triangle AEF$ and $\triangle BEG$ respectively. If the lines AH_1, BH_2 intersect at K, show that K lies on the semicircle.

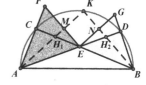

Insight.   Refer to the diagram on the right. Since $AF \cdot BG = AE \cdot BE$, one immediately sees that $\triangle AEF \sim \triangle BGE$, as $\angle EAF = \angle GBE$ (angles in the same arc).
Notice that the orthocenters and the diameter give right angles. In particular, $AK \perp EF$ and $BK \perp EG$. We are to show K lies on the semicircle. Hence, we **should** have $\angle AKB = 90°$ and $MENK$ **should** be a rectangle. Can you see it suffices to show $EF \perp EG$, i.e., $\angle BEG + \angle CEF = 90°$? This is easy because $\angle BEG = \angle AFE$ (since $\triangle AEF \sim \triangle BGE$) and $\angle AFE + \angle CEF = 90°$ (since $BC \perp AF$).
We leave it to the reader to write down the complete proof.

Note: F and G are constructed via $AF \cdot BG = AE \cdot BE$. This is not a commonly seen condition. Indeed, once we focus on this condition and see the similar triangles, it is not far away from the conclusion. Seeking clues from such an uncommon and useful condition is an effective strategy. We will discuss this further in Chapter 6.

Example 4.1.6   Let A be a point outside $\odot O$. AB, AC are tangent to $\odot O$ at B, C respectively. Let ℓ_1, ℓ_2 be two lines tangent to $\odot O$ and

$\ell_1 /\!/ \ell_2$. If the line AB intersects ℓ_1, ℓ_2 at D, E respectively, and the line AC intersects ℓ_1, ℓ_2 at F, G respectively. Show that $AD \cdot AG = AO^2$.

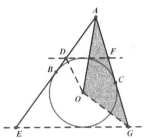

Insight. We are to show $\dfrac{AD}{AO} = \dfrac{AO}{AG}$.
Notice that $\odot O$ is tangent to the sides of $\triangle AEG$, i.e., it is the incircle of $\triangle AEG$ and O is the incenter. Hence, AO bisects $\angle BAC$. Refer to the diagram on the right.

Notice that we **should** have $\triangle AOD \sim \triangle AGO$. Can we show either $\angle ADO = \angle AOG$ or $\angle AOD = \angle AGO$? Notice that $\angle ADO$ and $\angle AOD$ can be expressed in terms of $\angle BAC$ and $\angle ADF$. How about $\angle AGO$ and $\angle AOG$? Recall that $\angle FOG = 90°$ (Example 3.2.5).

Proof. It is easy to see that AO bisects $\angle A$. In fact, O is the incenter of $\triangle AEG$.

Now $\angle AOG = 90° + \dfrac{1}{2}\angle AEG$ (Theorem 1.3.3)

$= 90° + \dfrac{1}{2}\angle ADF$ because $DF /\!/ EG$.

Since $\angle ADO = \angle ADF + \dfrac{1}{2}\left(180° - \angle ADF\right) = 90° + \dfrac{1}{2}\angle ADF = \angle AOG,$

we must have $\triangle AOD \sim \triangle AGO$. The conclusion follows. $\quad\square$

Note:
(1) If you cannot recall Theorem 1.3.3, simply calculate $\angle AOG$ by the fact that $\angle FOG = 90°$.

(2) One may also show $\angle AOD = \angle AGO$, where $\angle AGO = \dfrac{1}{2}\angle AGE$

$= \dfrac{1}{2}\angle AFD$ and $\angle AOD = \angle ODE - \dfrac{1}{2}\angle A.$

Example 4.1.7   Let O be the center of the semicircle where AB is the diameter. Draw a line $\ell \perp AB$ at B. Let D be a point on the semicircle and draw $DE \perp AB$ at E. Draw $OC \,/\!/ AD$, intersecting ℓ at C. If AC and DE intersect at P, show that $PD = PE$.

Insight.   Refer to the diagram below. It is easy to see that $DE \,/\!/ BC$ and hence, $\dfrac{PE}{BC} = \dfrac{AE}{AB}$.

However, one may not be able to relate this to PD in the diagram. What if we "fill up" the triangle by extending BC, intersecting AD extended at F?

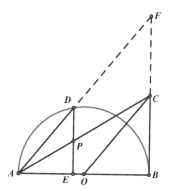

Notice that $\dfrac{PD}{CF} = \dfrac{AE}{AB} = \dfrac{PE}{BC}$. Now it suffices to show $BC = CF$.

Can you show it? (**Hint**: $OC \,/\!/ AD$ and $OA = OB$.) We leave the details to the reader.

Note: We did not construct any similar triangles, but simply applied the Intercept Theorem where $AD \,/\!/ CO$ and $DE \,/\!/ BC$. Notice that $OA = OB$ is an elementary property, but it could be overlooked occasionally.

Example 4.1.8   Refer to the diagram on the right. Given $\triangle ABC$ and its circumcircle Γ, MN is a line tangent to Γ at B such that MA, NC touch Γ at A, C respectively. Let P be a point on AC such that $BP \perp AC$. Show that BP bisects $\angle MPN$.

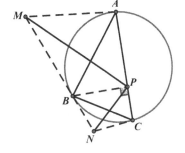

Insight.   We are only given a few tangent lines of the circle. Notice that P is **not** the center of the circle: there are no other given right angles in the diagram and it may be difficult to find concyclicity related to P.

Hence, showing $\angle BPM = \angle BPN$ by finding equal angles may not be an effective strategy.

If BP does bisect $\angle MPN$, we **should** have $\dfrac{PM}{PN} = \dfrac{BM}{BN}$ by the Angle Bisector Theorem. How could the tangent lines help us? Since we have $AM = BM$ and $BN = CN$, it suffices to show $\dfrac{PM}{PN} = \dfrac{AM}{CN}$. It *seems* that $\triangle AMP \sim \triangle CNP$ because $\angle PAM = \angle PCN$. (Can you see that the lines AM, CN are symmetric about the perpendicular bisector of AC?)

How can we show $\triangle AMP \sim \triangle CNP$? It seems we should find another pair of equal angles using the condition $BP \perp AC$, but this is equally difficult as showing the conclusion directly.

Notice that it is much easier to show the inverse: if we are given $\triangle AMP \sim \triangle CNP$, one could see that BP bisects $\angle MPN$ and $BP \perp AC$. Perhaps we should consider a proof by contradiction.

Proof. Choose P' on AC such that $\dfrac{AP'}{CP'} = \dfrac{AM}{CN}$. It is easy to see that $\angle P'AM = \angle P'CN$. Hence, $\triangle AMP' \sim \triangle CNP'$.

We have $\dfrac{P'M}{P'N} = \dfrac{AM}{CN}$ and $\angle AP'M = \angle CP'N$.   (1)

It is easy to see that $BM = AM$ and $BN = CN$ (equal tangent segments). Hence, $\dfrac{P'M}{P'N} = \dfrac{BM}{CN}$, which implies BP bisects $\angle MPN$ by the Angle Bisector Theorem. Now $\angle BP'M = \angle BP'C$.   (2)

(1) and (2) imply that $BP' \perp AC$, i.e., P and P' coincide. This completes the proof. $\square$

Note: One may still seek clues from $BP \perp AC$ and other right angles by introducing the center of Γ. Refer to the diagram on the right. It would be wise to erase unnecessary lines.

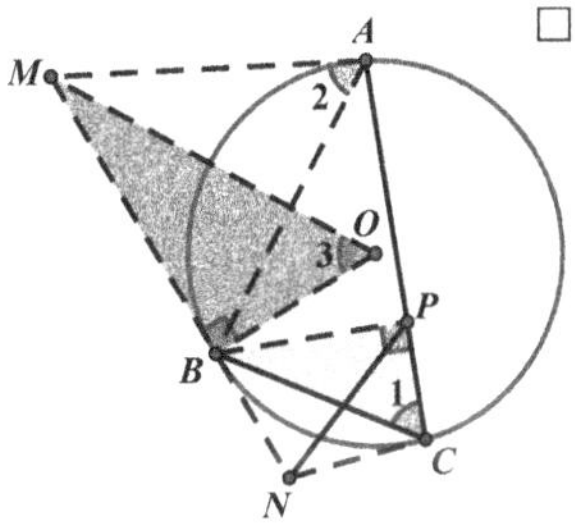

Can you see that $\triangle BOM \sim \triangle PCB$? (**Hint**: $\angle 3 = \frac{1}{2} \angle AOB = \angle 1$.)

Now $\triangle BOM \sim \triangle PCB$ implies $\dfrac{BM}{PB} = \dfrac{BO}{PC}$. (1)

A similar argument gives $\triangle BON \sim \triangle PAB$ and $\dfrac{BN}{PB} = \dfrac{BO}{PA}$. (2)

(1) and (2) imply that $\dfrac{AP}{CP} = \dfrac{BM}{BN} = \dfrac{AM}{CN}$ by equal tangent segments.

Now it is easy to see that $\triangle AMP \sim \triangle CNP$ and the conclusion follows. In fact, one familiar with angle properties in circle geometry may immediately see that $\angle 1 = \angle 2$ (Theorem 3.2.10) and $\angle 2 = \angle 3$ (because A, M, B, O are concyclic). Now it is easy to identify similar triangles and this alternative solution follows naturally.

In Chapter 2, we learnt Ceva's Theorem and Menelaus' Theorem, which are useful results solving problems on collinearity and concyclicity. When circles are introduced, one may find even more interesting results by applying Ceva's Theorem and Menelaus' Theorem, due to more equal angles and line segments. The following is a simple example.

Example 4.1.9   Given $\triangle ABC$ where $AB > AC$, its incircle $\odot I$ touches BC, AC, AB at D, E, F respectively. P is a point on BC extended. Draw a line PG tangent to $\odot I$ at G (distinct from D), intersecting AB, AC at M, N respectively. Let BG, DM intersect at Q and CG, DN intersect at R. Show that if P, E, F are collinear, then P, Q, R are collinear.

Insight.   Refer to the diagram on the right. We are to show collinearity and it seems we need to use either Ceva's Theorem or Menelaus' Theorem.

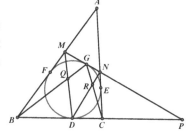

Now, which triangle should we start with?

Notice that we are given many tangent lines: those equal tangent segments could be helpful. If we choose the line PF intersecting $\triangle ABC$, Menelaus' Theorem gives $\dfrac{AF}{BF} \cdot \dfrac{BP}{CP} \cdot \dfrac{CE}{AE} = 1$.

Since $AE = AF$, $BF = BD$ and $CE = CD$, we have $\dfrac{BP}{CP} \cdot \dfrac{CD}{BD} = 1$. (1)

By applying Menelaus' Theorem to $\triangle BCG$, it suffices to show that

$\dfrac{GQ}{BQ} \cdot \dfrac{BP}{CP} \cdot \dfrac{CR}{GR} = 1$. However, this is not easy even if we use (1), because we do not know much about $\dfrac{BQ}{GQ}$ and $\dfrac{CR}{GR}$. Can we avoid these terms?

Notice that one may apply Ceva's Theorem instead: not only will we have more equal tangent segments, but also get rid of those line segments which are not preferred (i.e., those not along the tangent lines).

Proof.   Notice that $\dfrac{MF}{BF} \cdot \dfrac{BD}{PD} \cdot \dfrac{PG}{MG} = 1$ because $MF = MG$, $BD = BF$ and $PD = PG$ (equal tangent segments). By Ceva's Theorem applied to $\triangle BPM$, P, Q, F are collinear.

Similarly, we have $\dfrac{NE}{CE} \cdot \dfrac{CD}{PD} \cdot \dfrac{PG}{NG} = 1$. By Ceva's Theorem applied to $\triangle CPN$, P, E, R are collinear. (Notice that the points of division D, G are on the extension of PC, PN respectively.)

Since P, E, F are collinear, Q, R also lie on this line, i.e., P, Q, R are collinear. $\qquad\square$

4.2 Intersecting Chords Theorem and Tangent Secant Theorem

In most elementary geometry textbooks, Intersecting Chords Theorem and Tangent Secant Theorem are mentioned, but the application is not emphasized. Indeed, these are very useful results, with which we can show concyclicity **not** via equal angles.

Theorem 4.2.1 (Intersecting Chords Theorem) *Let AB and CD be two chords of a circle. If AB and CD intersect at E, we have* $AE \cdot BE = CE \cdot DE$.

Refer to the left diagram below. One sees the conclusion immediately from the fact that $\triangle ACE \sim \triangle DBE$.

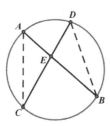

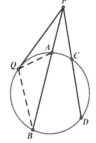

Intersecting Chords Theorem　　　　　　　Tangent Secant Theorem

Theorem 4.2.2 (Tangent Secant Theorem) *Let P be a point outside the circle and a line passing through P intersects the circle at A and B. If PQ touches the circle at Q, we must have* $PQ^2 = PA \cdot PB$.

Refer to the right diagram above. One may see the conclusion from the fact that $\triangle PAQ \sim \triangle PQB$ (because $\angle PQA = \angle PBQ$).

Note:
(1) An immediate corollary of the Tangent Secant Theorem is that if two lines passing through P intersect the circle at A, B and C, D respectively, we must have $PA \cdot PB = PC \cdot PD$, because both are equal to PQ^2.
(2) One easily sees that the inverse of the Intersecting Chords Theorem and the Tangent Secant Theorem hold. (Can you show it, say by

contradiction?) Hence, we may use these theorems, especially the inverse, to show concyclicity.

Example 4.2.3 In $\triangle ABC$, $AB = 9$, $BC = 8$ and $AC = 7$. Let M be the midpoint of BC. If AM extended intersects the circumcircle of $\triangle ABC$ at D, find MD.

Ans. Refer to the diagram on the right.

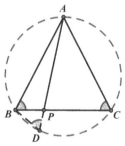

By Theorem 2.4.3, $AM^2 = \dfrac{1}{2} \cdot \left(9^2 + 7^2\right) - \dfrac{1}{4} \cdot 8^2$

$= 49$, i.e., $AM = 7$.

Now the Intersecting Chords Theorem gives $AM \cdot MD = BM \cdot CM$, where $BM = CM = 4$. Hence, $MD = \dfrac{4^2}{7} = \dfrac{16}{7}$.

Example 4.2.4 Let $\triangle ABC$ be an isosceles triangle where $AB = AC$ and P is a point on BC. Show that $\left(AB + AP\right)\left(AB - AP\right) = BP \cdot CP$.

Insight. From the first glance, it is not clear how the line segments are related to each other. In particular, it seems not easy to obtain $AB + AP$ or $AB - AP$. However, $BP \cdot CP$ reminds us of the Intersecting Chords Theorem, if we draw the circumcircle of $\triangle ABC$. Refer to the diagram on the right.

Let AP extended intersects the circumcircle at D. We immediately have $BP \cdot CP = AP \cdot PD$. Notice that $\left(AB + AP\right)\left(AB - AP\right) = AB^2 - AP^2$. Hence, it suffices to show that $AB^2 = AP^2 + AP \cdot PD = AP \cdot \left(AP + PD\right) = AP \cdot AD$. Now we **should** have $\triangle ABP \sim \triangle ADB$, which is not difficult to show.

Proof. Let AP extended intersect the circumcircle of $\triangle ABC$ at D. Since $\angle B = \angle C = \angle D$ (angles in the same arc), we have $\triangle ABP \sim \triangle ADB$. It follows that $\dfrac{AB}{AD} = \dfrac{AP}{AB}$, or $AB^2 = AP \cdot AD$.

Now $AB^2 = AP \cdot AD = AP \cdot (AP + PD) = AP^2 + AP \cdot PD$.

Hence, $AP \cdot PD = AB^2 - AP^2 = (AB + AP)(AB - AP)$. The conclusion follows as $AP \cdot PD = BP \cdot CP$ by the Intersecting Chords Theorem.   □

Example 4.2.5   Let X be a point inside $\triangle ABC$ and the lines AX, BX, CX intersect the circumcircle of $\triangle ABC$ at P, Q, R respectively. Let A' be a point on PX. Draw $A'B' /\!/ AB$ and $A'C' /\!/ AC$, where B', C' are on the lines QX, RX respectively. Show that B', C', R, Q are concyclic.

Insight.   Refer to the diagram on the right. Since X is an arbitrary point, the construction of the diagram seems symmetric, i.e., if we are to show B', C', R, Q are concyclic, we **might** have A', B', Q, P and A', C', R, P concyclic as well. **If** that is true, applying the Tangent Secant Theorem repeatedly gives $XB' \cdot XQ = XA' \cdot XP$ $= XC' \cdot XR$!

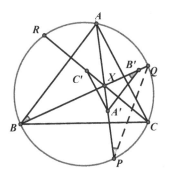

Proof.   Since $A'B' /\!/ AB$, we have $\angle A'B'X = \angle ABQ = \angle APQ$ (angles in the same arc). Hence, $A'B', Q, P$ are concyclic. By the Tangent Secant Theorem, $XA' \cdot XP = XB' \cdot XQ$. Similarly, A', C', R, P are concyclic and $XA' \cdot XP = XC' \cdot XR$. Now $XB' \cdot XQ = XC' \cdot XR$, which implies B', C', R, Q are concyclic.   □

Note: One might also show $B'C' /\!/ BC$ by applying the Intercept Theorem repeatedly, which also leads to the condition. We leave the details to the reader.

Example 4.2.6 *ABCD* is a quadrilateral inscribed in $\odot O$. *AB* extended and *DC* extended intersect at *P*. *AD* extended and *BC* extended intersect at *Q*. Draw *PE* tangent to $\odot O$ at *E* and *QF* tangent to $\odot O$ at *F*. Show that *PE*, *QF* and *PQ* give the sides of a right angled triangle.

Insight. Clearly, we should show that *PE*, *QF*, *PQ* satisfy Pythagoras' Theorem. Refer to the diagram on the right. What do we know about PE^2, QF^2 or PQ^2 ?

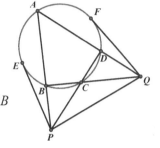

By the Tangent Secant Theorem, $PE^2 = PA \cdot PB$

$= PC \cdot PD$ and similarly, $QF^2 = QB \cdot QC$.

One sees that PQ^2 is related to those line segments above by Cosine Rule. However, it is difficult to use those line segments to express $\cos \angle A$ or $\cos \angle PCQ$.

Are there other methods to relate PE^2 and QF^2 to PQ^2? If the circumcircle of $\triangle CDQ$ intersects *PQ* at *X*, we must have $PE^2 = PC \cdot PD = PX \cdot PQ$.

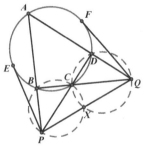

Hence, if $PQ^2 = PE^2 + QF^2$, we **should** have

$$QB \cdot QC = QF^2 = PQ^2 - PE^2 = PQ^2 - PX \cdot PQ = (PQ - PX) \cdot PQ$$
$$= QX \cdot PQ. \text{ Hence, } B, C, X, P \text{ should be concyclic. Can we prove it?}$$

Notice that we have $\angle CXP = \angle CDQ = \angle ABC$ by applying Corollary 3.1.5 repeatedly and hence, B, C, X, P are concyclic. (Can you see this is similar to the proof of the Simson's Line?) We leave it to the reader to write down the complete proof.

Note: One may *see* from the diagram that *PQ* is longer than *PE* and *QF*. (Drawing a reasonably accurate diagram would be helpful.) Even though this is not given, one should *aim* to show that $PQ^2 = PE^2 + QF^2$.

Example 4.2.7 (IMO 95)   The incircle of $\triangle ABC$ touches BC, AC, AB at D, E, F respectively. Let X be a point inside $\triangle ABC$ such that the incircle of $\triangle XBC$ touches BC, XB, XC at D, Y, Z respectively. Show that E, F, Y, Z are concyclic.

Insight. Refer to the diagram on the right. Apparently, there are very few conditions given: we only know that E, F, Y, Z are all points of tangency. Although there are incircles (i.e., angle bisectors), but E, F, Y, Z are not related to the incenter or any angle bisectors.

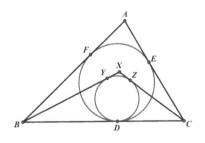

On the other hand, the diagram seems in an "upright" position because the two incircles share a common point of tangent. Do we have $YZ \parallel EF$? If yes, then perhaps we can show that $EFYZ$ is an isosceles trapezium.

Regrettably, this is not true. Refer to the diagram on the right where FE extended and YZ extended intersect. Can you see a clue in this diagram? Perhaps we could show that $PE \cdot PF = PY \cdot PZ$.

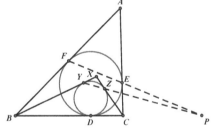

Since the two circles have one common point of tangency D, if P lies on BC extended, we would have $PD^2 = PE \cdot PF = PY \cdot PZ$.

How can we show that P lies on BC extended? In other words, if we let P be the intersection of BC extended and YZ extended, can we show that E, F, P are collinear? This looks like Menelaus' Theorem. Refer to the diagram below. Do we have $\dfrac{AF}{BF} \cdot \dfrac{BP}{CP} \cdot \dfrac{CE}{AE} = 1$? (1)

Notice that $AE = AF$. In fact, there are many equal tangent segments in this diagram.

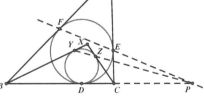

Perhaps we can also apply Menelaus' Theorem to $\triangle XBC$, which might give us sufficient equalities leading to (1).

Proof. Suppose BC extended and YZ extended intersect at P. Apply Menelaus' Theorem to $\triangle XBC$ and the line YZ: $\dfrac{XY}{BY}\cdot\dfrac{BP}{CP}\cdot\dfrac{CZ}{XZ}=1$.

Since $XY = XZ$ (equal tangent segments), we have $\dfrac{BP}{CP}=\dfrac{BY}{CZ}$. (*)

We claim that E,F,P are collinear, i.e., $\dfrac{AF}{BF}\cdot\dfrac{BP}{CP}\cdot\dfrac{CE}{AE}=1$.

Notice that $AF = AE$. By (*), we have

$$\frac{AF}{BF}\cdot\frac{BP}{CP}\cdot\frac{CE}{AE}=\frac{CE}{BF}\cdot\frac{BP}{CP}=\frac{CE}{BF}\cdot\frac{BY}{CZ}=1,\quad\text{because}\quad BF=BD=BY\quad\text{and}$$
$CE = CD = CZ$ (equal tangent segments). Hence, E,F,P are collinear.

Now by the Tangent Secant Theorem, $PE \cdot PF = PD^2 = PX \cdot PY$, which implies E,F,Y,Z are concyclic.

Notice that if BC and YZ do not intersect, i.e., $BC /\!/ YZ$, we must have $XB = XC$ (because $XY = XZ$) and hence, D is the midpoint of BC. Since $BF = BD = CD = CE$ and $AF = AE$, we have $AB = AF + BF = AE + CE = AE$. Now $\triangle ABC$ and $\triangle XBC$ are both isosceles triangles. Hence, the line AX is the perpendicular bisector of EF and YZ. It is easy to see that $EFYZ$ is an isosceles trapezium, which implies E,F,Y,Z are concyclic. $\square$

Example 4.2.8 (JPN 11) Given an acute angled triangle $\triangle ABC$ and its orthocenter H, M is the midpoint of BC. Draw $HP \perp AM$ at P. Show that $AM \cdot PM = BM^2$.

Insight. It seems AM, AP are not closely related to BM. However, given the orthocenter and the midpoints, one immediately sees $BM = DM = CM$, where $BD \perp AC$ at D.

Since we are to show $AM \cdot PM = DM^2$, we **should** have MD tangent to the circumcircle of $\triangle ADP$ by the Tangent Secant Theorem. Refer to the diagram on the right. It is easy to see H is on this circle as well. We have plenty of equal angles!

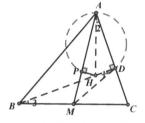

Proof. Let BD be the height on AC. In the right angled triangle $\triangle BCD$,

$$DM = \frac{1}{2}BC = BM.$$ Now it suffices to show $AM \cdot PM = DM^2$.

Since $\angle APH = \angle ADH = 90°$, A, D, H, P are concyclic. Notice that $\angle 2 = 90° - \angle C = \angle 3$ and $\angle 1 = \angle 3$ (because $BM = DM$). Hence, $\angle 1 = \angle 2$, which implies MD is tangent to the circumcircle of $\triangle ADP$ (Theorem 3.2.10).
By the Tangent Secant Theorem, $AM \cdot PM = DM^2$.   □

Example 4.2.9 (CMO 10)   Refer to the left diagram below. Two circles intersect at A and B. A line passing through B intersects the two circles at C, D respectively. Another line passing through B intersects the two circles at E, F respectively. CF intersects the two circles at P, Q respectively. Let M, N be the midpoints of arcs $\overparen{PB}, \overparen{QB}$ respectively. Show that if $CD = EF$, then C, M, N, F are concyclic.

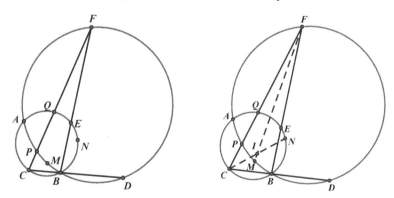

Insight.   Clearly, we must use the condition $CD = EF$ in the proof. How about $EF \cdot BF = FQ \cdot CF$ and $BC \cdot CD = CP \cdot CF$?

Since $CD = EF$, we have $\dfrac{FQ}{BF} = \dfrac{CP}{BC}$. Notice that all these line segments are in $\triangle BCF$. Perhaps we should focus on this triangle and see what we *may* discover.

Refer to the previous right diagram. How is $\triangle BCF$ related to the conclusion? Notice that CN and FM are the angle bisectors of $\triangle BCF$ (Corollary 3.3.3). Hence, they intersect at the incenter I of $\triangle BCF$. Since we are to show C, M, N, F concyclic, we **should** have $CI \cdot IN = FI \cdot IM$. Although we cannot apply the Intersecting Chords Theorem directly because these are chords in two different circles, there is a common chord AB! Since we are to show $CI \cdot IN = FI \cdot IM$, we **should** have AB passing through I. (Suppose otherwise, say BI extended intersects the two circles at A and A' respectively. By the Intersecting Chords Theorem, $AI \cdot IB = CI \cdot IN = FI \cdot IM = A'I \cdot IB$, which implies that A and A' coincide.)

Now it suffices to show that AB is the angle bisector of $\angle CBF$. Refer to the left diagram below. This is much simpler!

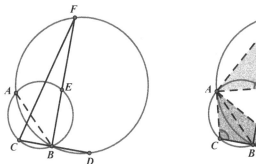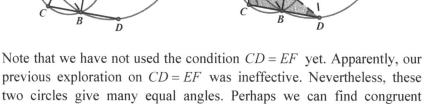

Note that we have not used the condition $CD = EF$ yet. Apparently, our previous exploration on $CD = EF$ was ineffective. Nevertheless, these two circles give many equal angles. Perhaps we can find congruent triangles.

Proof. Refer to the right diagram above. We have $\angle ADC = \angle AFE$ (angles in the same arc) and $\angle ACD = \angle AEF$ (Corollary 3.1.5).

Given $CD = EF$, we conclude that $\triangle ACD \cong \triangle AEF$ (A.A.S.) and hence, $AD = AF$. Now we have $\angle ABF = \angle ADF$ (angles in the same arc) $= \angle AFD$ (because $AD = AF$) $= \angle ABC$ (Corollary 3.1.5), i.e., BA is the angle bisector of $\angle CBF$.

Since M, N are the midpoints of arcs $\overset{\frown}{PB}, \overset{\frown}{QB}$ respectively, CN, FM are both angle bisectors of $\triangle CBF$ (Corollary 3.3.3). Let I be the incenter of $\triangle CBF$. We have $CI \cdot IN = AI \cdot IB = FI \cdot IM$ by the Intersecting Chords Theorem. Hence, C, M, N, F are concyclic. $\square$

Note: One sees many clues from the conditions given and hence, may explore in a wrong direction. For example, one may apply $\dfrac{FQ}{BF} = \dfrac{CP}{BC}$ and construct similar triangles, or seek angles in the same arc using the angle bisectors. Even though such (failed) attempts are not reflected in the final solution, these are inevitable during problem-solving and should not be considered a waste of effort. Indeed, beginners would learn much more from those attempts rather than merely reading the solution.

Example 4.2.10 (CGMO 10)   Refer to the diagram below. In an acute angled triangle $\triangle ABC$, M is the midpoint of BC. Let AP bisect the exterior angle of $\angle A$, intersecting BC extended at P. Draw $ME \perp AP$ at E and draw $MF \perp BC$, intersecting the line AP at F. Show that $BC^2 = 4PF \cdot AE$.

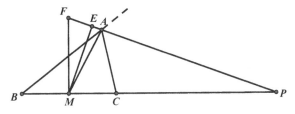

Insight.   It seems from the conclusion that the Intersecting Chords Theorem or the Tangent Secant Theorem should be applied, but where is the circle? Perhaps we can see concyclicity from the right angles given. Besides, we also have the angle bisector of the exterior angle. What does it remind you of? Recall that the angle bisectors of supplementary angles are perpendicular (Example 1.1.9)!

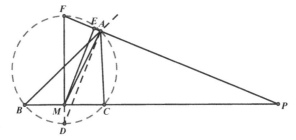

Refer to the diagram above. We draw the circumcircle of $\triangle ABC$ and AD which bisects $\angle A$, intersecting the circumcircle of $\triangle ABC$ at D. Notice that $AD \perp AP$. It *seems* that F lies on the circle as well. Can you prove it? (**Hint**: One may show that D, M, F are collinear and DF is indeed a diameter of the circle.)

Once we show that A, C, B, F are concyclic, by the Tangent Secant Theorem, $PB \cdot PC = PA \cdot PF$. How could we relate this to the conclusion $BC^2 = 4PF \cdot AE$? We have PF in both equations and $BC = PB - PC$. It is not clear at this stage how we should relate AE to the other line segments. Moreover, it seems the coefficient 4 does not appear naturally. Can we get rid of it?

Notice that M is the midpoint of BC, i.e., $BM = \dfrac{1}{2}BC$. Hence, it suffices to show $BM^2 = PF \cdot AE$.

We also note that $PB \cdot PC = (PM + BM)(PM - BM) = PM^2 - BM^2$, where $PM^2 = PE \cdot PF$ (Example 2.3.1). Apparently, we are very close to the conclusion.

Proof. Let D be the midpoint of the minor arc $\overset{\frown}{BC}$, i.e., $\overset{\frown}{BD} = \overset{\frown}{CD}$. It is easy to see that AD bisects $\angle BAC$ (Corollary 3.3.3). This implies that D lies on the perpendicular bisector of BC (because $BD = CD$). Since $MF \perp BC$, MF is also the perpendicular bisector of BC. It follows that D, M, F are collinear, the line of which passes through the center of the circumcircle of $\triangle ABC$.

Since $AD \perp AP$, we claim that F must lie on the circumcircle of $\triangle ABC$ as well. Otherwise, say the line MD intersects the circumcircle of $\triangle ABC$

at F', DF' must be a diameter of the circle and $\angle DAF' = 90°$, i.e., $AD \perp AF'$. This implies F' lies on the line AP, i.e., F and F' coincide.

Since A, C, B, F are concyclic, we have $PB \cdot PC = PA \cdot PF$, where $PB \cdot PC = (PM + BM)(PM - CM) = PM^2 - BM^2$ because $BM = CM$.

In the right angled triangle ΔPMF, $ME \perp PF$. Hence, $PM^2 = PE \cdot PF$.

It follows that $PA \cdot PF = PB \cdot PC = PM^2 BM^2 = PE \cdot PF - BM^2$, i.e., $BM^2 = PE \cdot PF - PA \cdot PF = (PE - PA) \cdot PF = AE \cdot PF$. The conclusion follows as $BM = \dfrac{1}{2} BC$.   □

Note: Once the circumcircle of ΔABC is drawn, it is easy to see that the line DM is a diameter of the circle, where AD bisects $\angle BAC$. Now the exterior angle bisector is used to construct right angles. Notice that applying the Angle Bisector Theorem may not be an effective strategy because AB, AC are not closely related to PF, AE.

4.3 Radical Axis

Given a circle, a straight line could intersect the circle at two points, or touch the circle at one point, i.e., a tangent line. Refer to the diagram on the right.

Can you show that no straight line intersects a circle at more than two points? (**Hint**: Suppose otherwise, say a line intersect $\odot O$ at A, B, C, we have $OA = OB = OC$, i.e., both ΔOAB and ΔOBC are isosceles triangles. Show that this is impossible by considering the base angles.)

Given a circle, another circle may intersect it at two points, or touch it at one point, in which case we say the circles are tangent to each other. Refer to the following diagrams.

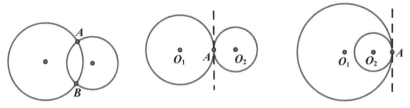

Circles Tangent to Each Other

Can you show that no two circles intersect at more than two points? (**Hint**: Suppose otherwise, say $\odot O_1$ and $\odot O_2$ intersect at A, B, C. It is easy to see that A, B, C cannot be collinear. Now consider the circumcircle of $\triangle ABC$.)

Given $\odot O_1$ and $\odot O_2$, if they intersect at A and B, then O_1O_2 must be the perpendicular bisector of AB (Theorem 3.1.20). In particular, if $\odot O_1$ and $\odot O_2$ touch each other at A, then O_1O_2 passes through A, i.e., O_1, O_2, A are collinear. Hence, one may consider two circles touching each other an extreme case of intersecting circles. Similarly, a tangent line of the circle is also an extreme case of a line intersecting the circle at two points, as reflected in the Tangent Secant Theorem. We may define radical axes when two or more circles intersect or touch each other.

Definition 4.3.1 If $\odot O_1$ and $\odot O_2$ intersect at A and B, we call the line AB the radical axis of $\odot O_1$ and $\odot O_2$. In particular, if $\odot O_1$ touches $\odot O_2$ at A, the radical axis of $\odot O_1$ and $\odot O_2$ is the common tangent of the two circles which passes through A.

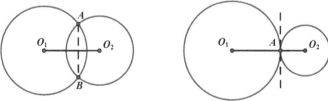

Radical axes

Note: One may also define a radical axis of two non-intersecting circles. However, we will only focus on radical axes of circles intersecting or tangent to each other, which are the most commonly seen applications in competitions.

Theorem 4.3.2   *If three circles are mutually intersecting each other, then the three radical axes are either parallel or concurrent.*

Proof.   Let the three circles be $\Gamma_1, \Gamma_2, \Gamma_3$ such that Γ_1, Γ_2 intersect at A, B, Γ_2, Γ_3 intersect at C, D and Γ_1, Γ_3 intersect at E, F. If the radical axes AB, CD, EF are parallel, there is nothing to prove. Refer to the left diagram below. Otherwise, say without loss of generality that AB and CD intersect at P. Extend PE, intersecting Γ_2 at X. We claim that X and F coincide. Refer to the right diagram below.

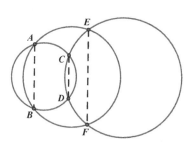
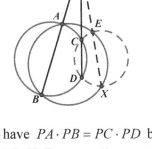

Since A, B, D, C are concyclic (on Γ_1), we have $PA \cdot PB = PC \cdot PD$ by the Tangent Secant Theorem. Similarly, A, B, X, E concyclic on Γ_2 implies $PA \cdot PB = PE \cdot PX$. It follows that $PC \cdot PD = PE \cdot PX$. Now C, D, X, E are concyclic and X must lie on the circumcircle of $\triangle CDE$, which is Γ_3. Since X lies on both Γ_2 and Γ_3, X and F coincide. This implies P, E, F are collinear, i.e., the radical axes are concurrent.   □

Note: This proof holds regardless of the relative positions of the three circles. Refer to the diagram on the right. Notice that $PA \cdot PB = PC \cdot PD = PE \cdot PF$ by the Intersecting Chords Theorem. Hence, we still have the radical axes AB, CD, EF concurrent.

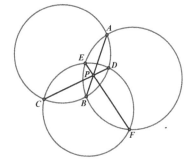

Example 4.3.3 Refer to the diagram on the right. $\odot O_1$, $\odot O_2$ and $\odot O_3$ are mutually tangent to each other at A, B, C respectively. Show that the circumcenter of $\triangle ABC$ is the incenter of $\triangle O_1 O_2 O_3$.

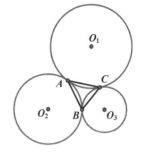

Insight. We see that $O_1 A = O_1 C$, $O_2 A = O_2 B$ and $O_3 B = O_3 C$.

What do we know about the incenter and the incircle of $\triangle O_1 O_2 O_3$? Refer to the left diagram below. It seems A, B, C **should** be the feet of the perpendiculars from the incenter of $\triangle O_1 O_2 O_3$. What if we draw perpendicular lines from A to $O_1 O_2$, B to $O_2 O_3$ and C to $O_3 O_1$? Can you see that the perpendicular from A to $O_1 O_2$ is indeed a common tangent of $\odot O_1$ and $\odot O_2$, and similarly for B and C? These common tangents are concurrent! (Can you show this by the Tangent Secant Theorem? Refer to Exercise 4.11.)

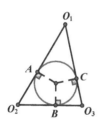
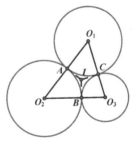

Proof. Refer to the right diagram above. Draw the perpendicular lines from A to $O_1 O_2$ and from B to $O_2 O_3$, intersecting at I. It is easy to show that $CI \perp O_1 O_3$ (Exercise 4.11). Notice that $AI = BI = CI$ (equal tangent segments). Hence, I is the circumcenter of $\triangle ABC$. Observe that $O_1 I$ bisects $\angle O_1$ since $\triangle O_1 AI \cong \triangle O_1 CI$ (H.L.). Similarly, $O_2 I$ bisects $\angle O_2$. Hence, I is the incenter of $\triangle O_1 O_2 O_3$. This completes the proof. □

Example 4.3.4   Refer to the diagram on the right. $\odot O_1$ and $\odot O_2$ intersect at P, Q. O_1A intersects $\odot O_2$ at B. O_2C intersects $\odot O_1$ at D. Given that A, C, B, D are concyclic, show that the circumcenter of $\triangle ABC$ lies on the line PQ.

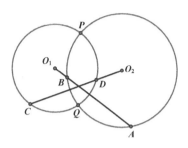

Insight.   Let us draw the circumcircle of $\triangle ABC$. Refer to the diagram on the right where A, B, C, D lie on the $\odot O$.

Can you see that the lines AB, CD, PQ are exactly the radical axes when $\odot O$, $\odot O_1$ and $\odot O_2$ intersect each other?

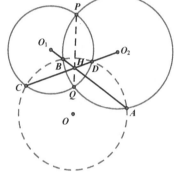

By Theorem 4.3.2, lines AB, CD, PQ must be concurrent, say at H.

Notice that $AB \perp OO_2$ and $CD \perp OO_1$ (Theorem 3.1.20). Can you see that H is the orthocenter $\triangle OO_1O_2$? Now can you see why O lies on the line PQ? (**Hint**: $OH \perp O_1O_2$ and $PH \perp O_1O_2$.) We leave it to the reader to complete the proof.

Note: Theorem 3.1.20 is an elementary but commonly used result. One may always apply it and seek clues when attempting questions with circles intersecting each other.

Definition 4.3.5   Let $\odot O$ be a circle centered at O with radius r. The power of a point P with respect to $\odot O$ is defined as $OP^2 - r^2$.

The concept of the power of a point with respect to a circle is closely related to the Intersecting Chords Theorem and the Tangent Secant Theorem. Refer to the following diagrams.

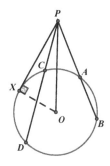 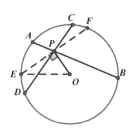

- If P is outside the circle where PX touches $\odot O$ at X, one sees that $PX^2 = OP^2 - OX^2$. By the Tangent Secant Theorem, we have $PA \cdot PB = PC \cdot PD = PX^2$, which equals the power of P with respect to $\odot O$.

- If P is inside the circle, draw $EF \perp OP$ at P, intersecting $\odot O$ at E and F. One sees that $OP^2 = OE^2 - PE^2$. By the Intersecting Chords Theorem, $PA \cdot PB = PC \cdot PD = PE \cdot PF = PE^2 = OE^2 - OP^2$, which is the negative of the power of P with respect to $\odot O$.

In conclusion, the power of a point P with respect to $\odot O$ is positive if P lies outside $\odot O$ and is negative if P lies inside $\odot O$. Clearly, the power of P is zero if it lies on $\odot O$.

Theorem 4.3.6 *Let $\odot O_1$ and $\odot O_2$ intersect at A, B. The power of a point P with respect to $\odot O_1$ and $\odot O_2$ is the same if and only if P lies on the line AB, which is also the radical axis of $\odot O_1$ and $\odot O_2$.*

Proof. Refer to the diagram on the right.

Let P be any point. Suppose the line PA intersects $\odot O_1$ and $\odot O_2$ at C and D respectively. Notice that the power of P with respect to $\odot O_1$ is $PA \cdot PC$, and the power with respect to $\odot O_2$ is $PA \cdot PD$.

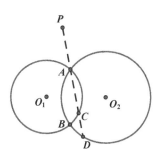

One sees that the power of P with respect to $\odot O_1$ and $\odot O_2$ is the same if and only if $PC = PD$, i.e., C, D coincide with B, the line PA passes through B and P lies on the radical axis AB.

Notice that this proof still holds if P lies inside $\odot O_1$ and $\odot O_2$. Now the power of P with respect to $\odot O_1$ and $\odot O_2$ are $-PA \cdot PC$ and $-PA \cdot PD$ respectively. Hence, the power of P with respect to $\odot O_1$ and $\odot O_2$ is the same if and only if $PC = PD$, i.e., if and only if P lies on the radical axis AB.   □

Note: One may easily show Theorem 4.3.2 by applying Theorem 4.3.6: If a point P lies on the radical axis of $\odot O_1$ and $\odot O_2$ and the radical axis of $\odot O_2$ and $\odot O_3$, its power with respect to $\odot O_1$, $\odot O_2$ and $\odot O_3$ is the same. Hence, P must also lie on the radical axis of $\odot O_1$ and $\odot O_3$.

Example 4.3.7 (RUS 13)   Given an acute angled triangle $\triangle ABC$, draw squares $BCDE$ and $ACFG$ outwards from BC, AC respectively. Let DC extended intersect AG at P and FC extended intersect BE at Q. X is a point inside $\triangle ABC$ which lies on the circumcircles of both $\triangle PDG$ and $\triangle QEF$. If M is the midpoint of AB, show that $\angle ACM = \angle BCX$.

Insight.   Refer to the diagram on the right. It seems that the properties of $\angle ACM$ and $\angle BCX$ are not clear. Let the two circles intersect at X and Z. There are many right angles in the diagram and hence, a lot of concyclicity. In particular, one sees that the lines DE and FG intersect at Z. (Can you show it?)

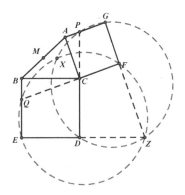

Observe the diagram. It *seems* that C lies on XZ, the common chord (and the radical axis) of the two circles. This is not difficult to show, by calculating the power of C with respect to the two circles, because we have many equal lengths in the diagram due to the squares.

It follows that $\angle BCX = \angle CZD$ (because C lies on XZ). One sees that C, D, Z, F are concyclic and hence, $\angle CZD = \angle CFD$. It suffices to show that $\angle ACM = \angle CFD$, where M is the midpoint of AB. Refer to the diagram on the right. It is much simpler! Does the diagram look familiar? (Refer to Example 1.2.11.)

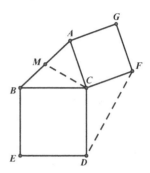

Proof. Let the lines FG and DE intersect at Z. Since $\angle PDZ = \angle PGZ = 90°$, P, D, Z, G are concyclic. Similarly, Q, E, Z, F are concyclic because $\angle QEZ = \angle QFZ = 90°$. Let Γ_1, Γ_2 denote the circumcircles of $\triangle PDG$ and $\triangle QEF$ respectively. We see that Z lies on both Γ_1 and Γ_2, i.e., XZ is the common chord of Γ_1 and Γ_2.

Notice that the power of C with respect to Γ_1 is $-PC \cdot CD$ and the power of C with respect to Γ_2 is $-QC \cdot CF$. Observe that $PC \cdot CD = PC \cdot BC$ and $QC \cdot CF = QC \cdot AC$.

It is easy to see that $\triangle BCQ \sim \triangle ACP$ because $\angle ACP = 90° - \angle ACB = \angle BCQ$. Hence, $\dfrac{PC}{AC} = \dfrac{QC}{BC}$, i.e., $PC \cdot BC = QC \cdot AC$. This implies the power of C with respect to Γ_1 and Γ_2 is the same. By Theorem 4.3.6, C lies on XZ, the radical axis of Γ_1 and Γ_2.

Now $\angle BCX = \angle CZD$ (because $BC \, // \, DE$)$= \angle CFD$ (angles in the same arc). Refer to the diagram on the right. Extend CM to C' such that $CM = C'M$. One sees that $\triangle ACC' \cong \triangle CFD$ (Example 1.2.11, or simply by S.A.S.). Now we have $\angle ACM = \angle CFD = \angle BCX$.

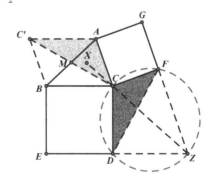

$\square$

Example 4.3.8 (IMO 85)   In a non-isosceles acute angled triangle $\triangle ABC$, D, E are points on AC, AP respectively such that B, C, D, E are concyclic on $\odot O$. Let the circumcircles of $\triangle ABC$ and $\triangle ADE$ intersect at A and P. Show that $AP \perp OP$.

Insight.   Refer to the diagram below. It is easy to see that BC and DE are not parallel. Since there are three circles, we immediately see that the radical axes are concurrent, say at X.

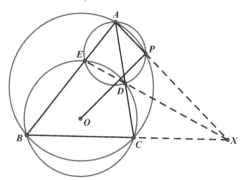

Notice that AO, OX (or more precisely, AO^2, OX^2) are closely related to the power of points A and X with respect to $\odot O$. One may also express the power of A with respect to $\odot O$ as $AD \cdot AC$ and the power of X with respect to $\odot O$ as $XB \cdot XC$. Since X lies on all radical axes (or by the Tangent Secant Theorem), we have $XB \cdot XC = XA \cdot XP$. How are these line segments helpful? Perhaps we can show $AP \perp OP$ by calculating $AO^2 - AP^2$ and $OX^2 - PX^2$. (Recall Theorem 2.1.9: $AP \perp OP$ if and only if $AO^2 - AP^2 = OX^2 - PX^2$.)

Proof.   If $DE \, / / \, BC$, $BCDE$ must be an isosceles trapezium (Exercise 3.1). Now $AB = AC$, which contradicts the fact that $\triangle ABC$ is non-isosceles. Hence, DE and BC are not parallel, say intersecting at X.

We conclude that the radical axes BC, DE, AP are concurrent at X (Theorem 4.3.2). Let the radius of $\odot O$ be R. Now the power of X with respect to $\odot O$ is $OX^2 - R^2 = BX \cdot CX = AX \cdot PX$ and the power of A

with respect to $\odot O$ is $AO^2 - R^2 = AD \cdot AC$. It follows that $AO^2 - OX^2 = AD \cdot AC - AX \cdot PX$. (1)

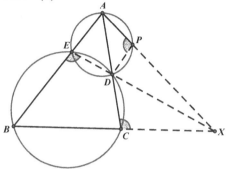

Refer to the diagram above. We have $\angle APD = \angle BED = \angle ACX$ (Corollary 3.1.5) and hence, C, D, P, X are concyclic. By the Tangent Secant Theorem, $AC \cdot AD = AP \cdot AX$. (2)

(1) and (2) imply that $AO^2 - OX^2 = AP \cdot AX - AX \cdot PX$

$$= AX \cdot (AP - PX) = (AP + PX)(AP - PX) = AP^2 - PX^2$$

In conclusion, $AO^2 - AP^2 = OX^2 - PX^2$, which implies $AP \perp OP$ (Theorem 2.1.9). $\square$

Note:

(1) One may see (2) from (1) and reverse engineering: Since we are to show $AO^2 - OX^2 = AP^2 - PX^2$, we **should** have

$$AC \cdot AD - AX \cdot PX = AP^2 - PX^2, \text{ or equivalently,}$$

$$AC \cdot AD = AX \cdot PX + (AP^2 - PX^2) = AP^2 + PX \cdot (AX - PX)$$

$$= AP^2 + PX \cdot AP = AP \cdot (AP + PX) = AP \cdot PX.$$

Hence, C, D, P, X **should** be concyclic. Once we see the necessity of this intermediate step, the proof is not difficult.

(2) One may also show the conclusion by angles. First, we show that E, D, X are collinear and C, D, P, X are concyclic as in the proof above. Now $\angle APE = \angle ADE = \angle CDX = \angle CPX$ (Corollary 3.1.3). Refer to the diagram below. We are to show $\angle OPA = \angle OPX = 90°$. Hence, it suffices to show that OP bisects $\angle CPE$.

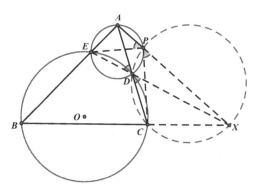

Consider $\triangle OEP$ and $\triangle OCP$. Refer to the diagram below. We have $OE = OC$ and we **should** have $\angle OPE = \angle OPC$. However, it seems that $\triangle OEP$ and $\triangle OCP$ are **not** congruent. Hence, we should have $\angle OEP + \angle OCP = 180°$! (Refer to the remarks before Example 3.3.6.) This implies C, O, E, P **should** be concyclic. Can we show it?

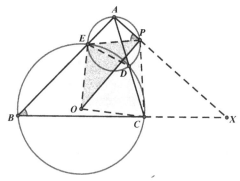

Notice that $\angle APE = \angle ADE = \angle B$ (Corollary 3.1.5). Since we have $\angle APE = \angle CPX$, it follows that $180° - \angle CPE = \angle APE + \angle CPX$ $= 2\angle B = \angle COE$ (Theorem 3.1.1), i.e., $\angle CPE + \angle COE = 180°$. Hence, C, O, E, P are concyclic.

Now $\angle OEP + \angle OCP = 180°$ and hence, $\sin \angle OEP = \sin \angle OCP$.

By Sine Rule, $\dfrac{OE}{\sin \angle OPE} = \dfrac{OP}{\sin \angle OEP} = \dfrac{OP}{\sin \angle OCP} = \dfrac{OC}{\sin \angle OPC}$.

Since $OC = OE$, we must have $\sin \angle OPE = \sin \angle OPC$.
Clearly, $\angle OPE + \angle OPC = \angle CPE < 180°$.
It follows that $\angle OPE = \angle OPC$, which completes the proof.

4.4 Ptolemy's Theorem

Besides the Intersecting Chords Theorem and the Tangent Secant Theorem, Ptolemy's Theorem provides another way to determine concyclicity *without* finding equal angles. Moreover, it gives useful identity regarding the sides and diagonals of a cyclic quadrilateral.

Theorem 4.4.1 (Ptolemy's Theorem)   *In a quadrilateral ABCD, $AB \cdot CD + BC \cdot AD \geq AC \cdot BD$ and the equality holds if and only if ABCD is cyclic.*

Proof.   Refer to the diagram on the right. Choose P such that $\angle ABD = \angle CBP$ and $\angle ADB = \angle BCP$, i.e., we construct similar triangles $\triangle ABD \sim \triangle PBC$.

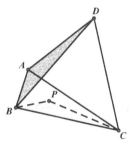

Hence, $\dfrac{AB}{BD} = \dfrac{BP}{BC}$ (1), and

$\dfrac{AD}{BD} = \dfrac{PC}{BC}$, i.e., $AD \cdot BC = BD \cdot PC$.   (2)

(1) implies that there is another pair of similar triangles: $\triangle ABP \sim \triangle DBC$. This is because the angles between the corresponding sides are the same: $\angle ABP = \angle ABD + \angle PBD = \angle CBP + \angle PBD = \angle CBD$.

Refer to diagram on the right. We have

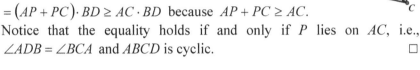

$$\frac{AB}{AP} = \frac{BD}{CD}, \text{ i.e., } AB \cdot CD = AP \cdot BD. \quad (3)$$

(2) and (3) give that $AB \cdot CD + BC \cdot AD$
$= AP \cdot BD + BD \cdot PC$
$= (AP + PC) \cdot BD \geq AC \cdot BD$ because $AP + PC \geq AC$.
Notice that the equality holds if and only if P lies on AC, i.e.,
$\angle ADB = \angle BCA$ and $ABCD$ is cyclic.   □

Ptolemy's Theorem is useful when solving problems regarding sides and diagonals about cyclic quadrilaterals. Refer to Example 3.1.10. One may see the conclusion immediately by applying Ptolemy's Theorem.

Example 4.4.2   Refer to the diagram on the right. $ABCD$ is a cyclic quadrilateral. Show that:

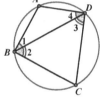

$\sin(\angle 1 + \angle 2) \cdot \sin(\angle 2 + \angle 3) \cdot \sin(\angle 3 + \angle 4) \cdot \sin(\angle 4 + \angle 1)$
$\geq 4 \sin \angle 1 \cdot \sin \angle 2 \cdot \sin \angle 3 \cdot \sin \angle 4$.

Insight.   One could see that $\sin(\angle 1 + \angle 2) = \sin \angle B = \sin \angle D$ because $\angle B + \angle D = 180°$ (Corollary 3.1.4). Hence, $\sin(\angle 1 + \angle 2) = \sin(\angle 3 + \angle 4)$
.
Similarly, $\sin(\angle 2 + \angle 3) = \sin(180° - \angle C) = \sin \angle A = \sin(\angle 1 + \angle 4)$.
Now it suffices to show that
$(\sin \angle A \cdot \sin \angle B)^2 \geq 4 \sin \angle 1 \cdot \sin \angle 2 \cdot \sin \angle 3 \cdot \sin \angle 4$.   (*)
However, it seems not easy to show (*) directly because we do not know how the *product* of $\sin \angle 1, \sin \angle 2, \sin \angle 3, \sin \angle 4$ is related to $\sin \angle A$ and $\sin \angle B$. Perhaps we should consider another strategy.
Notice that each of these angles (on the circumference) corresponds to a line segment in $ABCD$ by Sine Rule. For example, $AB = 2R \sin \angle 4$, $BD = 2R \sin \angle C = 2R \sin(\angle 2 + \angle 3)$, etc., where R is the radius of the circle.
Now (*) is equivalent to $(BD \cdot AC)^2 \geq 4AD \cdot CD \cdot BC \cdot AB$. We have all the four sides and the two diagonals of $ABCD$. Perhaps we can apply Ptolemy's Theorem.

Proof. By Sine Rule, $\sin(\angle 1 + \angle 2) = \dfrac{AC}{2R} = \sin(\angle 3 + \angle 4)$, where R is the radius of the circle. Similarly, we have $\sin(\angle 2 + \angle 3) = \sin \angle C = \sin(\angle 1 + \angle 4) = \dfrac{BD}{2R}$, $\sin \angle 1 = \dfrac{AD}{2R}$, $\sin \angle 2 = \dfrac{CD}{2R}$, $\sin \angle 3 = \dfrac{BC}{2R}$ and $\sin \angle 4 = \dfrac{AB}{2R}$. Now it suffices to show $(BD \cdot AC)^2 \geq 4AD \cdot CD \cdot BC \cdot AB$.

Ptolemy's Theorem gives $BD \cdot AC = AB \cdot CD + BC \cdot AD$. Hence, it suffices to show that $AB \cdot CD + BC \cdot AD \geq 2\sqrt{AD \cdot CD \cdot BC \cdot AB}$. (1)

Notice that (1) follows from the inequality $x^2 + y^2 \geq 2xy$, where $x = \sqrt{AB \cdot CD}$ and $y = \sqrt{AD \cdot BC}$. This completes the proof. □

Note: $x^2 + y^2 \geq 2xy$ because $x^2 + y^2 - 2xy = (x-y)^2 \geq 0$. Even though this is a commonly known fact and could be found in any elementary algebra textbook, one may not be able to recognize it immediately when it takes the form of (1).

Example 4.4.3 Given a parallelogram $ABCD$ where $\angle A > 90°$, a circle passing through B intersects AB, BC, BD at P, Q, R respectively. Show that $BP \cdot AB + BQ \cdot BC = BR \cdot BD$.

Insight. One notices that the conclusion looks like Ptolemy's Theorem. Refer to the diagram on the right.

In fact, we are given a circle, even though applying Ptolemy's Theorem on that circle directly does not give the conclusion. Instead, we have $BP \cdot QR + BQ \cdot PR = BR \cdot PQ$. Notice that AB, BC, BD are replaced by QR, PR, PQ respectively.

Are these line segments in ratio? If they are, i.e., $\dfrac{AB}{QR} = \dfrac{BC}{PR} = \dfrac{BD}{PQ}$, then we immediately have the conclusion. Do we have any pair of similar triangles which leads to such equal ratio? Considering the line segments involved, it must be $\triangle PQR$ and another triangle.

Can you see $\triangle PQR \sim \triangle BDC$? It should not be difficult to show, by circle properties and parallel lines, that the corresponding angles of these triangles are all equal. For example, $\angle PRQ = 180° - \angle ABC = \angle BCD$. We leave it to the reader to complete the proof.

Example 4.4.4   Given an acute angled triangle $\triangle ABC$ where O is the circumcenter, M, N are on AB, AC respectively such that O lies on MN. Let D, E, F be the midpoints of MN, BN, CM respectively. Show that O, D, E, F are concyclic.

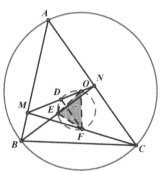

Insight.   Refer to the diagram on the right. Since we are to show that O, D, E, F are concyclic, it is natural to consider angles. Can we show that $\angle EDF = \angle EOF$? Since D, E, F are midpoints, we must have $DE \parallel AB$ and $DF \parallel AC$. Hence, $\angle EDF = \angle A$. Can we show $\angle EOF = \angle A$?   (1)
Similarly, $AC \parallel DF$ gives $\angle ANM = \angle ODF$.
Can we show $\angle ODF = \angle ANM = \angle OEF$?   (2)
Since O, E, F, D **should** be concyclic, (1) and (2) **should** be true, i.e., we **should** have $\triangle OEF \sim \triangle ANM$. Can we show this, say by the ratio of corresponding sides? Although DE, DF are related to BM, CN, we cannot apply Ptolemy's Theorem because we have **not** shown O, E, F, D are concyclic.

Apparently, there are many clues, but none of them is useful unless O, E, F, D are concyclic. Perhaps we can draw the circumcircle of $\triangle DEF$, which intersects MN at O' and show that O' coincide with O. By applying Ptolemy's Theorem to $O'DEF$ and replacing the lengths by those in $\triangle ABC$ (by similar triangles or the Midpoint Theorem), we might be close to the conclusion.

Proof. Let the circumcircle of $\triangle DEF$ intersect MN at O'. It is easy to see that $DE \parallel AB$ and $DF \parallel AC$. Hence, $\angle A = \angle EDF = \angle EO'F$ and $\angle O'EF = \angle O'DF = \angle ANM$, which imply $\triangle O'EF \sim \triangle ANM$. Since O', D, E, F are concyclic, Ptolemy's Theorem gives

$$O'D \cdot EF + O'F \cdot DE = O'E \cdot DF \text{ , or } O'D = \frac{O'E}{EF} \cdot DF - \frac{O'F}{EF} \cdot DE . \quad (1)$$

Since $\triangle O'EF \sim \triangle ANM$, $\dfrac{O'E}{EF} = \dfrac{AN}{MN}$ and $\dfrac{O'F}{EF} = \dfrac{AM}{MN}$.

Now $O'D = \dfrac{1}{MN}\big(AN \cdot DF - AM \cdot DE\big) = \dfrac{1}{2MN}\big(AN \cdot CN - AM \cdot BM\big),$ where $DE = \dfrac{1}{2}BM$ and $DF = \dfrac{1}{2}CN.$ (3)

Notice that $AN \cdot CN$ and $AM \cdot BM$ are the negative of the power of N, M with respect to the circumcircle of $\triangle ABC$ respectively. Hence, we have $AN \cdot CN = R^2 - NO^2$ and $AM \cdot CM = R^2 - MO^2$, where R denotes the radius of the circumcircle of $\triangle ABC$.

Now $O'D = \dfrac{1}{2MN}\big(MO^2 - NO^2\big) = \dfrac{1}{2MN}\big(MO + NO\big)\big(MO - NO\big)$ where $MO + NO = MN$. It follows that $O'D = \dfrac{MO - NO}{2} = OD$. (*)

This implies O and O' coincide. $\qquad \square$

Note:
(1) One sees that (*) holds regardless of the positions of O and O' on MN, i.e., if $MO < NO$, both $O'D$ and OD are negative, which means O and O' lie between M and D. If $MO > NO$, O and O' lie between N and D.
(2) Considering the power of M, N upon (3) is expected: We have not used the condition that O is the circumcenter of $\triangle ABC$ and we are to remove A, B, C in the expression of $O'D$!

Example 4.4.5   Given $\triangle ABC$, E, F are on AC, AB respectively such that BE, CF bisect $\angle B, \angle C$ respectively. P, Q are on the minor arc $\overarc{AC}$ of the circumcircle of $\triangle ABC$ such that $AC /\!/ PQ$ and $BQ /\!/ EF$. Show that $PA + PB = PC$.

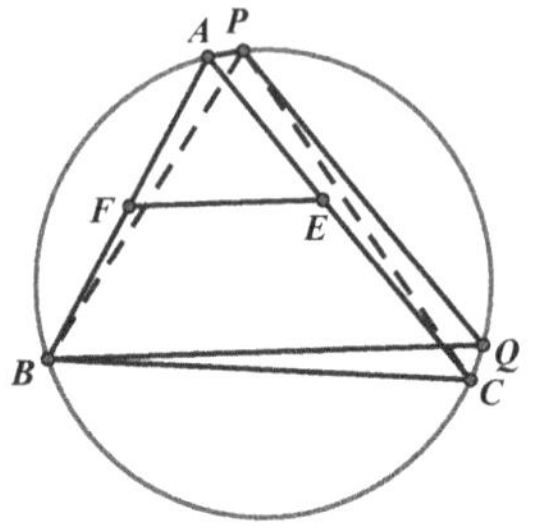

Insight.   Refer to the diagram on the right. We are to show the relationship among PA, PB and PC, which lie in the quadrilateral $PABC$. On the other hand, we might obtain equal angles from the circle and parallel lines.

For example, we have $\angle BAC = \angle BPQ$ and $\angle AEF = \angle PQB = \angle PCB$. It follows that $\triangle AEF \sim \triangle PCB$.

In the quadrilateral $PABC$, PA, PB, PC are related to AB, BC, AC by Ptolemy's Theorem. PB, PC are also related to AE, AF by similar triangles. Since AE, AF are angle bisectors, they can be expressed in terms of AB, BC, AC (Example 2.3.8). It seems that we are close to the conclusion.

Proof.   We have $\angle AEF = \angle PQB$ (since $EF /\!/ BQ$ and $AC /\!/ PQ$) $= \angle PCB$ (angles in the same arc). Since $\angle BAC = \angle BPC$ (angles in the same arc), we have $\triangle AFE \sim \triangle PBC$. Hence, $\dfrac{PB}{PC} = \dfrac{AF}{AE}$. (1)

Let $BC = a$, $AC = b$ and $AB = c$. We see $AE = \dfrac{bc}{a+c}$ and $AF = \dfrac{bc}{a+b}$ (Example 2.3.8). It follows that $\dfrac{AF}{AE} = \dfrac{a+c}{a+b}$. (2)

We are to show $PA + PB = PC$. By (1) and (2), it suffices to show that

$$PA + PB = PB \cdot \frac{a+b}{a+c}, \text{ or } PA = PB \cdot \left(\frac{a+b}{a+c} - 1 \right) = PB \cdot \frac{b-c}{a+c}.$$

Ptolemy's Theorem implies $PA \cdot a + PC \cdot c = PB \cdot b$.

Hence, $PA \cdot a + \left(PB \cdot \dfrac{a+b}{a+c} \right) \cdot c = PB \cdot b$, which gives

$PA \cdot a = PB \cdot \left(b - \dfrac{a+b}{a+c} \cdot c \right) = PB \cdot \left(\dfrac{b-c}{a+c} \cdot a \right)$. It follows that

$PA = PB \cdot \dfrac{b-c}{a+c}$, which completes the proof. □

4.5 Exercises

1. Given $\triangle ABC$ and its circumcircle $\odot O$, P is a point outside $\odot O$ such that $\odot P$ touches $\odot O$ at C. AC extended intersects $\odot P$ at D and BC extended intersects $\odot P$ at E. Show that if A, B, D, E are concyclic, then $AC \cdot CP = OC \cdot CE$.

2. Let AB be the diameter of a semicircle centered at O. $BP \perp AB$ at B and AP intersects the semicircle at C. Let D be the midpoint of BP. If $ACDO$ is a parallelogram, find $\sin \angle PAD$.

3. Given a cyclic quadrilateral $ABCD$, E is a point on AB such that $DE \perp AC$. Draw $BF /\!/ DE$, intersecting AD extended at F. Show that if $\angle B = 90°$, then $\dfrac{AB^3}{AD^3} = \dfrac{AF}{AE}$.

4. (CHN 96) In a quadrilateral $ABCD$, its diagonals AC and BD intersect at M. Draw a line $EF /\!/ AD$ passing through M, intersecting AB, CD at E, F respectively. Let EF extended intersect BC extended at O. Draw a circle centered at O with radius OM and P is a point on this circle. Show that $\angle OPF = \angle OEP$.

5. Given a circle and a point P outside the circle, draw PA, PB touching the circle at A, B respectively. C is a point on the minor arc $\overgroup{AB}$ and PC extended intersects the circle at D. Let E be a point on AC

extended and F a point on AD such that $EF \parallel PA$. If EF intersect AB at Q, show that $QE = QF$.

6. Let P be a point outside $\odot O$ and PA, PB touch $\odot O$ at A, B respectively. C is a point on the minor arc $\overset{\frown}{AB}$ and PC extended intersects $\odot O$ at D. If M is the midpoint of AB, show that O, M, C, D are concyclic.

7. Given a semicircle with the diameter AB, C is a point on the semicircle and D is the midpoint of the minor arc $\overset{\frown}{BC}$. Let AD, BC intersect at E. If $CE = 3$ and $BD = 2\sqrt{5}$, find AB.

8. Refer to the diagram on the right. Let $\odot O$ be the circumcircle of $\triangle ABC$. D is a point on the line BC such that the line AD touches $\odot O$ at A. E is a point on the line AC such that the line BE touches $\odot O$ at B. F is a point on the line AB such that the line CF touches $\odot O$ at C. Show that D, E, F are collinear.

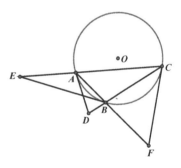

9. (CGMO 05) Given an acute angled triangle $\triangle ABC$ and its circumcircle, P is a point on the minor arc $\overset{\frown}{BC}$. AB extended intersects CP extended at E. AC extended intersects BP extended at F. If the perpendicular bisector of AC intersects AB at J and the perpendicular bisector of AB intersects AC at K, show that $\dfrac{CE^2}{BF^2} = \dfrac{AJ \cdot JE}{AK \cdot KF}$.

10. An acute angled triangle $\triangle ABC$ is inscribed inside $\odot O$. BO extended intersects AC at D. CO extended intersects AB at E. If the line DE intersects $\odot O$ at P, Q respectively and it is given that $AP = AQ$, show that $DE \, /\!/ \, BC$.

11. Refer to the diagram on the right. We have $\odot O_1$, $\odot O_2$ and $\odot O_3$ mutually tangent to each other at A, B, C respectively, while ℓ_1, ℓ_2, ℓ_3 are the common tangents passing through A, B, C respectively. Show that ℓ_1, ℓ_2, ℓ_3 are concurrent.

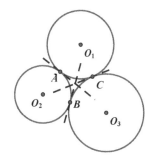

12. Let O_1, O_2 be two points inside $\odot O$. Draw $\odot O_1$ and $\odot O_2$, which touch $\odot O$ at A, B respectively. If $\odot O_1$ and $\odot O_2$ intersect at C, D and A, B, C are collinear, show that $OD \perp CD$.

13. In $\triangle ABC$, $\angle B = 2\angle C$, show that $AC^2 = AB \cdot (AB + BC)$.

14. $\odot O_1$ is tangent to two parallel lines ℓ_1, ℓ_2. Let O_2 be a point outside $\odot O_1$. $\odot O_2$ is tangent to $\odot O_1$ and ℓ_1 at A, B respectively. Let O_3 be a point outside $\odot O_1$ and $\odot O_2$. $\odot O_3$ is tangent to $\odot O_1$, ℓ_2 and $\odot O_2$ at C, D, E respectively. Show that the intersection of AD and BC is the circumcenter of $\triangle ACE$.

15. Let O be the circumcenter of $\triangle ABC$. P, Q are points on AC, AB respectively. Let M, N, L be the midpoints of BP, CQ, PQ respectively. Show that if PQ is tangent to the circumcircle of $\triangle MNL$, then we have $OP = OQ$.

Chapter 5

Basic Facts and Techniques in Geometry

5.1 Basic Facts

We have learnt a number of theorems and corollaries through the previous chapters. Besides those well-known results, we have also seen many examples, some of which are indeed commonly used facts in geometry. One familiar with these basic facts could find it significantly more effective when seeking clues and insights during problem-solving. Hence, we shall have a summary of these basic facts in this section.

❖ Most Commonly Used Facts

The following are standard results which could be used directly in problem solving, i.e., one may simply state these results without proof.

- In an acute angled triangle $\triangle ABC$, BD, CE are heights. We have $\angle ABD = \angle ACE$.

  Moreover, B, C, D, E are concyclic and A, D, H, E are concyclic, where H is the orthocenter of $\triangle ABC$.

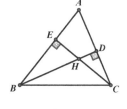

- In $\triangle ABC$, $\angle A = 90°$ and $AD \perp BC$ at D. We have $\angle BAD = \angle C$ and $\angle CAD = \angle B$.

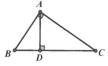

  Moreover, we have $AB^2 = BD \cdot BC$ and $AD^2 = BD \cdot CD$.

- Given $\triangle ABC$ where M is the midpoint of BC, we have $AM = \dfrac{1}{2}BC$ if and only if $\angle A = 90°$.

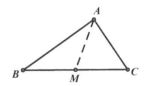

- Angle bisectors of neighboring supplementary angles are perpendicular to each other.

  Hence, if D, E are the ex-centers opposite B, C respectively in $\triangle ABC$, then DE passes through A and is perpendicular to the angle bisector of $\angle A$.

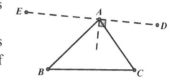

- Let $ABCD$ be a trapezium where $AD \,/\!/\, BC$. The angle bisectors of $\angle A$ and $\angle B$ are perpendicular to each other.

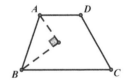

- In a right angled isosceles triangle $\triangle ABC$ where $\angle A = 90°$ and $AB = AC$, we have $\dfrac{AB}{BC} = \dfrac{1}{\sqrt{2}}$.

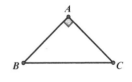

  In a right angled triangle $\triangle ABC$ where $\angle A = 90°$ and $\angle B = 30°$, we have $AC : AB : BC = 1 : \sqrt{3} : 2$.

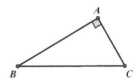

- In $\triangle ABC$, D is a point on BC and P is a point on AD. We have $\dfrac{[\triangle ABP]}{[\triangle ACP]} = \dfrac{BD}{CD}$.

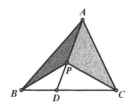

- Given $\triangle ABC$ and D is on AC such that $\angle ABD = \angle C$, we have $\triangle ABD \sim \triangle ACB$. Refer to the left diagram below.

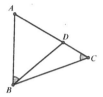 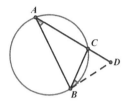

In particular, given $\triangle ABC$, if D is a point on AC extended and BD touches the circumcircle of $\triangle ABC$ at B, then $\triangle ABD \sim \triangle BCD$. Refer to the right diagram above.

- Let AD be the angle bisector of $\angle A$ in $\triangle ABC$. We have $BD = BC \cdot \dfrac{AB}{AB + AC}$.

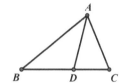

- Let AD, BE, CF be the heights of an acute angled triangle $\triangle ABC$. We have $\angle ABE = \angle ADF = \angle ADE = \angle ACF$.

  In particular, the orthocenter of $\triangle ABC$ is the incenter of $\triangle DEF$.

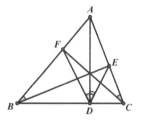

- Let H, I, O be the orthocenter, incenter and circumcenter of an acute angled triangle $\triangle ABC$ respectively. We have $\angle BHC = 180° - \angle A$, $\angle BIC = 90° + \dfrac{1}{2}\angle A$ and $\angle BOC = 2\angle A$.

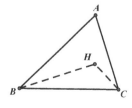 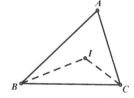 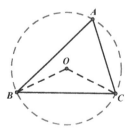

- Given an acute angled triangle $\triangle ABC$ and its circumcenter O, we must have $\angle A + \angle OBC = 90°$.

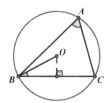

- Refer to the diagram on the right. Given a circle with $\overset{\frown}{AB} = \overset{\frown}{CD}$, we must have $AD \,/\!/\, BC$.

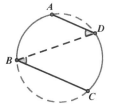

  Proof. Notice that $\overset{\frown}{AB} = \overset{\frown}{CD}$ implies $\angle ADB = \angle CBD$. Hence, $AD \,/\!/\, BC$.   □

- Refer to the left diagram below. AB, CD are two chords in $\odot O$ and AB, CD intersect at P. We have $\angle AOD + \angle BOC = 2\angle APD$.

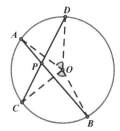
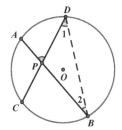

  Proof. Refer to the right diagram above. We have $\angle BOC = 2\angle 1$ and $\angle AOD = 2\angle 2$ (Theorem 3.1.1).
  Now $\angle APD = \angle 1 + \angle 2$ and the conclusion follows.   □

- Given $\triangle ABC$ and its circumcircle, D, E, F are the midpoints of arcs $\overset{\frown}{BC}$, $\overset{\frown}{AC}$, $\overset{\frown}{AB}$ respectively. Refer to the diagram on the right.

  We have AD, BE, CF the angle bisectors of $\triangle ABC$ and hence, concurrent at I, the incenter of $\triangle ABC$.

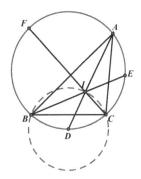

Notice that D, E, F are the circumcenters of $\triangle BCI, \triangle ACI, \triangle ABI$ respectively (Example 3.4.2).

- Given a right angled triangle $\triangle ABC$ with $\angle C = 90°$, we have $r = \dfrac{1}{2}(AC + BC - AB)$, where r is the radius of the incircle of $\triangle ABC$.

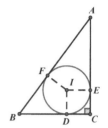

Proof. Let I be the incenter of $\triangle ABC$.
Suppose the incircle of $\triangle ABC$ touches BC, AC, AB at D, E, F respectively. Refer to the diagram above.
It is easy to see that $AE = AF$, $BD = BF$ and $CDIE$ is a square.

It follows that $r = CD = \dfrac{1}{2}(CD + CE) = \dfrac{1}{2}(AC + BC - AB)$.   □

Note: Let $BC = a$, $AC = b$, $AB = c$. We have $r = \dfrac{1}{2}(a + b - c)$.

By Theorem 3.2.9, we have $r = \dfrac{2S}{a + b + c}$ where $S = [\triangle ABC]$.

Indeed, $2r \cdot (a + b + c) = (a + b - c)(a + b + c) = (a + b)^2 - c^2$

$= (a + b)^2 - (a^2 + b^2) = 2ab = 4S$.

❖ Useful Facts

One familiar with the following facts may see clues and intermediate steps in problem-solving quickly, which might tremendously simplify the conclusion to be shown. While experienced contestants simply state these well-known facts during competitions, beginners are recommended not to omit any necessary proof to these results (which were illustrated in the previous chapters).

Occasionally, one may derive an intermediate step, but find it irrelevant to the problem given. If it seems not a useful clue, one should put it aside and refrain from wasting time exploring that piece further.

- Let H be the orthocenter of an acute angled triangle $\triangle ABC$. We have $AH = \dfrac{BC}{\tan A}$.

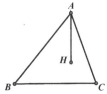

- Let $ABCD$ be a cyclic quadrilateral. E, F are on AB, CD respectively such that $BC /\!/ EF$. We must have A, E, F, D concyclic.

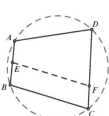

- In a quadrilateral $ABCD$, $AB = AD$ and $BC \neq CD$, if AC bisects $\angle C$, then $ABCD$ is cyclic.

- Let O be the circumcenter of an acute angled triangle $\triangle ABC$ and AD is a height. We have $\angle CAD = \angle BAO$.

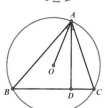

- Let I be the incenter of $\triangle ABC$. If BI extended intersects the circumcircle of $\triangle ABC$ at P, we have $AP = PI = CP$.

Hence, P is the circumcenter of $\triangle AIC$.

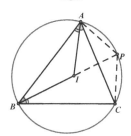

- Let H be the orthocenter of an acute angled triangle $\triangle ABC$. $AD \perp BC$ at D and AD extended intersects the circumcircle of $\triangle ABC$ at E. We have $DE = DH$. Refer to the left diagram below.

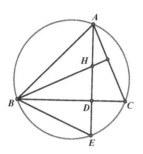
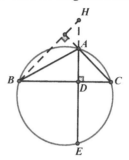

Notice that the conclusion still holds if $\triangle ABC$ is a right angled triangle (i.e., A is the orthocenter and BC is the diameter of the circumcircle) or an obtuse angled triangle. Refer to the right diagram above. The proof is similar and we leave it to the reader.

- Let H be the orthocenter of $\triangle ABC$ and M be the midpoint of AB. Let HM extended intersect the circumcircle of $\triangle ABC$ at D. We have that $BDCH$ is a parallelogram and hence, AD is a diameter of the circumcircle of $\triangle ABC$. Refer to the left diagram below.

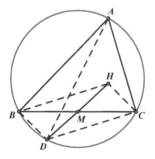
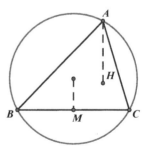

Moreover, we have $AH = 2OM$, where O is the circumcenter of $\triangle ABC$. Refer to the right diagram above.

- Let BD, CE be the heights of an acute angled triangle $\triangle ABC$. If the line DE intersects the circumcircle of $\triangle ABC$ at P, Q respectively, we have $AP = AQ$.

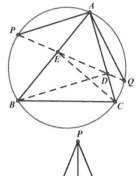

- Let P be a point outside $\odot O$ and PA, PB touch $\odot O$ at A, B respectively. We have that the incenter of $\triangle PAB$ is the midpoint of the minor arc $\overset{\frown}{AB}$.

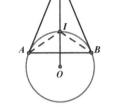

We shall see how these facts (together with theorems and standard results) could be helpful in problem-solving.

Example 5.1.1   In an acute angled triangle $\triangle ABC$, AD, BE, CF are heights. Let the incircle of $\triangle DEF$ touch EF, DF, DE at P, Q, R respectively. Show that $\triangle PQR \sim \triangle ABC$.

Proof. Refer to the diagram on the right. Let H be the orthocenter of $\triangle ABC$. It is well-known that H is also the incenter of $\triangle DEF$. In particular, DH bisects $\angle EDF$ and we have

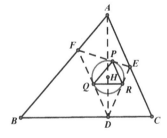

$$\angle BDF = \angle CDE = \frac{1}{2}\left(180° - \angle EDF\right).$$

Since $DR = DQ$, we have $\angle DQR = \frac{1}{2}\left(180° - \angle EDF\right) = \angle BDF$, which implies $BC // QR$. Similarly, $PQ // AB$ and $PR // AC$.

We must have $\angle A = \angle P$, $\angle B = \angle Q$ and hence the conclusion.   □

Note: One may attempt to show $\angle A = \angle P$ by observing that $\angle P = \angle DQR$ and $\angle A = \angle CDE$. Can we show that $\angle DQR = \angle DRQ = \angle CDE$? Notice that **if** we have $\angle DRQ = \angle CDE$, it follows immediately that $BC \, // \, QR$ and similarly, $PQ \, // \, AB$ and $PR \, // \, AC$. Indeed, we **should** have these parallel lines.

Example 5.1.2 (CHN 10)   Let ℓ be a straight line and P is a point which does not lie on ℓ. A, B, C are distinct points on ℓ. Let the circumcenters of ΔPAB, ΔPBC, ΔPCA be O_1, O_2, O_3 respectively. Show that P, O_1, O_2, O_3 are concyclic.

Insight.   Refer to the diagram on the right. This is indeed a complicated diagram and if we draw all the perpendicular bisectors explicitly, it will be unreadable!

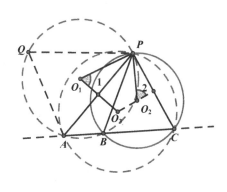

Since we are asked to show P, O_1, O_2, O_3 are concyclic, it is natural to search for equal angles. For example, can we show $\angle 1 = \angle 2$?

Notice that both $\angle 1$ and $\angle 2$ are at the center of a circle. Moreover, it is easy to see that $O_1O_3 \perp PA$ and $O_2O_3 \perp PC$, i.e., $\angle 1 = \dfrac{1}{2} \angle AO_1P$ and $\angle 2 = \dfrac{1}{2} \angle CO_2P$. Can we show $\angle AO_1P = \angle CO_2P$?

Notice that $\angle CO_2P = 2\angle CBP$, where $\angle CBP = \angle AQP$ (Corollary 3.1.5) $= \dfrac{1}{2} \angle AO_1P$ (Theorem 3.1.1). This completes the proof.

Alternatively, one may also show that P, O_1, O_2, O_3 are concyclic via $\angle PO_1O_2 = \angle PO_3O_2$: Since $O_1O_2 \perp PB$, we have

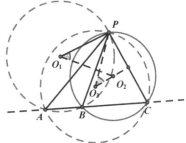

$$\angle PO_1O_2 = \frac{1}{2}\angle PO_1B = \angle PAB.$$

Meanwhile, $\angle PO_3O_2 = \frac{1}{2}\angle PO_3C = \angle PAC.$ This completes the proof.

Note: This could be considered a very easy problem if one is familiar with the basic properties in circle geometry, including recognizing the angles needed while disregarding the unnecessary line segments. Indeed, if one decides to show the concyclicity via angle properties, it is natural to consider either of the approaches above.

Example 5.1.3   Given $\triangle ABC$ and its incenter I, the circumcircles of $\triangle AIB$ and $\triangle AIC$ intersect BC at D, E respectively. Show that $DE = AB + AC - BC$.

Insight.   One recalls that the circumcenter of $\triangle AIB$ is indeed the intersection of CI extended with the circumcircle of $\triangle ABC$. Refer to the diagram on the right. Let CI extended intersect the circumcircle of $\triangle ABC$ at P. We have $PA = PB = PI = PD$. Notice that CI bisects $\angle C$.

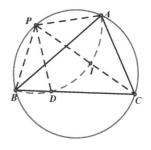

It is not difficult to see $\triangle ACP \cong \triangle DCP$. (Can you show it?) Hence, $AC = CD$. Similarly, $AB = BE$. Now it is easy to see the conclusion because $BE + CD - DE = BC$. We leave the details to the reader.

Warning: One should **not** conclude $\triangle ACP \cong \triangle DCP$ via $PA = PD$, $\angle DCP = \angle ACP$ and $PC = PC$. This is NOT S.A.S.! Instead, one may show that $\angle PAC = 180° - \angle PBC = 180° - \angle PDB = \angle PDC$ and apply A.A.S.

Example 5.1.4 (CMO 11) Let P be a point inside $\triangle ABC$ such that $\angle PBA = \angle PCA$. Draw $PD \perp AB$ at D and $PE \perp AC$ at E. Show that the perpendicular bisector of DE passes through the midpoint of BC.

Insight. Refer to the diagram on the right. It seems the conclusion is easy to show **if** P is the orthocenter of $\triangle ABC$ (i.e., when BD and CE intersect at P), in which case we have $ME = MD = \dfrac{1}{2} BC$ where M is the midpoint of BC. The conclusion follows immediately.

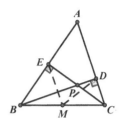

Of course, P may not be the orthocenter of $\triangle ABC$, but we **should** still have $MD = ME$. How can we show it? We cannot apply the previous argument since M is not the midpoint of the hypotenuse in a right angled triangle. What if we construct one, say the midpoint of BP?

Proof. Refer to the diagram on the right. Let M be the midpoint of BC. Let F, G be the midpoints of BP, CP respectively.

In the right angled triangle $\triangle BEP$, $EF = \dfrac{1}{2} BP$.

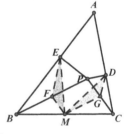

In $\triangle BCP$, MG is a midline and hence, $MG = \dfrac{1}{2} BP$ and $MG /\!/ BP$. It follows that $EF = MG$.

Similarly, $FM /\!/ CP$ and $FM = DG$. Now $FPGM$ is a parallelogram. Notice that $\angle EFM = \angle EFP + \angle PFM = 2\angle PBA + \angle PFM$. Similarly, $\angle MGD = 2\angle PCA + \angle PGM$. Since $\angle PFM = \angle PGM$ (in the parallelogram $FPGM$) and given that $\angle PBA = \angle PCA$, we must have $\angle EFM = \angle MGD$.

Now $\triangle EFM \cong \triangle MGD$ (S.A.S.), which implies $MD = ME$. It follows that M lies on the perpendicular bisector of DE. $\qquad\square$

Note: The condition $\angle PBA = \angle PCA$ seems not easy to apply at first. We leave it aside. Once we see that $\triangle EFM$ and $\triangle MGD$ **should** be congruent, it becomes natural to show equal angles using this condition.

Example 5.1.5   Let I be the incenter of $\triangle ABC$. AI extended intersects the circumcircle of $\triangle ABC$ at P. Draw $ID \perp BP$ at D and $IE \perp CP$ at E. Show that $ID + IE = AP \sin \angle BAC$.

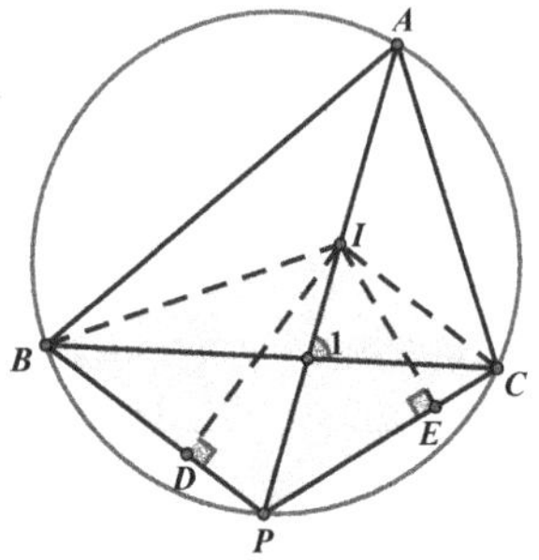

Insight.   Refer to the diagram on the right. We immediately recall that $PB = PC = PI$. However, one may find it difficult to construct a line segment equal to $ID + IE$. Since ID, IE are heights, perhaps we could use the area method. Notice that

$$[BPCI] = [\triangle BPI] + [\triangle CPI] = \frac{1}{2} BP \cdot ID + \frac{1}{2} CP \cdot IE = \frac{1}{2} BP \cdot (ID + IE).$$

On the other hand, $[BPCI] = \dfrac{1}{2} BC \cdot PI \sin \angle 1 = \dfrac{1}{2} BC \cdot BP \sin \angle 1$.   (*)

It follows that $ID + IE = BC \sin \angle 1$. Now it suffices to show that $BC \sin \angle 1 = AP \sin \angle BAC$.

Is it reminiscent of Sine Rule? Shall we show that $\dfrac{BC}{\sin \angle BAC} = \dfrac{AP}{\sin \angle 1}$?

Indeed, applying Sine Rule repeatedly gives $\dfrac{BC}{\sin \angle BAC} = \dfrac{AB}{\sin \angle ACB}$

$= \dfrac{AB}{\sin \angle APB} = \dfrac{AP}{\sin \angle ABP}$.

One sees the conclusion by showing $\angle ABP = \angle 1$. We leave it to the reader to complete the proof. (**Hint:** $\angle PBC = \angle PAC = \angle PAB$.)

Note: We use the fact that $[BPCI] = \frac{1}{2}BC \cdot PI \sin \angle 1$ in (*). Indeed, this holds for a general quadrilateral. Refer to the left diagram below where AC, BD intersect at P. We must have $[ABCD] = \frac{1}{2}AC \cdot BD \sin \angle 1$.

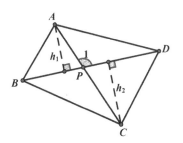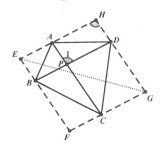

Notice that $[ABCD] = [\triangle ABD] + [\triangle BCD] = \frac{1}{2}BD \cdot h_1 + \frac{1}{2}BD \cdot h_2$, where h_1, h_2 are heights from A, C to BD respectively. Notice that $h_1 = AP \sin \angle 1$ and $h_2 = CP \sin \angle 1$ because $\sin \angle 1 = \sin \angle APB$. Hence,

$$[ABCD] = \frac{1}{2}BD \cdot (h_1 + h_2) = \frac{1}{2}BD \cdot (AP + CP) \cdot \sin \angle 1 = \frac{1}{2}AC \cdot BD \sin \angle 1.$$

Alternatively, one may also draw lines passing through A, C and parallel to BD, and lines passing through B, D and parallel to AC. Refer to the right diagram above. Notice that $EFGH$ is a parallelogram. One sees that $[ABCD] = \frac{1}{2}EFGH = [\triangle EGH] = \frac{1}{2}EH \cdot GH \sin \angle H$. Hence, we still have

$[ABCD] = \frac{1}{2}AC \cdot BD \sin \angle 1$ since $GH = AC$, $EH = BD$ and $\angle 1 = \angle H$.

Example 5.1.6 (CWMO 12) In an acute angled triangle $\triangle ABC$, D is on BC such that $AD \perp BC$. Let O and H be the circumcenter and orthocenter of $\triangle ABC$ respectively. The perpendicular bisector of AO intersects BC extended at E. Show that the midpoint of OH is on the circumcircle of $\triangle ADE$.

Insight. Refer to the diagram on
the right. Let N be the midpoint of
OH. One sees that M lies on the
circumcircle of $\triangle ADE$. i.e.,
A, M, D, E are concyclic since
$\angle AME = \angle ADE = 90°$. We are to
show N also lies on this circle.
It seems easier to show the
concyclicity involving M instead
of E, as we know more about M
than E.

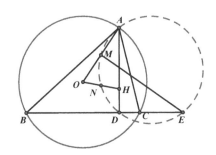

Can we show that A, M, N, D are concyclic? Notice that M, N are both
midpoints and we have $MN \,/\!/\, AH$. Hence, we **should** have $MNDA$ an
isosceles trapezium. How can we show it?

Notice that we have used the condition
about M and the perpendicular bisector
of AO, the midpoint N and the
orthocenter H, but we have not used the
condition about O and the circumcircle.
How could we relate O and the
circumcircle to A, M, N and D?

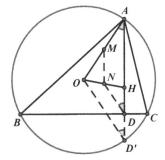

Recall that if AD extended intersects the circumcircle at D'. we have
$DH = DD'$. Refer to the diagram above.
It follows that $DN \,/\!/\, OD'$. This implies $\angle ADN = \angle AD'O = \angle OAD$.
Hence, $ADNM$ is an isosceles trapezium and the conclusion follows.

Example 5.1.7   Given a non-isosceles
acute angled triangle $\triangle ABC$, O is its
circumcenter. P is a point on AO
extended such that $\angle BPA = \angle CPA$.
Refer to the diagram on the right. Draw
$PQ \perp AB$ at Q, $PR \perp AC$ at R and
$AD \perp BC$ at D. Show that $PQDR$ is a
parallelogram.

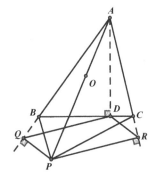

Insight. It is natural to consider showing $PQ // DR$ and $PR // DQ$. Given that $PQ \perp AB$ and $PR \perp AC$, it suffices to show $DQ \perp AC$ and $DR \perp AB$. Let us focus on one of them, say $DQ \perp AC$: most probably a similar argument applies for the other. How can we show that $\angle BAC + \angle AQD = 90°$?

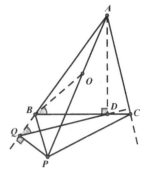

Recall $\angle BAC = 90° - \angle OBC$, i.e., it suffices to show $\angle AQD = \angle OBC$. Refer to the diagram above. How are these two angles related? It seems not very clear.

On the other hand, we are given $\angle BPA = \angle CPA$. How can we use this condition? Can you see that O, B, P, C are concyclic (Example 3.1.11)? Now we have $\angle OBC = \angle APC$. Can we show $\angle APC = \angle AQD$? One may also notice that $\angle CAO = \angle BAD$. It seems that we **should** have $\triangle PAC \sim \triangle QAD$. Refer to the diagram below.

Proof. It is easy to see $\angle CAD = \angle BAO$ (Example 3.4.1). Hence, $\triangle AQP \sim \triangle ADC$

and $\dfrac{AQ}{AP} = \dfrac{AD}{AC}$.

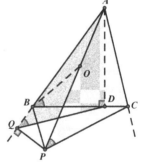

Notice that $\angle CAO = \angle BAD$. We must have $\triangle QAD \sim \triangle PAC$. It follows that $\angle AQD = \angle APC$.

Since $\triangle ABC$ is non-isosceles, we must have $PB \neq PC$. Otherwise OP is the perpendicular bisector of BC, which implies $AB = AC$. It follows that O, B, P, C are concyclic (Example 3.1.11), which implies that $\angle OBC = \angle APC = \angle AQD$. Notice that $\angle OBC + \angle BAC = 90°$ because O is the circumcenter of $\triangle ABC$. It follows that $\angle AQD + \angle BAC = 90°$, which implies that $DQ \perp AC$, i.e., $DQ // PR$. Similarly, $DR // PQ$ and the conclusion follows. □

Note: One sees that familiarity with basic facts in geometry is important in solving this problem.

Example 5.1.8   In an equilateral triangle $\triangle ABC$, D is a point on BC. Let O_1, I_1 be the circumcenter and incenter of $\triangle ABD$ respectively, and O_2, I_2 be the circumcenter and incenter of $\triangle ACD$ respectively. If the lines $O_1 I_1$ and $O_2 I_2$ intersect at P, show that D is the circumcenter of $\triangle O_1 P O_2$.

Insight.   Apparently, the construction of the diagram is not simple. Perhaps we shall consider the circumcenters and incenters separately.

Refer to the following diagrams. Can you see $\angle I_1 A I_2 = 30°$? Can you see $\odot O_1$ and $\odot O_2$ have the same radius (by Sine Rule) and hence, $A O_1 D O_2$ is a rhombus?

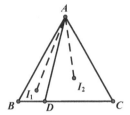 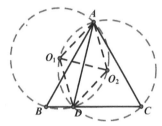

If we focus on one triangle, say $\triangle ACD$ with its incenter and circumcenter, we have $\angle AO_2 D = 2\angle C$ and $\angle AI_2 D = 90° + \dfrac{1}{2}\angle C$.

But these two angles are the same since $\angle C = 60°$! This implies A, O_2, I_2, D are concyclic. Refer to the left diagram below.

   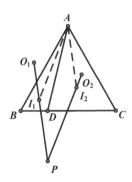

One sees that $\Delta O_1 O_2 D$ is an equilateral triangle. Hence it suffices to show that $\angle P = 30°$. Refer to the previous right diagram. We **should** have $\angle P = \angle I_1 A I_2$. It *seems* that $A I_1 P I_2$ is a parallelogram. Can we show it? (We have not used the concyclicity of A, O_2, I_2, D.)

Proof. Since O_2, I_2 are the circumcenter and incenter of ΔACD respectively, we have $\angle A O_2 D = 2\angle C$ and $\angle A I_2 D = 90° + \dfrac{1}{2}\angle C$.

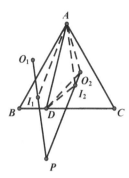

It follows that $\angle A O_2 D = \angle A I_2 D = 120°$ because $\angle C = 60°$. Hence, A, O_2, I_2, D are concyclic and we have $\angle A I_2 O_2 = \angle A D O_2 = 30°$ (because $A O_2 = D O_2$ and $\angle A O_2 D = 120°$).

Notice that $\angle I_1 A I_2 = \dfrac{1}{2}\angle BAC = 30°$, which implies $\angle I_1 A I_2 = \angle A I_2 O_2$, i.e., $A I_1 \,/\!/\, O_2 P$. Similarly, $A I_2 \,/\!/\, O_1 P$ and $A I_1 P I_2$ is a parallelogram.

In particular, $\angle P = \angle I_1 A I_2 = 30°$.

On the other hand, let the circumradius of ΔABD and ΔACD be r_1, r_2 respectively. By Sine Rule, $\dfrac{AB}{\sin \angle ADB} = 2r_1$ and $\dfrac{AC}{\sin \angle ADC} = 2r_2$.

Notice that $\sin \angle ADB = \sin \angle ADC$ (because $\angle ADB = 180° - \angle ADC$) and $AB = AC$. It follows that $r_1 = r_2$ and hence, $A O_1 D O_2$ is a rhombus. In particular, $\Delta O_1 O_2 D$ is an equilateral triangle.

Now $\angle P = 30° = \dfrac{1}{2}\angle O_1 D O_2$ implies that P lies on the circle centered at D with radius $O_1 D$. This completes the proof. $\qquad\square$

Example 5.1.9 (EGMO 12)   Given an acute angled triangle $\triangle ABC$, its circumcircle Γ and orthocenter H, K is a point on the minor arc $\overset{\frown}{BC}$. Let L be the reflection of K about the line AB and M be the reflection of K about the line BC. The circumcircle of $\triangle BLM$ intersects Γ at B and E. Show that the lines KH, EM and BC are concurrent.

Insight.   Refer to the diagram on the right. It seems not easy to show the concurrency using Ceva's Theorem. However, we notice that H and D are symmetric about BC, where D is the intersection of AH extended and Γ. On the other hand, it is given that M and K are symmetric about BC.

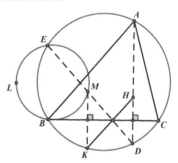

Now it is easy to see that MD, KH and BC are concurrent, because BC is the perpendicular bisector of HD and MK, where $HD \,/\!/\, MK$.

Since we are to show the lines KH, EM and BC are concurrent, it suffices to show that E, M, D are collinear. Notice that there are many equal angles in the diagram due to the two circles and the symmetry of K, L and K, M. Is there any angle related to say the point E?

How about $\angle BEM$? One sees immediately that $\angle BEM = \angle BLM$. Refer to the diagram on the right. Since L, M are reflections of K about AB, BC respectively, we have $BK = BM = BL$. It follows that $\angle BLM = \angle BML$. Can we show that $\angle BAD = \angle BED = \angle BEM$?

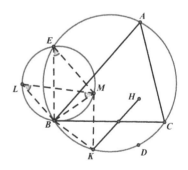

Unfortunately, neither $\angle BLM$ nor $\angle BML$ seems directly related to $\angle BAD$. Perhaps we can write $\angle BLM = 90° - \dfrac{1}{2}\angle MBL$. Notice that $\angle MBL = \angle ABL + \angle ABM$ and these angles, after applying the reflections, might be related to $\angle BAD$.

Proof. Let AH extended intersect Γ at D. We know that D is the reflection of H about BC. Since M is the reflection of K about BC, BC is the perpendicular bisector of both MK and HD. Hence, $MK \parallel HD$. Now $DHMK$ is an isosceles trapezium and it is easy to see that KH, DM, BC are concurrent. We claim that E, M, D are collinear.

Since L, M are reflections of K about AB, BC respectively, one sees that $BK = BM = BL$, which implies $\angle BLM = \angle BML$.

Now we have $\angle BEM = \angle BLM = 90° - \dfrac{1}{2} \angle MBL$

$$= 90° - \frac{1}{2}\left(\angle ABL + \angle ABM\right) = 90° - \frac{1}{2}\left[\angle ABK + \left(\angle ABC - \angle CBM\right)\right]$$

$$= 90° - \frac{1}{2}\left[\left(\angle ABC + \angle CBK\right) + \left(\angle ABC - \angle CBK\right)\right]$$

$= 90° - \angle ABC = \angle BAD = \angle BED$. Hence, E, M, D are collinear. We conclude that KH, EM and BC are concurrent. $\square$

Note: One may find it difficult to show that E, M, D are collinear by $\angle BME + \angle BMD = 180°$. Indeed, we do not know much about $\angle BMD$ or $\angle BKH$ because K is an arbitrary point.

Example 5.1.10 (USA 12) Let $ABCD$ be a cyclic quadrilateral whose diagonals AC, BD intersect at P. Draw $PE \perp AB$ at E and $PF \perp CD$ at F. BF and CE intersect at Q. Show that $PQ \perp EF$.

Insight. Refer to the diagram on the right. Apparently, the construction of the diagram is straightforward, but it is not clear how we could show $PQ \perp EF$. Even if we extend QP, intersecting EF, it seems difficult to find the angles at the intersection. Perhaps we shall leave the conclusion aside and study the diagram further.

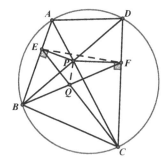

Notice that E and F are introduced by perpendicular lines. $ABCD$ is cyclic. If we introduce more perpendicular lines, we should obtain more concyclicity by the right angles. Refer to the left diagram below.

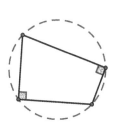 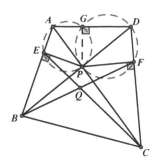

Let us draw say $PG \perp AD$ at G. Refer to the right diagram above. We immediately obtain two circles, i.e., A, E, P, G and D, F, P, G are concyclic. Even though this seems not directly related to our conclusion, it gives us an inspiration: what if we draw $PG \perp EF$ at G instead? Perhaps we could still obtain concyclicity and it would suffice to show that P, G, Q are collinear, or PG passes through Q. Refer to the diagram below.

Since BF, CE intersect at Q, it suffices to show PG, BF, CE are concurrent. Is it reminiscent of radical axes? Let Γ_1, Γ_2 denote the circumcircles of $\triangle EGP$ and $\triangle FGP$ respectively. We see that PG is the radical axis of Γ_1, Γ_2.

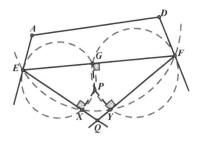

If we can find another circle Γ_3 such that EQ, FQ are the radical axes of Γ_1, Γ_3 and Γ_2, Γ_3 respectively, the conclusion follows. Let Γ_1 intersect the line EQ at X and Γ_2 intersect the line FQ at Y. It is easy to see that X, Y are the feet of the perpendiculars from P to EQ, FQ respectively. Can we show that E, X, Y, F are concyclic? This should not be difficult to as we have an abundance of concyclicity in the diagram (for example, B, E, P, Y are concyclic because $\angle BEP = \angle BYP = 90°$) and hence, many pairs of equal angles.

Proof. Draw $PG \perp EF$ at G, $PX \perp CE$ at X and $PY \perp BF$ at Y. Clearly, E,G,P,X are concyclic and F,G,P,Y are concyclic. We claim that E,X,Y,F are concyclic, which implies the radical axes EX, FY and PG are concurrent at Q (Theorem 4.3.2) and hence, leads to the conclusion. Refer to the left diagram below.

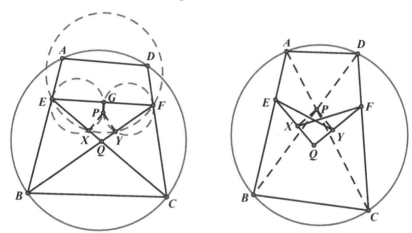

It suffices to show $\angle EXF = \angle EYF$. Since $\angle PXE = \angle PYF = 90°$, it suffices to show $\angle PXF = \angle PYE$. Refer to the right diagram above. Notice that $\angle BEP = \angle BYP = 90°$. Hence, B,E,P,Y are concyclic and $\angle PYE = \angle ABD$. Similarly, $\angle PXF = \angle ACD$. This completes the proof as $\angle ABD = \angle ACD$ (angles in the same arc). $\qquad\square$

5.2 Basic Techniques

Knowing the basic facts and important theorems well is important for solving geometry problems, but is still insufficient. In fact, it is common to see beginners who diligently learn many theorems, but do not know how to apply those results and solve geometry problems. Indeed, many beginners are not aware of the commonly used *techniques* (instead of theorems), which are not found in most textbooks.

The following is an elementary example: **NO** advanced knowledge is required to solve this problem. Can you see the clues without referring to the solution?

Example 5.2.1   Given a quadrilateral $ABCD$ where $AD = BC$ and $\angle BAC + \angle ACD = 180°$, show that $\angle B = \angle D$.

Insight.   It seems not easy to apply the condition $\angle BAC + \angle ACD = 180°$ since the angles are far apart. Can we put them together? If we extend the line CD, say the lines AB and CD intersect at E, can you see that we obtain an isosceles triangle?

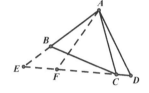

If $\angle BAC = \angle ACD = 90°$, it is easy to see that $ABCD$ is a parallelogram and we have $\angle B = \angle D$ immediately. Otherwise, say without loss of generality that $\angle BAC < 90°$, AB extended and DC extended intersect at E. Refer to the diagram above. We have $AE = CE$. It seems not clear how $AD = BC$ leads to the conclusion because they are far apart. Can we put them together? If we draw $AF = BC$, where F is on DC extended, we obtain an isosceles trapezium!

Proof.   If $\angle BAC = \angle ACD = 90°$, we have $\triangle BAC \cong \triangle DCA$ (H.L.) and hence, $ABCD$ is a parallelogram and $\angle B = \angle D$.

Suppose $\angle BAC < 90°$. Let DC extended and AB extended intersect at E. Since $\angle BAC + \angle ACD = 180°$, we have $\angle BAC = \angle ECA$ and $AE = CE$. Choose F on the line CD such that $AF = AD$. We have $\angle D = \angle AFD$. Now $BC = AD = AF$ gives $\triangle ABC \cong \triangle CFA$ (S.A.S.). It follows that $\angle B = \angle AFD = \angle D$.

If $\angle BAC > 90°$, the lines AB and CD intersect at the other side of AC and a similar argument applies.   □

Note: We used "cut and paste" to find clues in this problem: since $\angle BAC$ and $\angle ACD$ are supplementary, if we put them together, a straight line is obtained. We also put the line segments AD, BC together, which gives an isosceles trapezium. Notice that simply applying any theorem directly to this problem will not give the conclusion.

Basic and commonly used techniques in solving geometry problems include the following:

- Cut and paste

  When given equal line segments, equal or supplementary angles, and sum of angles or line segments which are far apart, one may cut and paste, moving those angles or line segments together. This technique may give straight lines, isosceles triangles or congruent triangles.

- Construct congruent and similar triangles.

  One strategy to show equal angles or line segments is to place them in congruent or similar triangles. If no such triangles exist in the diagram, consider drawing auxiliary lines and construct one! Notice that any other angles or line segments known to be equal may give inspiration on which triangles **could** be congruent or similar.

- Reflection about an angle bisector

  When given an angle bisector, it is naturally a line of symmetry. Reflecting about the angle bisector may bring angles and line segments together and hence, it may be an effective technique besides "cut and paste".

- Double the median.

  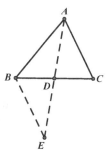

  Refer to the diagram on the right. Given $\triangle ABC$ and its median AD, extending AD to E with $AD = DE$ gives $\triangle ABE$ where $BE = AC$ and $\angle ABE = 180° - \angle A$.

  Hence, $\sin \angle A = \sin \angle ABE$ and $[\triangle ABC] = [\triangle ABE]$.

Moreover, (twice) the median of $\triangle ABC$ becomes a side of $\triangle ABE$. This may be a useful technique when constructing congruent and similar triangles.

- Midpoints and Midpoint Theorem

When midpoints are given, it is natural to apply the Midpoint Theorem, which not only gives parallel lines, but also moves the (halved) line segments around. In particular, if connecting the midpoints does not give a midline of the triangle, one may choose more midpoints and draw the midlines. Refer to the diagram below.

Given a quadrilateral $ABCD$ where M, N are the midpoints of AD, BC respectively, simply connecting MN does not give any conclusion. If we choose P, the midpoint of BD, then

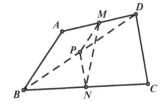

$$PM = \frac{1}{2}AB \text{ and } PN = \frac{1}{2}CD.$$

If we know more about AB and CD, say $AB = CD$, then we conclude that $\triangle PMN$ is an isosceles triangle.

On the other hand, if midpoints are given together with right angled triangles, one may consider the median on the hypotenuse. Example 5.1.4 illustrates this technique.

- Angle bisector plus parallel lines

One may easily see an isosceles triangle from an angle bisector plus parallel lines. Refer to the diagram on the right. If AD bisects $\angle A$, we have $\angle 1 = \angle 2$. If $AC /\!/ BD$, $\angle 2 = \angle 3$. It follows that $AB = BD$.

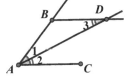

Notice that this technique could also be applied reversely. In the diagram above, if we know $AB = BD$, then by showing $AC // BD$, we conclude that AD bisects $\angle A$.

- Similar triangles sharing a common vertex

A pair of similar triangles sharing a common vertex may immediately give another pair of similar triangles. Refer to the following diagrams where $\triangle ABC \sim \triangle AB'C'$.

Since $\dfrac{AB}{AB'} = \dfrac{AC}{AC'}$ and $\angle BAC = \angle B'AC'$, by subtracting $\angle B'AC$, we see that $\angle BAB' = \angle CAC'$. It follows that $\triangle ABB' \sim \triangle ACC'$.

Notice that this technique applies for the inverse as well. If we have $\triangle ABB' \sim \triangle ACC'$, we may also conclude that $\triangle ABC \sim \triangle AB'C'$.

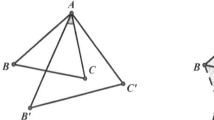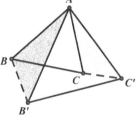

One may recall that we applied this technique in the proof of Ptolemy's Theorem, as well as in Example 5.1.7.

- Angle-chasing

This is an elementary but effective technique when we explore angles related to a circle, especially when an incircle or circumcircle of a triangle is given (because the incenter and circumcenter give us even more equal angles). If more than one circle is given, it is a basic technique to apply the angle properties repeatedly and identify equal angles far apart or apparently unrelated. Indeed, experienced contestants are very familiar with the angle properties and are sharp in observing and catching equal angles. (For example, can you write down the proof of Simson's Line quickly?)

However, one should avoid long-winded angle-chasing which leads nowhere. If that happens, one may seek clues from the line segments instead, say identifying similar triangles, or applying the Intersecting Chords Theorem and the Tangent Secant Theorem.

- Watch out for right angles.

When right angles are given, it is worthwhile to spend time and effort digging out more information about them, because right angles may lead to a number of approaches:
(1)   If a right angled triangle with a height on the hypotenuse is given, we will have similar triangles.
(2)   If there are other heights or the orthocenter of a triangle, we may find parallel lines.
(3)   One may see concyclicity when a few right angles are given.
(4)   If a right angle is extended on the circumference of a circle, it corresponds to a diameter of the circle.

One should always refer to the context of the problem and determine which approach might be effective.

- Perpendicular bisector of a chord

Introducing a perpendicular from the center of a circle to a chord is a simple technique but occasionally, it may be decisively useful. Notice that the perpendicular bisector gives both right angles and the midpoint of the chord.

- Draw a line connecting the centers of two intersecting circles.

This is a very basic technique where the line connecting the centers of the two circles is a line of symmetry.

Refer to the diagram on the right. Notice that $O_1O_2 \perp AB$ and O_1O_2 is the angle bisector of both $\angle AO_1B$ and $\angle AO_2B$. Even though this is an elementary result, one may apply it to solve difficult problems.

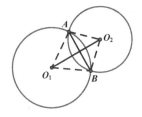

Example 3.2.13 illustrates this technique. It is noteworthy that beginners tend to overlook this elementary property during problem-solving, especially when the diagram is complicated.

- Relay: Tangent Secant Theorem and Intersecting Chords Theorem

When more than one circle is given and there is a common chord or concurrency, one may apply the Tangent Secant Theorem or the Intersecting Chords Theorem repeatedly to acquire more concyclicity. Refer to the diagrams below. Can you see C, D, E, F are concyclic in both diagrams? Can you see that $PC \cdot PD = PA \cdot PB = PE \cdot PF$?

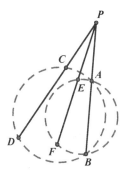

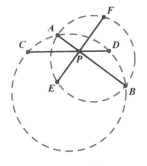

Refer to the diagram on the right. If A, B, C, D are concyclic, C, D, E, F are concyclic and E, F, G, H are concyclic, can you see that A, B, G, H are concyclic? (**Hint:** $PA \cdot PB = PC \cdot PD = PE \cdot PF = PG \cdot PH$.)

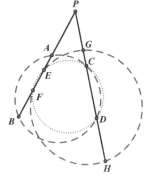

One may recall that we applied these basic techniques extensively when solving problems in the previous chapters. We shall illustrate these techniques with more examples in this section.

Example 5.2.2 (ITA 11)   Given a quadrilateral $ABCD$, the external angle bisectors of $\angle CAD, \angle CBD$ intersect at P. Show that if $AD + AC = BC + BD$, then $\angle APD = \angle BPC$.

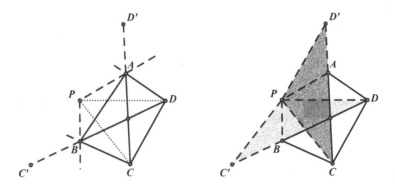

Insight.   Refer to the left diagram above, where AP, BP are the external angle bisectors of $\angle CAD, \angle CBD$ respectively. How can we apply the condition $AD + AC = BC + BD$? Cut and paste!

Extend DB to C' such that $BC' = BC$ and extend CA to D' such that $AD = AD'$. Can you see that C, C' are symmetric about the line PB, and D, D' are symmetric about the line PA? (**Hint**: $\triangle BCC'$ is an isosceles triangle and PB is the perpendicular bisector of CC'.) Now $BC = BC'$ and $AD = AD'$. Refer to the right diagram above. Can you see that $AD + AC = BC + BD$ implies $CD' = C'D$? Can you see that $PC = PC'$, $PD = PD'$ and hence, $\triangle PC'D \cong \triangle PCD'$?

Now $\angle C'PD = \angle CPD'$ and the conclusion follows. We leave the details to the reader.

Example 5.2.3 (GER 08)   Given an acute angled triangle $\triangle ABC$, AD is the angle bisector of $\angle A$, BE is a median and CF is a height. Show that AD, BE, CF are concurrent if and only if F lies on the perpendicular bisector of AD.

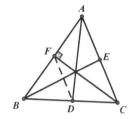

Insight. We are to show $AF = DF$ if and only if AD, BE, CF are concyclic. Since AD bisects $\angle A$, the isosceles triangle $\triangle ADF$ gives $AC /\!/ DF$. Refer to the diagram on the right. How can we show the concurrency?

What if we apply Ceva's Theorem to the height CF, the median BE and the angle bisector AD? By the Angle Bisector Theorem and $AE = CE$, we may obtain the ratio of line segments leading to $AC /\!/ DF$.

Proof. By Ceva's Theorem, AD, BE, CF are concurrent if and only if $\dfrac{AF}{BF} \cdot \dfrac{BD}{CD} \cdot \dfrac{CE}{AE} = 1$. Since BE is a median, it is equivalent to $\dfrac{AF}{BF} = \dfrac{CD}{BD}$, or $DF /\!/ AC$.

We claim that $DF /\!/ AC$ if and only if $AF = DF$. In fact, since AD bisects $\angle A$, $DF /\!/ AC$ if and only if $\angle ADF = \angle CAD = \angle BAD$, which is equivalent to $AF = DF$.

In conclusion, AD, BE, CF are concurrent if and only if $AF = DF$, i.e., F lies on the perpendicular bisector of AD. $\qquad\square$

Example 5.2.4 (BRA 08) Given a quadrilateral $ABCD$ inscribed inside $\odot O$, draw lines ℓ_1, ℓ_2 such that ℓ_1 and the line AB is symmetric about the angle bisector of $\angle CAD$, and ℓ_2 and the line AB is symmetric about the angle bisector of $\angle CBD$. If ℓ_1 and ℓ_2 intersect at M, show that $OM \perp CD$.

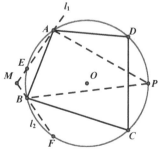

Insight. It is easy to see that the angle bisectors of $\angle CAD$ and $\angle CBD$ meet at the midpoint of the arc $\overset{\frown}{CD}$, say P. Refer to the diagram on the right. Notice that the reflections ℓ_1, ℓ_2 and the circle give a lot of equal angles.

How can we show $OM \perp CD$?
It may not be wise to find the angle directly because we do not know where OM and CD intersect. Shall we explore the angles around the

circles and seek more clues? If for example OM is the perpendicular bisector of AF (i.e., $AM = FM$), then it suffices to show $AF \parallel CD$.

Proof. Let P be the midpoint of $\overset{\frown}{CD}$. Clearly, AP, BP are the angle bisectors of $\angle CAD, \angle CBD$ respectively.

Let ℓ_1 and ℓ_2 intersect $\odot O$ at A, E and B, F respectively.

Since ℓ_1 and AB are symmetric about AP, we must have $\angle BAP = 180° - \angle EAP = \angle ECP$ (because A, E, C, P are concyclic). (1)

Since P is the midpoint of $\overset{\frown}{CD}$, we have $\angle PCD = \angle PAC$. (2)

(1) and (2) imply that $\angle BAP - \angle PAC = \angle ECP - \angle PCD$, which gives $\angle BAC = \angle DCE$, i.e., $\overset{\frown}{BC}$ and $\overset{\frown}{DE}$ extend the same angle on the circumference. This implies $BC = DE$ and hence, $BCDE$ is an isosceles trapezium with $BE \parallel CD$.

Since ℓ_2 and AB are symmetric about BP, a similar argument applies which gives $AF \parallel CD$ and $ADCF$ is an isosceles trapezium. Now it is easy to see that $AEBF$ is also an isosceles trapezium. Notice that $AM = MF$ and hence, OM is the perpendicular bisector of AF. Since $AF \parallel CD$, we must have $OM \perp CD$. □

Note: If the diagram becomes complicated and the angles on the circumference do not give clear insight, it might be easier to consider the corresponding arcs. Notice that we showed $\overset{\frown}{BC} = \overset{\frown}{DE}$ in the proof above, which simplifies the argument. In fact, one would easily see that isosceles trapeziums via equal arc lengths.

Example 5.2.5   Let I be the incenter of $\triangle ABC$. M, N are the midpoints of AB, AC respectively. NM extended and CI extended intersect at P. Draw $QP \perp MN$ at P such that $QN \parallel BI$. Show that $QI \perp AC$.

Insight. Refer to the following diagram. Notice that the angle bisector CI and the parallel lines $BC \, / \! / \, MN$ give $PN = CN$.

Since $CN = \dfrac{1}{2} AC$, we must have $\angle APC = 90°$.

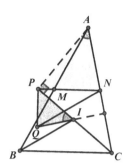

Given $PQ \perp PN$, i.e., $\angle QPN = 90° = \angle APC$, one immediately sees that $\angle QPI = \angle APN$. Since we are to show $QI \perp AC$, we **should** have $\angle PIQ = 90° - \angle ACI = \angle PAC$.

Hence, we **should** have $\triangle APN \sim \triangle IPQ$.

Can we show it, say via $\dfrac{AP}{PN} = \dfrac{IP}{PQ}$? Notice that AP, PN, IP, PQ are the sides of the right angled triangles $\triangle API$ and $\triangle PQN$. Indeed, if we can show that $\triangle API \sim \triangle NPQ$, it follows that $\triangle APN \sim \triangle IPQ$.

We have not used the condition $QN \, / \! / \, BI$ yet. Perhaps this could help us to find an equal pair of angles in $\triangle API$ and $\triangle NPQ$.

Proof. Since CI bisects $\angle C$ and $BC \, / \! / \, MN$, we have $\angle NCP = \angle BCP = \angle NPC$, i.e., $PN = CN$. Since N is the midpoint of AC, we have $PN = AN = CN$ and hence, $\angle APC = 90°$ (Example 1.1.8).

Since I is the incenter of $\triangle ABC$, we have $\angle AIC = 90° + \dfrac{1}{2} \angle ABC$ and

hence, $\angle AIP = 180° - \angle AIC = 90° - \dfrac{1}{2} \angle ABC = 90° - \angle CBI$.

Notice that $\angle CBI = \angle PNQ$ (because $MN \, / \! / \, BC$ and $BI \, / \! / \, QN$). Hence, $\angle AIP = 90° - \angle PNQ = \angle PQN$. Since $\angle APC = \angle QPN = 90°$, we must have $\triangle API \sim \triangle NPQ$. Refer to the diagram below.

Now we have $\dfrac{AP}{PN} = \dfrac{IP}{PQ}$ and $\angle QPI = \angle APN$.

It follows that $\triangle APN \sim \triangle IPQ$.

Let QI extended intersect AC at D. We have $\angle CID = \angle PIQ = \angle PAC = 90° - \angle ACI$, i.e., $\angle CDI = 90°$. This completes the proof.   □

Example 5.2.6 (HEL 09)   Let O, G denote the circumcenter and the centroid of $\triangle ABC$ respectively. Let the perpendicular bisectors of AG, BG, CG intersect mutually at D, E, F respectively. Show that O is the centroid of $\triangle DEF$.

Insight.   Refer to the diagram below. What can we say about O and $\triangle DEF$? O is the circumcenter of $\triangle ABC$ where $\triangle DEF$ is constructed by the perpendicular bisectors of AG, BG, CG. Can you see the *link* between perpendicular bisectors and circumcenters? Indeed, one immediately concludes that D, E, F are the circumcenters of $\triangle BCG, \triangle ACG, \triangle ABG$ respectively.

How can we show that O is the centroid of $\triangle DEF$? Let DO extended intersect EF at P. If we can show that $EP = FP$ (which **should** be true), perhaps it is similar to show that EO, FO pass through the midpoints of DF, DE respectively. Notice that we have many right angles in the diagram, which give a lot of concyclicity.

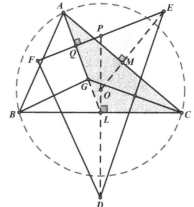

Let L be the midpoint of BC. Can you see that $\angle CAL = \angle OEP$ and $\angle ACB = \angle EOP$? What can you say about $\triangle ACL$ and $\triangle EOP$? How about $\triangle ABC$ and $\triangle DEF$?

Proof.   It is easy to see that D, E, F are the circumcenters of $\triangle BCG, \triangle ACG, \triangle ABG$ respectively. Let L, M, N be the midpoints of BC, AC, AB respectively. Notice that the lines DL, EM, FN are the perpendicular bisectors of BC, AC, AB respectively and hence, intersect at O. Let DL extended intersect EF at P. We claim that P is the midpoint of EF.

Let AG intersect EF at Q. Since $AG \perp EF$ and $EM \perp AC$, A, E, M, Q are concyclic and hence, $\angle CAL = \angle OEP$. (1)

Since $\angle CLO + \angle CMO = 180°$, we also have C, L, O, M concyclic and hence, $\angle ACL = \angle EOP$. (2)

(1) and (2) give $\triangle ACL \sim \triangle EOP$ and hence, $\dfrac{EP}{AL} = \dfrac{OP}{CL}$. (3)

Similarly, one sees that $\triangle ABL \sim \triangle FOP$ and hence, $\dfrac{FP}{AL} = \dfrac{OP}{BL}$. (4)

(3) and (4) imply $EP = FP$, i.e., DO extended passes through the midpoint of EF. Similarly, EO extended and FO extended pass through the midpoints of DF and DE respectively. We conclude that O is the centroid of $\triangle DEF$. $\qquad\square$

Note: Even though we did not explicitly double the median in the proof above, it is essentially the technique we applied. Refer to Example 1.2.11, where $\triangle ABC$ and $\triangle AEF$ are related in a similar manner as $\triangle ABC$ and $\triangle OEF$ in this example. Refer to the diagram on the right. If we extend AL to A' such that $AL = AL'$, can you see $\triangle ACA' \sim \triangle EOF$? Notice that P and L are corresponding points because $\angle ACL = \angle EOP$.

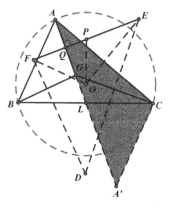

One may also show that $\triangle DEP \sim \triangle CGL$ and $\triangle DFP \sim \triangle BGL$, which also leads to the conclusion.

Example 5.2.7 Let $\Gamma_1, \Gamma_2, \Gamma_3$ be three circles such that Γ_1, Γ_2 intersect at A and P, Γ_2, Γ_3 intersect at C and P, and Γ_1, Γ_3 intersect at B and P.

Refer to the following diagram. If AP extended intersects Γ_3 at D, BP extended intersects Γ_2 at E and CP extended intersects Γ_1 at F, show that $\dfrac{AP}{AD} + \dfrac{BP}{BE} + \dfrac{CP}{CF} = 1$.

Insight. We focus on $\dfrac{AP}{AD}$ first. Since we do not have much information about the line segments, we may consider re-writing the ratio by areas of triangles.

For example, $\dfrac{AP}{AD} = \dfrac{[\triangle APF]}{[\triangle ADF]}$.

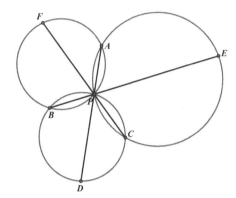

However, applying this to $\dfrac{BP}{BE}$ and $\dfrac{CP}{CF}$ gives ratios of no common denominator and hence, it is not easy to calculate the sum.

Perhaps we should use the triangles independent of AD, BE, CF. Notice that AP, BP, CP are common chords of circles. How about connecting the centers of the circles? It gives us the perpendicular bisector of the common chords. Refer to the diagram below, where we denote the centers of $\Gamma_1, \Gamma_2, \Gamma_3$ by O_1, O_2, O_3 respectively. If we draw $O_3H \perp AD$, it is the perpendicular bisector of DP. Hence, $\dfrac{AP}{AD} = \dfrac{MP}{HP}$. This seems closely related to $\triangle O_1O_2O_3$.

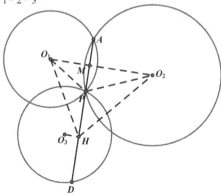

Proof. Let O_1, O_2, O_3 denote the centers of $\Gamma_1, \Gamma_2, \Gamma_3$ respectively. Let O_1O_2 intersect AP at M. Clearly, $AM = PM$. Draw $O_3 \perp DP$ at H.

It is easy to see that $DH = PH$. Hence, $MH = \dfrac{1}{2}(AP + DP) = \dfrac{1}{2}AD$.

Now $\dfrac{AP}{AD} = \dfrac{\dfrac{1}{2}AP}{\dfrac{1}{2}AD} = \dfrac{PM}{HM} = \dfrac{[\Delta O_1O_2P]}{[\Delta O_1O_2H]}$.

Notice that $[\Delta O_1O_2H] = [\Delta O_1O_2O_3] = \dfrac{1}{2}O_1O_2 \cdot MH$, because $O_1O_2 \perp AD$ and $O_3H \perp AD$, i.e., $O_1O_2 \, /\!/ \, O_3H$. It follows that $\dfrac{AP}{AD} = \dfrac{[\Delta O_1O_2P]}{[\Delta O_1O_2O_3]}$.

Similarly, $\dfrac{BP}{BE} = \dfrac{[\Delta O_1O_3P]}{[\Delta O_1O_2O_3]}$ and $\dfrac{CP}{CF} = \dfrac{[\Delta O_2O_3P]}{[\Delta O_1O_2O_3]}$.

Refer to the diagram below.

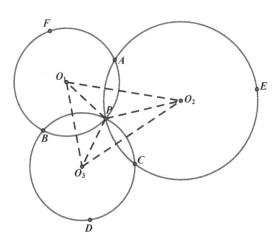

The conclusion follows as $\dfrac{[\Delta O_1O_2P] + [\Delta O_1O_3P] + [\Delta O_2O_3P]}{[\Delta O_1O_2O_3]} = 1$. $\square$

Example 5.2.8 (CHN 07)   Let AB be the diameter of a semicircle centered at O. Given two points C, D on the semicircle, BP is tangent to the circle, intersecting CD extended at P. If the line PO intersects CA extended and AD extended at E, F respectively, show that $OE = OF$.

Insight.   Clearly, $AO = BO$. One sees that $AEBF$ **should** be a parallelogram. How can we show it? Refer to the left diagram below. Perhaps the most straightforward way is to show $\angle ABE = \angle BAF$.

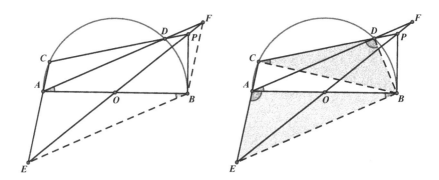

By applying circle properties, we obtain many equal angles, for example $\angle BAD = \angle BCD$ and $\angle BDC = \angle BAE$. It seems that we **should** have $\triangle BDC \sim \triangle EAB$. Refer to the right diagram above. Can we show it by considering the sides, say $\dfrac{BD}{CD} = \dfrac{AE}{AB}$? Unfortunately, this is not easy because we do not know much about CD or AE.

On the other hand, we have not used the condition $BP \perp AB$. This is when drawing a perpendicular to the chord becomes handy: we bisect CD and obtain a right angle as well. Notice that the midpoint of CD and O **should** be corresponding points in $\triangle BDC$ and $\triangle EAB$.

Proof.   Draw $OM \perp CD$ at M. We have $CM = DM$. Since $BP \perp AB$, we have B, O, M, P concyclic and hence, $\angle BMP = \angle BOP = \angle AOE$. (1) Since A, B, D, C are concyclic, we have $\angle BDC = \angle BAE$. (2)

(1) and (2) imply that $\triangle BDM \sim \triangle EAO$ and hence, $\dfrac{AE}{AO} = \dfrac{BD}{DM}$. Refer to the following diagram.

Since O and M are midpoints of AB, CD respectively, we have

$$\frac{AE}{AB} = \frac{AE}{2AO} = \frac{BD}{2DM} = \frac{BD}{CD}. \quad (3)$$

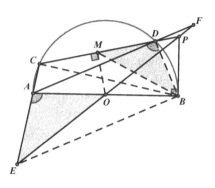

(2) and (3) imply that $\triangle BDC \sim \triangle EAB$. Hence, $\angle BCD = \angle ABE$.

Since $\angle BCD = \angle BAD$, we must have $\angle BAD = \angle ABE$.

One sees that $\triangle AOF \cong \triangle BOE$ (A.A.S.) and hence, $OE = OF$. $\qquad\square$

Example 5.2.9 (IRN 09) Given an acute angled triangle $\triangle ABC$ where AD, BE, CF are heights, draw $FP \perp DE$ at P. Let Q be the point on DE such that $QA = QB$. Show that $\angle PAQ = \angle PBQ = \angle PFC$.

Insight. Refer to the diagram on the right. Clearly, $\angle PAQ = \angle PBQ$ if and only if A, B, Q, P are concyclic. We are given many perpendicular lines, but we should **not** draw all the lines explicitly: otherwise, the diagram will be in a mess. Notice that there are a few concyclicity due to the right angles. For example, A, B, D, E are concyclic.

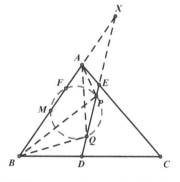

Can you see that the circumcircle of $\triangle PQF$ passes through M, the midpoint of AB? (**Hint:** QM is the perpendicular bisector of AB.) Can you see that that the circumcircle of $\triangle DEF$ pass through M as well? (**Hint**: Consider the nine-point circle of $\triangle ABC$.) Suppose BA extended and DE extended intersect at X. Perhaps we can apply the Tangent Secant Theorem repeatedly and show that A, B, Q, P are concyclic.

How about $\angle PFC$? Can you see that $\angle PFC = \angle X$, because $FP \perp DE$ and $CF \perp AB$?

Proof. Clearly, Q lies on the perpendicular bisector of AB. Let M be the midpoint of AB. We must have $QM \perp AB$. Since $FP \perp DE$, F, M, Q, P are concyclic. Let the lines AB and DE intersect at X. By the Tangent Secant Theorem, $XP \cdot XQ = XF \cdot XM$. (1)

It is well known that A, B, D, E are concyclic and hence, we have

$XA \cdot XB = XD \cdot XE$. (2)

Notice that D, E, F, M are concyclic because they lie on the nine-point circle of $\triangle ABC$.

Hence, $XD \cdot XE = XF \cdot XM$. (3)

Refer to the diagram on the right.

(1), (2) and (3) give $XA \cdot XB = XP \cdot XQ$.

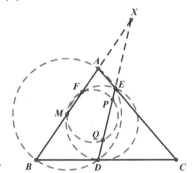

Hence, A, B, Q, P are concyclic and $\angle PAQ = \angle PBQ$.

Let H denote the orthocenter of $\triangle ABC$. Consider the right angled triangle $\triangle FHX$. Since $FP \perp HX$, we have $\angle PFC = \angle X$. Refer to the left diagram below. It suffices to show $\angle X = \angle PAQ$.

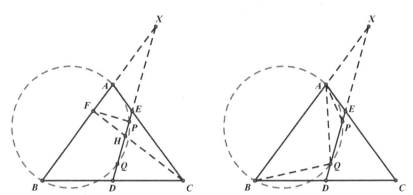

Notice that $\angle X = \angle PAB - \angle APX$, where $\angle APX = \angle ABQ = \angle BAQ$. It follows that $\angle X = \angle PAB - \angle BAQ = \angle PAQ$. Refer to the right diagram above. This completes the proof. □

5.3 Constructing a Diagram

Most geometry problems in competitions held recently were presented in descriptive sentences without any diagram. Contestants are expected to construct the diagram on their own, usually with a straightedge and a compass allowed. Indeed, a well-constructed diagram is very important, if not indispensable, for solving a geometry problem: it not only helps in seeking geometric insight (for example, catching equal angles around a circle), but also gives inspiration on what **could** or **should** be true.

Constructing a diagram with only a straightedge and a compass involves a lot of skills. For example, given $\odot O$ and a point P outside the circle, do you know how to introduce tangent lines from P to $\odot O$ accurately? (**Hint**: Draw a circle Γ where OP is a diameter. Let $\odot O$ and Γ intersect at A, B. Can you see that PA, PB are the tangent lines from P to $\odot O$, because the diameter OP extends right angles on the circumference of Γ? Refer to the diagram on the right.)

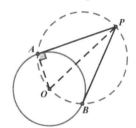

In this section, we shall introduce a few techniques (related to the diagram) which one may find useful.

- Turn the paper around.

If one thinks there might be symmetry in the diagram constructed but cannot see it clearly, a wise strategy is rotating the diagram (by turning the paper around) to the *upright* position, for example, with respect to the angle bisector, the perpendicular bisector or the line connecting the centers of two intersecting circles. Usually, the symmetry would become clearer in this view.

This technique is also helpful for beginners to catch the geometric insight. It is common that beginners cannot identify similar triangles or equal tangent segments if a (complicated) diagram is drawn in an *oblique* manner. Hence, by turning the paper around, one may observe the diagram more thoroughly and find clues more easily.

- Coincidence and equivalent conclusions

Occasionally, finding a direct proof could be difficult (or infeasible due to technical difficulties). Hence, one may consider showing an equivalent conclusion instead by coincidence. For example, if showing that a line ℓ passes through a specific point X on a circle Γ is difficult, one may let ℓ intersect Γ at X' and show that X and X' coincide. In fact, this technique is often applied when showing collinearity and concurrency, and is also illustrated in Example 1.4.3.

- Uniquely determined points

It is an advanced technique to examine the diagram and check **how** it could be constructed and which points (and angles, line segments, etc.) are uniquely determined by the given conditions. For example, given a circle Γ and a point O outside Γ, if we are to construct $\odot O$ which touches Γ, then it is easy to see that the point of tangency, called P, is uniquely determined. In fact, P lies on the line connecting O and the center of Γ. Notice that OP, the radius of $\odot O$, is also uniquely determined.

Although this technique may not help the problem-solving directly, it gives clues on how the diagram **could** vary and which points and line segments are more closely related. Acquiring such insight may greatly help us understand the diagram, identify the links and design an effective strategy leading to the solution.

Example 5.3.1 (RUS 09)   Let $\odot O$ be the circumcircle of $\triangle ABC$. D is on AC such that BD bisects $\angle B$. Let BD extended intersect $\odot O$ at E. Draw a circle Γ with a diameter DE, intersecting $\odot O$ at E and F. Draw a line ℓ such that the line BF and ℓ are symmetric about the line BD. Show that ℓ passes through the midpoint of AC.

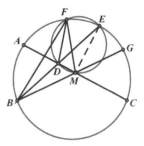

Insight. Refer to the diagram on the right. Notice that there are a few symmetries in the diagram due to the angle bisector. Suppose ℓ intersects AC at M (which **should** be the midpoint of AC). It *seems* from the diagram that M lies on Γ as well! Can we show it?

On the other hand, it may not be easy to show $AM = CM$ directly because we do not know much about the point M. How about choosing M as the midpoint of AC? Would it be easier to show BD bisects $\angle MBF$? (We can probably apply the angle properties about $\odot O$ and Γ.)

Notice that E is the midpoint of the arc $\overset{\frown}{AC}$ and hence, EM is the perpendicular bisector of AC.

Proof. Let M be the midpoint of AC. Let BM extended intersect $\odot O$ at G. Since BE bisects $\angle ABC$, E must be the midpoint of $\overset{\frown}{AC}$. Hence, EM is the perpendicular bisector of AC. We claim that BM coincides with ℓ, i.e., BE bisects $\angle FBG$. Notice that it suffices to show that F and G are symmetric about EM, or equivalently, $\angle EFM = \angle G$. Refer to the left diagram below.

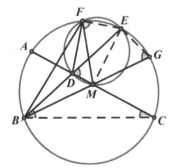
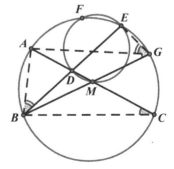

Since $EM \perp AC$, M must lie on Γ where DE is a diameter. It follows that $\angle EFM = \angle EDM = \angle CBD + \angle C = \angle ABD + \angle C$.

Refer to the right diagram above. Notice that $\angle ABD = \angle AGE$ and $\angle C = \angle AGB$. It follows that $\angle ABD + \angle C = \angle AGE + \angle AGB = \angle BGE$. This completes the proof. $\square$

Note:

(1) Given $\overset{\frown}{AC}$ and $\odot O$, E and M are determined regardless of the choice of B. By choosing D, other points including B, F and G are determined. Hence, it is a wise strategy to explore the properties of angles around D.

(2) By rotating the diagram, one may see the symmetry about the line EM. Refer to the diagram on the right. Let FM extended intersects $\odot O$ at B'. Notice that BG and $B'F$ are symmetric about the line EM.

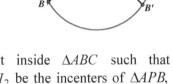

Example 5.3.2   Let P be a point inside $\triangle ABC$ such that $\angle APB - \angle ACB = \angle APC - \angle ABC$. Let I_1, I_2 be the incenters of $\triangle APB$, $\triangle APC$ respectively. Show that AP, BI_1, CI_2 are concurrent.

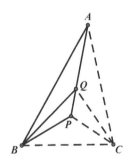

Insight.   Apparently, the conditions given are unusual, not easy to apply and unrelated to the conclusion. In fact, we do not even know **how** to construct such a diagram. Let us focus on the conclusion: we are to show AP, BI_1, CI_2 are concurrent. Since BI_1, CI_2 are angle bisectors of $\angle ABP, \angle ACP$ respectively, it suffices to show that these angle bisectors intersect AP at the same position.

Refer to the left diagram above. Let us draw $\triangle ABP$ first where BQ is the angle bisector of $\angle ABP$. We shall find a point C such that CQ bisects $\angle ACP$. What conditions must C satisfy? For example, we must have $\dfrac{AC}{CP} = \dfrac{AQ}{PQ}$. In this case, we see that it suffices to show $\dfrac{AB}{BP} = \dfrac{AC}{CP}$, which leads to the conclusion.

Now we are to apply the condition $\angle APB - \angle ACB = \angle APC - \angle ABC$. Notice that these angles are far apart. Can we bring them together? Refer to the left diagram below. Notice that $\angle APB - \angle ACB = \angle 1 + \angle 2$ and $\angle APC - \angle ABC = \angle 3 + \angle 4$. Hence, $\angle 1 + \angle 2 = \angle 3 + \angle 4$.

 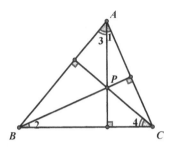

It seems these angles are still far apart. Recall that **if** P is the orthocenter, then we have $\angle 1 = \angle 2$ and $\angle 3 = \angle 4$. Refer to the right diagram above. This is because the perpendicular lines imply concyclicity and give equal angles. For a general P, there are no perpendicular lines given, but perhaps we can introduce some!

Proof. Refer to the diagram on the right. Since $\angle APB - \angle ACB = \angle 1 + \angle 2$ and $\angle APC - \angle ABC = \angle 3 + \angle 4$, we have $\angle 1 + \angle 2 = \angle 3 + \angle 4$.
Let D, E, F be the feet of the perpendiculars from P to BC, AC, AB respectively. Since $\angle AFP = \angle AEP = 90°$, A, F, P, E are concyclic and $\angle 1 = \angle EFP$. Similarly, $\angle 2 = \angle DFP$. It follows that $\angle 1 + \angle 2 = \angle EFP + \angle DFP = \angle DFE$.

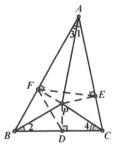

A similar argument gives $\angle 3 + \angle 4 = \angle DEF$. Now $\angle DEF = \angle DFE$ and we must have $DE = DF$.

By Sine Rule, $\dfrac{DF}{\sin \angle ABC} = BP$ (since BP is a diameter of the circumcircle of $\triangle BDF$). Similarly, $\dfrac{DE}{\sin \angle ACB} = CP$. Since $DE = DF$, we have $\dfrac{BP}{CP} = \dfrac{\sin \angle ACB}{\sin \angle ABC} = \dfrac{AB}{AC}$ by applying Sine Rule to $\triangle ABC$.

Hence, $\dfrac{AB}{BP} = \dfrac{AC}{CP}$. By the Angle Bisector Theorem, the angle bisectors of $\angle ABP$ and $\angle ACP$ must intersect AP at the same point. (Otherwise, say they intersect AP at X, Y respectively, we have $\dfrac{AX}{PX} = \dfrac{AP}{BP} = \dfrac{AC}{CP}$ $= \dfrac{AY}{PY}$, which implies X, Y coincide.) This completes the proof.   □

Example 5.3.3   Refer to the diagram on the right. $\odot O_1$ and $\odot O_2$ intersect at A and B. $\odot O_3$ touches $\odot O_1$ and $\odot O_2$ at C, D respectively. A common tangent of $\odot O_1$ and $\odot O_2$ touches the two circles at E, F respectively.

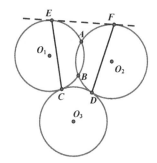

If the lines CE and DF intersect at P, show that P lies on the line AB.

Insight.   One sees that AB is the radical axis of $\odot O_1$ and $\odot O_2$. Hence, it suffices to show that the powers of P with respect to $\odot O_1$ and $\odot O_2$ are the same, i.e., $PC \cdot PE = PD \cdot PF$ (or by the Tangent Secant Theorem if one is not familiar with the power of a point with respect to circles).

However, the difficulty is that we do not know the position of P and hence, we cannot calculate PC, PD, PE, PF directly.

Refer to the diagram on the right. (We omit A, B to have a clearer view of the angles.) It *seems* from the diagram that P, the intersection of the lines CE and DF, lies on $\odot O_3$. Can we prove it?

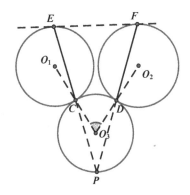

Since $\odot O_3$ is tangent to $\odot O_1$ and $\odot O_2$, the line connecting the centers of the circles must pass through the point of tangency, i.e., O_1O_3 passes through C and O_2O_3 passes through D. Notice that $\angle CO_3D$ is an angle at the center of $\odot O_3$. Now P lies on $\odot O_3$ if and only if $\angle CO_3D = 2\angle CPD$. Can we show this?

Notice that $\angle CO_3D$ could be calculated via the pentagon $O_1O_3O_2FE$ (which has two right angles) and $\angle CPD$ could be calculated via $\triangle EPF$.

If we denote $\angle O_1EC = \alpha$ and $\angle O_2FD = \beta$, all the interior angles in the pentagon $O_1O_3O_2FE$ and $\triangle EPF$ could be expressed in α, β (using the fact that $\triangle O_1CE$ and $\triangle O_2DF$ are isosceles triangles). Refer to the diagram on the right.

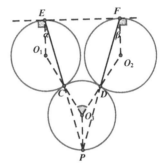

On the other hand, if P indeed lies on $\odot O_3$, we have similar isosceles triangles $\triangle O_1CE \sim \triangle O_3CP$ and $\triangle O_2DF \sim \triangle O_3DP$. Now $\dfrac{PC}{CE}$ and $\dfrac{PD}{DF}$ could be expressed using the radii of $\odot O_1$, $\odot O_2$ and $\odot O_3$. We should not be far away from the conclusion.

Proof. First, we claim that P lies on $\odot O_3$. Let $\angle O_1EC = \alpha$ and $\angle O_2FD = \beta$. Consider $\triangle EPF$. We have

$$\angle CPD = 180° - \angle CEF - \angle DFE = 180° - (90° - \alpha) - (90° - \beta) = \alpha + \beta.$$

On the other hand, by considering the pentagon $O_1O_3O_2FE$, we have

$$\angle CO_3D = 540° - \angle O_1EF - \angle O_2FE - \angle CO_1E - \angle DO_2F$$

$$= 540° - 90° - 90° - (180° - 2\alpha) - (180° - 2\beta) = 2(\alpha + \beta) = 2\angle CPD.$$

Hence, P lies on $\odot O_3$ (Theorem 3.1.1).

Now it is easy to see that $\triangle O_1CE \sim \triangle O_3CP$ since both are isosceles triangles and $\angle O_1CE = \angle O_3CP = \alpha$. Similarly, $\triangle O_2DF \sim \triangle O_3DP$.

Let the radii of $\odot O_1$, $\odot O_2$ and $\odot O_3$ be a, b, c respectively. Let $CE = x$ and $DF = y$. We have $\dfrac{PC}{CE} = \dfrac{c}{a}$, i.e., $PC = \dfrac{xc}{a}$. Similarly, $PD = \dfrac{yc}{b}$.

Consider $\triangle O_1CE$. We have $\dfrac{x}{a} = 2 \cdot \dfrac{x/2}{a} = 2\cos\alpha$. Similarly, $\dfrac{y}{b} = 2\cos\beta$.

Now $\dfrac{PC}{PD} = \dfrac{2\cos\alpha \cdot c}{2\cos\beta \cdot c} = \dfrac{\cos\alpha}{\cos\beta}$. On the other hand, applying Sine Rule in $\triangle PEF$ gives $\dfrac{PE}{PF} = \dfrac{\sin\angle PFE}{\sin\angle PEF} = \dfrac{\cos\beta}{\cos\alpha}$.

It follows that $\dfrac{PC}{PD} = \dfrac{PF}{PE}$, or $PC \cdot PE = PD \cdot PF$.

Now the power of P with respect to $\odot O_1$ and $\odot O_2$ are the same, which implies that P lies on the line AB, the radical axis of $\odot O_1$ and $\odot O_2$. $\square$

Example 5.3.4 (CHN 10)   Refer to the diagram on the right. $\odot O$ is tangent to AB at H. Draw a semicircle with the diameter AB, touching $\odot O$ at E. C is a point on the semicircle such that $CD \perp AB$ at D and CD touches $\odot O$ at F. Show that $CH^2 = 2DH \cdot BH$.

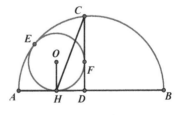

Insight.   Clearly, $\angle BCH$ is not 90°, but if it were, we could have concluded $CH^2 = DH \cdot BH$. Now we are to show $CH^2 = 2DH \cdot BH$.

Hence, if one draws $PC \perp CH$ at C, intersecting AB extended at P, we **should** have $PH = 2BH$, i.e., $BH = BP$. Can we show it?

Refer to the diagram on the right. It seems we do not have many clues about BH and BP, although there are many points of tangency given in the diagram.

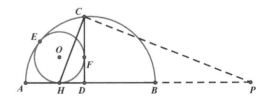

One may find equal angles or apply the Tangent Secant Theorem, but those are not directly related to BH or BP. Perhaps we should study the diagram more carefully and see how it **could** be constructed.

Suppose we are given $\odot O$. Notice that if we choose AB casually, the semicircle may not touch $\odot O$. In fact, once the center of the semicircle, called O_1, is chosen, the positions of A, B, E (and C, F) are uniquely determined. Refer to the left diagram below. Can you see that $OFDH$ is a square?

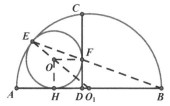 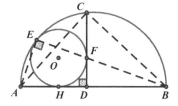

Since E is the point of tangency, O, O_1, E are collinear. Since $OF \parallel AB$, the isosceles triangles $\triangle OEF$ and $\triangle O_1EB$ are similar, which implies B, E, F are collinear! Now we have plenty of clues to apply the Tangent Secant Theorem. Refer to the right diagram above. One sees that $BE \cdot BF = BH^2$. Can you see that $BE \cdot BF = BA \cdot BD$ because A, D, F, E are concyclic? Can you see that $BA \cdot BD = BC^2$?

It follows that $BC = BH$. This is almost what we want. Refer to the diagram on the right. Can you see why $BH = BP$?

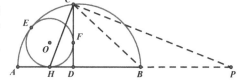

Proof. Let O_1 be the midpoint of AB. Extend AB to P such that $BH = BP$. Connect OF and O_1E. It is easy to see that O_1E passes through O. Refer to the diagram below.

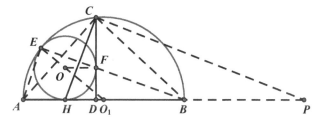

Since $OF \perp CD$ and $AB \perp CD$, we must have $OF // AB$ and hence, $\angle EOF = \angle EO_1 B$. Since $\triangle OEF$ and $\triangle O_1 EB$ are both isosceles triangles, we have $\angle OFE = \angle O_1 BE$. It follows that B, E, F are collinear.

Connect AC, BC and AE. Notice that $\angle ACB = \angle AEB = \angle ADC = 90°$. Hence, A, D, F, E are concyclic. Now $BC^2 = BD \cdot BA$ (Example 2.3.1) $= BE \cdot BF = BH^2$ (Tangent Secant Theorem). Hence, $BC = BH$.

Notice that $BC = BP = BH$ implies $CH \perp CP$ (Example 1.1.8). Now we have $CH^2 = DH \cdot PH = DH \cdot 2BH$. This completes the proof.   □

Example 5.3.5   Let P be a point inside the cyclic quadrilateral $ABCD$ such that $\angle BPC = \angle BAP + \angle CDP$. Draw $PE \perp AB$ at E, $PF \perp AD$ at F and $PG \perp CD$ at G. Show that $\triangle FEG \sim \triangle PBC$.

Insight.   Refer to the diagram on the right. It seems $\angle BPC = \angle BAP + \angle CDP$ is not a straightforward condition. How can we show $\triangle FEG \sim \triangle PBC$? It should be via equal angles or sides of equal ratio. One easily sees that A, E, P, F are concyclic and D, F, P, G are concyclic. Can you see that $\angle BPC = \angle EFG$?

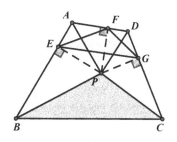

What else can we derive from $\angle BPC = \angle BAP + \angle CDP$? Even though this condition is not straightforward, it seems the only source for us to understand the diagram. (Notice that E, F, G could be obtained simply by drawing circles using AP, DP as diameters.) Hence, we shall explore further about this condition.

One sees that $\angle BPC, \angle BAP, \angle CDP$ are either an angle around P, or an angle inside $ABCD$, both of whom might give $360°$:

$$\angle APB + \angle CPD = 360° - \angle BPC - \angle APD \quad (1)$$

$$\angle APB + \angle CPD = \left(180° - \angle ABP - \angle BAP\right) + \left(180° - \angle CDP - \angle DCP\right) \quad (2)$$

(1) and (2) give $\angle APD = \angle ABP + \angle DCP$. This is a symmetric version of what is given. Is it useful?

Perhaps we shall examine the construction of our diagram, i.e., **how** can we locate a point P such that $\angle BPC = \angle BAP + \angle CDP$? By taking $\angle BPX = \angle BAP$, we must have PX tangent to the circumcircle of $\triangle ABP$ at P (Theorem 3.2.10). Refer to the diagram on the right.

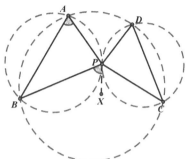

Now we construct another circle tangent to the circumcircle of $\triangle ABP$ at P (which is simple because the line connecting the centers of the two circles must be perpendicular to PX). This circle intersects the circumcircle of $\triangle ABC$ at D because $\angle CDP = \angle CPX$.

In conclusion, given $\triangle ABC$ and P, D is uniquely determined and PX **should** be a common tangent of the circumcircles of $\triangle ABP$ and $\triangle CDP$.

Since $ABCD$ is cyclic, we now have three circles, whose radical axes should be concurrent (Theorem 4.3.2). Refer to the left diagram below. Can you see similar triangles in this diagram involving BP and CP, for example, $\triangle QAP \sim \triangle QPB$ and $\triangle QDP \sim \triangle QPC$? Recall that we are to show $\dfrac{EF}{BP} = \dfrac{FG}{CP}$. What do we know about EF and FG?

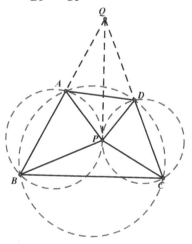

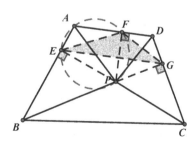

Refer to the right diagram above. Since A, E, P, F are concyclic where AP is a diameter, we have $EF = AP \sin \angle BAD$ (Sine Rule). Similarly, $FG = DP \sin \angle ADC$. Now AP, BP, CP, DP are related by similar triangles. It seems we have gathered all the links!

Please note that in the formal proof, one should also consider the case if $AB \,/\!/\, CD$ or if AB, CD intersect at the other side of line AD, i.e., our argument should not depend on the diagram.

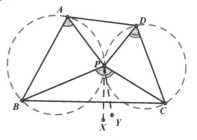

Proof. First, we claim that the circumcircles of $\triangle ABP$ and $\triangle CDP$ touch at P. Let PX be tangent to the circumcircle of $\triangle ABP$ at P. We have $\angle BPX = \angle BAP$. Refer to the diagram on the right.

We also draw PY tangent to the circumcircle of $\triangle CDP$ at P, which implies $\angle CDP = \angle CPY$. It follows that $\angle BPC = \angle BAP + \angle CDP = \angle BPX + \angle CPY$, i.e., P, X, Y are collinear. This is only possible if the circumcircles of $\triangle ABP$ and $\triangle CDP$ are tangent at P.

Consider the lines AB and CD.

Case I: BA extended and CD extended intersect at Q.

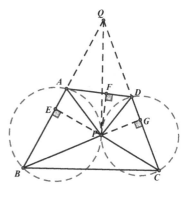

Refer to the diagram on the right. Since $ABCD$ is cyclic, $QA \cdot QB = QC \cdot QD$, i.e., the power of Q with respect to the circumcircles of $\triangle ABP$ and $\triangle CDP$ are the same. Hence, QP must be tangent to both circles.

It is easy to see that $\triangle QAP \sim \triangle QPB$.

Hence, $\dfrac{AQ}{AP} = \dfrac{QP}{BP}$. (1)

Similarly, $\triangle QDP \sim \triangle QPC$ and we have $\dfrac{DQ}{DP} = \dfrac{QP}{CP}$. (2)

Since $\angle AEP = \angle AFP = 90°$, A, E, P, F are concyclic where AP is a diameter. By Sine Rule, $\dfrac{EF}{\sin \angle BAD} = AP$, i.e., $EF = AP \sin \angle BAD$.

Similarly, $FG = DP \sin \angle ADC$.

It follows that $\dfrac{EF}{FG} = \dfrac{AP \sin \angle BAD}{DP \sin \angle ADC} = \dfrac{AP}{DP} \cdot \dfrac{\sin \angle QAD}{\sin \angle QDA} = \dfrac{AP}{DP} \cdot \dfrac{DQ}{AQ}$.

By (1) and (2), we have $\dfrac{EF}{FG} = \dfrac{BP}{PQ} \cdot \dfrac{PQ}{CP} = \dfrac{BP}{CP}$.

Case II: AB extended and DC extended intersect at Q.

Refer to the left diagram below. A similar argument applies and we still have $\dfrac{EF}{FG} = \dfrac{BP}{CP}$.

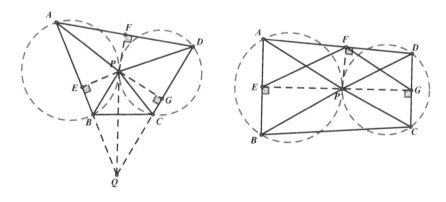

Case III: $AB /\!/ CD$

Refer to the right diagram above. We see that $ABCD$ is an isosceles trapezium (Exercise 3.1) and E, P, G are collinear (Example 1.1.11). We still have the circumcircles of $\triangle ABP$ and $\triangle CDP$ tangent at P. Now the radical axes, two of which are AB, CD, must be parallel (Theorem 4.3.2). Hence, AB, CD are perpendicular to the line connecting the circumcenters of $\triangle ABP$ and $\triangle CDP$. It follows that P lies on the perpendicular bisectors of AB and CD, which implies $AP = BP$ and $CP = DP$.

Since A, E, P, F are concyclic where AP is a diameter, we still have $EF = AP \sin \angle BAD$ and similarly, $FG = DP \sin \angle ADC$.

Since $AB \parallel CD$, we have $\sin \angle BAD = \sin \angle ADC$ because $\angle BAD$ and $\angle ADC$ are supplementary. Now $\dfrac{EF}{FG} = \dfrac{AP}{DP} = \dfrac{BP}{CP}$.

In conclusion, $\dfrac{EF}{FG} = \dfrac{BP}{CP}$ holds in all cases.

Now $\angle BPC = \angle BAP + \angle CDP = \angle EFP + \angle GFP$ (angles in the same arc) $= \angle EFG$. We conclude that $\triangle FEG \sim \triangle PBC$. $\qquad\square$

5.4 Exercises

1. Given an acute angled triangle $\triangle ABC$ and its circumcenter O, BD, CE are heights. Show that $AO \perp DE$.

2. Given a semicircle centered at O whose diameter is AB, draw $OP \perp AB$, intersecting the semicircle at P. Let M be the midpoint of AP. Draw $PH \perp BM$ at H. Show that $PH^2 = AH \cdot OH$.

3. (IND 94) Let I be the incenter of $\triangle ABC$ and the incircle of $\triangle ABC$ touches BC, AC at D, E respectively. If BI extended and DE extended intersect at P, show that $AP \perp BP$.

4. (AUT 09) Given an acute angled triangle $\triangle ABC$ where D, E, F are the midpoints of BC, AC, AB respectively and AP, BQ, CR are heights. Let X, Y, Z be the midpoints of QR, PR, PQ respectively. Show that DX, EY, FZ are concurrent.

5. Given a non-isosceles acute angled triangle $\triangle ABC$ and its circumcircle $\odot O$, H is the orthocenter of $\triangle ABC$ and M, N are the midpoints of AB, BC respectively. If MH extended and NH extended intersect $\odot O$ at P, Q respectively, show that P, Q, M, N are concyclic.

6. Given a right angled triangle $\triangle ABC$ where $\angle A = 90°$, $AD \perp BC$ at D. Let the radii of the incircles of $\triangle ABC, \triangle ABD, \triangle ACD$ be r, r_1, r_2 respectively. Show that $r + r_1 + r_2 = AD$.

7. Given a rectangle $ABCD$ where $AB = 1$ and $BC = 2$, P, Q are on BD, BC respectively. Find the smallest possible value of $CP + PQ$.

8. Given an acute angled triangle $\triangle ABC$ and its orthocenter H, M is the midpoint of BC. Draw a line ℓ passing through H and perpendicular to MH, intersecting AB, AC at P, Q respectively. Show that H is the midpoint of PQ.

9. Let P be a point outside $\odot O$ and PA, PB touch $\odot O$ at A, B respectively. C is a point on AB and the circumcircle of $\triangle BCP$ intersects $\odot O$ at B and D. Let Q be a point on PA extended such that $OP = OQ$. Show that $AD \parallel CQ$.

10. Given an acute angled triangle $\triangle ABC$ and its circumcircle, AD, BE are heights. X lies on the minor arc $\overset{\frown}{AC}$. If the lines BX and AD intersect at P, and the lines AX and BE intersect at Q, show that DE passes through the midpoint of the line segment PQ.

11. Given a right angled triangle $\triangle ABC$ where $\angle A = 90°$ and its circumcircle Γ, P is a point on Γ and $PH \perp BC$ at H. D, E are points on Γ such that $PD = PE = PH$. Show that DE bisects PH.

12. Let AB be a diameter of $\odot O$. P, Q are points outside the circle such that PA intersects $\odot O$ at C, PB extended intersects $\odot O$ at D and

QC, QD touch $\odot O$ at C, D respectively. If AD extended and PQ extended intersect at E, show that B, C, E are collinear.

13. (CHN 12)   Let Γ be the circumcircle of $\triangle ABC$ and I be the incenter of $\triangle ABC$. Let AI extended and BI extended intersect Γ at D, E respectively. Draw a line ℓ_1 passing through I such that $\ell_1 \parallel AB$. Draw a line ℓ_2 tangent to Γ at C. If ℓ_1, ℓ_2 intersect at F, show that D, E, F are collinear.

14. In $\triangle ABC$, $\angle A = 90°$. D, E are on AC, AB respectively such that BD, CE bisect $\angle B, \angle C$ respectively. Draw $AP \perp DE$, intersecting BC at P. Show that $AB - AC = BP - CP$.

15. Given a parallelogram $ABCD$, the circumcircle of $\triangle ABD$ intersects AC extended at E. P is a point on BD such that $\angle BCP = \angle ACD$. Show that $\angle AED = \angle BEP$.

16. Let CD be a diameter of $\odot O$. Points A, B on $\odot O$ are on opposite sides of CD. PC is tangent to $\odot O$ at C, intersecting the line AB at P. If the lines BD and OP intersect at E, show that $AC \perp CE$.

17. Refer to the diagram on the right. Given a cyclic quadrilateral $ABCD$, where BA extended and CD extended intersect at P, E, F lie on CD. Let G, H denote the circumcenters of $\triangle ADE$ and $\triangle BCF$ respectively. Show that if A, B, F, E are concyclic, then P, G, H are collinear.

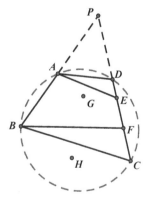

Chapter 6

Geometry Problems in Competitions

We have included a number of geometry problems from competitions in the previous chapters as examples. One may see that those problems are generally much harder than the standard exercises: simply applying a known theorem will not be an effective strategy. It could be difficult to even relate the conclusion to the conditions given. Indeed, this is a major obstacle encountered by the beginners: how to *start* problem solving? On the other hand, reading the solutions provided does not seem to be inspiring. Those solutions are usually written in an elegant and splendid manner, but do not show the beginners *how* one can think of such a solution.

One definitely finds it useful to be familiar with the basic skills and commonly used techniques illustrated in the previous chapters. Besides, we will introduce a few strategies in this chapter to tackle challenging geometry problems, while elaborating these strategies with examples from various competitions in the past years. Our focus is to seek clues and insights for each problem and hence, carry out the strategy which gradually leads to the solution.

6.1 Reverse Engineering

Not all competition questions are unreasonably difficult. Indeed, for those (relatively) easy questions, one simple but effective strategy is reverse engineering. This includes the following:

- Expect what the last step of the solution *could* be.

For example, if we are to show concyclicity, it could be concluded by equal angles, supplementary angles or line segments which compose of the Tangent Secant Theorem or the Intersecting Chords Theorem. If we are to show collinearity, it could be concluded by either Menelaus' Theorem or supplementary angles. If we are to show equal line segments, it could be concluded by isosceles, congruent or similar triangles.

Knowing the sketch of (the last part of) the proof gives inspiration on what intermediate steps one may expect and attempt to show.

- Discover what **should** be true by assuming the conclusion is true.

Of course, the conclusion to be shown should be true. Hence, by assuming this extra *condition*, we may discover what **should** be true (but is yet to be shown). For example, if we are to show equal angles and we assume they are, we may find a pair of triangles which **should** be similar. Now showing the similar triangles (say by line segments in ratio) leads to the conclusion!

- Simplify the conclusion ("It suffices to show…").

One shall always attempt to *link* the conclusion to the conditions given. Writing down "it suffices to show…" could transform the conclusion, moving it *towards* the given conditions.

Unfortunately, there is often more than one way to approach the conclusion or the intermediate steps, while most approaches will **not** lead to a complete proof. Be resilient and do not give up easily! It is common for even the most experienced contestants to have a few failed attempts before reaching a valid proof.

Example 6.1.1 (HRV 09) Given a quadrilateral $ABCD$, the circumcircle of $\triangle ABC$ intersects CD, AD at E, F respectively, and the circumcircle of $\triangle ACD$ intersects AB, BC at P, Q respectively. If BE, BF intersect PQ at X, Y respectively, show that E, F, Y, X are concyclic.

Insight. We are not given much information besides the two circles. Hence, it is natural to expect a proof by the angle properties. Refer to the diagram on the right.
Since we do not know much about $\angle EXF$ or $\angle EYX$, can we show that $\angle FYP = \angle FEX$?

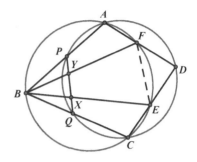

$\angle FEC$ is on the circumference of a circle, but $\angle FYP$ is not. However, one may write $\angle FYP = \angle PBY + \angle BPY$.

Proof. Refer to the diagram on the right. Since A, B, E, F are concyclic, we have $\angle BEF = 180° - \angle BAF$ $= \angle ABF + \angle AFB$.
Notice that $\angle AFB = \angle ACB = \angle BPY$ (because A, C, Q, P are concyclic).
Hence, $\angle BEF = \angle ABF + \angle BPY$ $= \angle FYP$. It follows that E, F, Y, X are concyclic. $\quad\square$

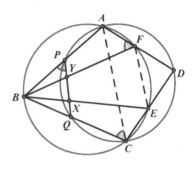

Note: There is more than one way to solve this problem. For example, one sees that $\angle FYP = \angle PBY + \angle BPY = \angle ACF + \angle ACB$ $= \angle BCF = \angle BEF$, which also leads to the conclusion. Indeed, it is an effective strategy to apply reverse engineering for this problem, i.e., repeatedly simplifying the conclusion by writing down "it suffices to show..." which eventually leads to a clear fact (about angles) and completes the proof.

Example 6.1.2 (SVN 08) $ABCD$ is a trapezium where $BC \parallel AD$ and $AB \perp BC$. It is also given that $AC \perp BD$. Draw $AE \perp CD$, intersecting CD extended at E. Show that $\dfrac{BE}{CE} = \dfrac{AD \cdot BD}{BD^2 - AD^2}$.

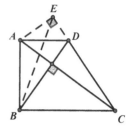

Insight. Refer to the diagram on the right. One immediately sees $BD^2 - AD^2 = AB^2$ and hence, we are to show $\dfrac{BE}{CE} = \dfrac{AD \cdot BD}{AB^2}$.

Can we simplify $\dfrac{AD \cdot BD}{AB^2}$? If yes, the problem may be solved by similar triangles.

How is AB related to AD and BD? Can you see that $AB^2 = AD \cdot BC$?

Proof. It is easy to see $\triangle ABD \sim \triangle BCA$. Hence, we have $\dfrac{AB}{BC} = \dfrac{AD}{AB}$, or $AB^2 = AD \cdot BC$. It follows that $\dfrac{AD \cdot BD}{BD^2 - AD^2} = \dfrac{AD \cdot BD}{AB^2} = \dfrac{AD \cdot BD}{AD \cdot BC}$ $= \dfrac{BD}{BC}$. Now it suffices to show $\dfrac{BE}{CE} = \dfrac{BD}{BC}$, or $\triangle BCD \sim \triangle ECB$.

Since $\angle AEC = \angle ABC = 90°$, we have A, B, C, E concyclic and hence, $\angle CBD = \angle BAC = \angle CEB$. Now $\triangle BCD \sim \triangle ECB$ and the conclusion follows.   □

Note: If one writes $AB^2 = BD \cdot BF$ where AC and BD intersect at F, it may not be easy to show $\dfrac{BE}{CE} = \dfrac{AD}{BF}$ because it is not clear how AD is related to BE or CE. Hence, AD should be cancelled out, i.e., we shall write $AB^2 = AD \cdot *$. Now it is easy to see that $*$ is BC.

Example 6.1.3 (CGMO 12) Let I be the incenter of $\triangle ABC$ whose incircle touches AB, AC at D, E respectively. If O is the circumcenter of $\triangle BCI$, show that $\angle ODB = \angle OEC$.

Insight. Refer to the left diagram below. Even though BD, CE are tangent to the incircles, it is not clear which angle on the circumference is equal to $\angle ODB$ or $\angle OEC$, as we do not know where OD, OE intersect $\odot I$. How about the supplement of these angles? Can we show $\angle ADO = \angle AEO$? At least we know $\angle ADI = \angle AEI = 90°$. Can we show $\angle ODI = \angle OEI$?

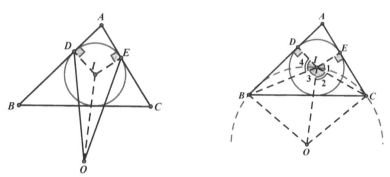

Since $DI = EI$, we **should** have $\triangle ODI \cong \triangle OEI$. How can we show these triangles are congruent? Can we show $\angle OID = \angle OIE$?

Proof. Refer to the right diagram above. We write $\angle OIE = \angle 1 + \angle 2$ and $\angle OID = \angle 3 + \angle 4$.

Notice that $\angle 1 = 90° - \angle ECI = 90° - \dfrac{1}{2}\angle ACB$ and since $OC = OI$, $\angle 2 = 90° - \dfrac{1}{2}\angle COI$. Notice that $\angle COI = 2\angle CBI$ (Theorem 3.1.1). Hence, $\angle COI = \angle ABC$ and we have $\angle 2 = 90° - \dfrac{1}{2}\angle ABC$.

Similarly, $\angle 3 = 90° - \dfrac{1}{2}\angle ACB$ and $\angle 4 = 90° - \dfrac{1}{2}\angle ABC$.

It follows that $\angle 1 + \angle 2 = \angle 3 + \angle 4$, i.e., $\angle OID = \angle OIE$. This implies $\triangle ODI \cong \triangle OEI$ (S.A.S.). Now we have $\angle ODI = \angle OEI$ and hence the conclusion. $\square$

Note: One familiar with the basic facts about the incenter and the circumcircle easily sees that AI extended intersects the circumcircle of $\triangle ABC$ at O, the circumcenter of $\triangle BIC$ (Example 3.4.2 and Exercise 3.14). Since O lies on AI, the perpendicular bisector of DE, the conclusion follows immediately.

Example 6.1.4 (APMO 13)   Given an acute angled triangle $\triangle ABC$ and its circumcenter O, AD, BE, CF are heights. Show that the line segments OA, OF, OB, OD, OC, OE dissect $\triangle ABC$ into three pairs of triangles that have equal areas.

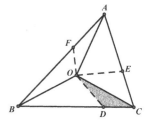

Insight.   First, we shall decide which of the triangles **could** be of equal area. Refer to the diagram on the right. Since F is not the midpoint of AB, $[\triangle AOF] \neq [\triangle BOF]$. Observe that we shall **not** have $[\triangle AOF] = [\triangle AOE]$. (Consider the case when $\angle C$ is almost $90°$.)

Nor shall we have $[\triangle AOF] = [\triangle COE]$. Otherwise the triangles cannot be paired up in a *symmetric* manner. It seems that we should show $[\triangle AOF] = [\triangle COD]$, $[\triangle AOE] = [\triangle BOD]$ and $[\triangle BOF] = [\triangle COE]$.

Apparently, these triangles are not congruent. Notice that

$$[\triangle AOF] = \frac{1}{2} AO \cdot AF \sin \angle OAF \text{ and } [\triangle COD] = \frac{1}{2} CO \cdot CD \sin \angle OCD.$$

Since $AO = CO$, it suffices to show $AF \sin \angle OAF = CD \sin \angle OCD$. This should not be difficult since we have the right angled triangles (heights) and the circumcircle.

Proof.   Refer to the diagram on the right. We have $AF = AC \cos \angle A$ and $CD = AC \cos \angle C$.

Hence, $\dfrac{AF}{CD} = \dfrac{\cos \angle A}{\cos \angle C}$.   (1)

Notice that $\angle OAF = 90° - \dfrac{1}{2} \angle AOB = 90° - \angle C$.

Similarly, $\angle OCD = 90° - \angle A$.

Now $\dfrac{[\Delta AOF]}{[\Delta COD]} = \dfrac{\frac{1}{2}AF \cdot AO\sin\angle OAF}{\frac{1}{2}CD \cdot CO\sin\angle OCD} = \dfrac{AF}{CD} \cdot \dfrac{\sin(90° - \angle C)}{\sin(90° - \angle A)}$

$= \dfrac{AF}{CD} \cdot \dfrac{\cos\angle C}{\cos\angle A} = 1$ by (1).

Similarly, $[\Delta AOE] = [\Delta BOD]$ and $[\Delta BOF] = [\Delta COE]$. □

Example 6.1.5 (IMO 98)   In a cyclic quadrilateral $ABCD$, $AC \perp BD$ and AB, CD are not parallel. If the perpendicular bisectors of AB and CD intersect at P, show that $[\Delta ABP] = [\Delta CDP]$.

Insight.   One notices that ΔABP and ΔCDP are isosceles triangles. Hence, we are to show

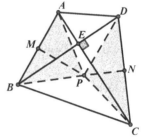

$$\dfrac{[\Delta ABP]}{[\Delta CDP]} = \dfrac{\frac{1}{2}AP^2 \sin\angle APB}{\frac{1}{2}CP^2 \sin\angle CPD} = 1.$$

How are AP and CP related? Since $ABCD$ is cyclic, say inscribed inside the circle Γ, the center of Γ must lie on the perpendicular bisectors of AB, CD. Indeed, P is the center of Γ and we have $AP = CP$. Refer to the diagram above.

Now it suffices to show $\sin\angle APB = \sin\angle CPD$. It seems from the diagram $\angle APB \neq \angle CPD$. Can we show $\angle APB = 180° - \angle CPD$ instead? (**Hint**: Can you see $\angle APB = 2\angle ACB$?)

Proof.   Since $ABCD$ is cyclic, one sees that P is the center of the circumcircle of $ABCD$. Hence, $PA = PB = PC = PD$.

Since $[\Delta ABP] = \frac{1}{2} PA^2 \sin \angle APB$ and $[\Delta CDP] = \frac{1}{2} PC^2 \sin \angle CPD$, it suffices to show $\sin \angle APB = \sin \angle CPD$.

We claim that $\angle APB + \angle CPD = 180°$. In fact, since P is the centre of the circumcircle of $ABCD$, $\angle APB = 2\angle ACB$ (Theorem 3.1.1). Similarly, $\angle CPD = 2\angle CBD$. Since $\angle ACB + \angle CBD = 90°$, we must have $\angle APB + \angle CPD = 180°$. This completes the proof. □

Note: One may also solve the problem by considering

$$[\Delta ABP] = \frac{1}{2} AB \cdot PM \text{ and } [\Delta CDP] = \frac{1}{2} CD \cdot PN.$$

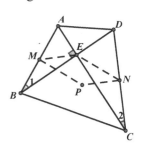

It suffices to show $\dfrac{AB}{PN} = \dfrac{CD}{PM}$.

Notice that in the right angled triangle ΔABE, $AB = 2EM$ because M is the midpoint of AB.

Similarly, $CD = 2EN$. Now it suffices to show $\dfrac{EM}{PN} = \dfrac{EN}{PM}$. (1)

In fact, we claim that $EMPN$ is a parallelogram. Refer to the diagram above on the right. We have $\angle BEM = \angle 1 = \angle 2 = \angle CEN$. Now
$$\angle EMP = 90° - \angle AME = 90° - 2\angle 1 \qquad (2)$$
$$\angle MEN = 90° + \angle BEM + \angle CEN = 90° + 2\angle 1 \qquad (3)$$
(2) and (3) give $\angle EMP + \angle MEN = 180°$ and hence, $EM \, /\!/ \, PN$. Similarly, $EN \, /\!/ \, PM$ and $EMPN$ is a parallelogram. This implies (1) and the conclusion follows.

Example 6.1.6 (HEL 11) In an acute angled triangle ΔABC, $AB < AC$, $AD \perp BC$ at D and AD extended intersects the circumcircle of ΔABC at E. The perpendicular bisector of AB intersects AD at L. BL extended intersects AC at M and intersects the circumcircle of ΔABC at N. EN and the perpendicular bisector of AB intersect at Z. Show that if $AC = BC$, then $MZ \perp BC$.

Insight. Refer to the diagram below. Since $AD \perp BC$, we **should** have $MZ \parallel AD$. Can we show $\angle 1 = \angle 2$? We **should** have $\angle 2 = \angle MCN$ and hence, C, N, M, Z **should** be concyclic.

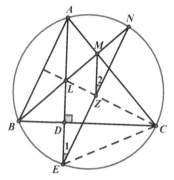

Notice that there are many equal angles in the diagram. In fact, the isosceles triangle $\triangle ABC$ is symmetric about the perpendicular bisector of AB. One may also notice that L is the orthocenter of $\triangle ABC$. It should not be difficult to show the concyclicity by angle-chasing.

Proof. Since $AC = BC$, it is easy to see that $\triangle ABC$ is symmetric about the perpendicular bisector of AB. Hence, C, Z, L are collinear, which gives the angle bisector of $\angle ACB$ (and the perpendicular bisector of AB). In particular, L is the orthocenter of $\triangle ABC$.

Recall that E and L are symmetric about BC (Example 3.4.3). It follows that $\angle N = \angle BCE = \angle BCL = \angle ACL$. Hence, C, N, M, Z are concyclic. Now $\angle 2 = \angle MCN = \angle 1$, which implies $AE \parallel MZ$, i.e., $MZ \perp BC$. □

Example 6.1.7 (USA 07) Refer to the diagram on the right. Γ_1, Γ_2 are circles intersecting at P, Q. AC, BD are chords in Γ_1, Γ_2 respectively such that AB intersects CD extended at P. AC intersects BD extended at X. Let Y, Z be on Γ_1, Γ_2 respectively such that $PY \parallel BD$ and $PZ \parallel AC$. Show that Y, X, Q, Z are collinear.

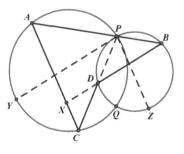

Insight. It seems that Y and Z are constructed in a symmetric manner. If we can show that X, Q, Z are collinear, perhaps a similar argument applies for X, Q, Y. Refer to the following diagram.

We are to show $\angle 1 = 180° - \angle DQZ = \angle 2$. Since $PZ \, / \! / \, AC$, we have $\angle 2 = \angle 3$ and hence, it suffices to show $\angle 1 = \angle 3$. Hence, C, Q, D, X **should** be concyclic. Can we show $\angle DCQ = \angle DXQ$?

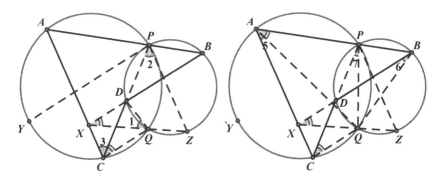

Refer to the right diagram above. Clearly, $\angle DCQ = \angle PAQ$ and hence, we **should** have $\angle BAQ = \angle BXQ$ and A, B, Q, X **should** be concyclic. Hence, it suffices to show $\angle 5 = \angle 6$.

Can you see that $\angle 5 = \angle 7 = \angle 6$ by concyclicity? We leave it to the reader to complete the proof. (**Hint**: One may conclude that Q, X, Y are collinear by observing that $\angle CQX = \angle CQY$, where $\angle CQX = \angle CDX = \angle CPY = \angle CQY$ by concyclicity and $PY \, / \! / \, BD$.)

6.2 Recognizing a Relevant Theorem

Occasionally, one may encounter a geometry problem in the competition where the construction seems closely related to a particular theorem (or a well-known fact). It might be a wise strategy to apply the theorem and find the missing links during the process. If you are successful, there is a high chance that the proof is almost complete.

Example 6.2.1 (CHN 08) Given a convex quadrilateral $ABCD$ where $\angle B + \angle D < 180°$, P is an arbitrary point in $\triangle ACD$ and we define $f(P) = PA \cdot BC + PD \cdot AC + PC \cdot AB$. Show that when $f(P)$ attains the minimal value, B, P, D are collinear.

Insight. One should recognize that $f(P)$ is closely related to Ptolemy's Theorem. In fact, we have $PA \cdot BC + PC \cdot AB \geq PB \cdot AC$ and the equality holds if and only if P lies on the circumcircle of $\triangle ABC$. Refer to the diagram on the right. Now it is easy to complete the proof.

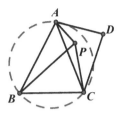

Proof. Notice that $f(P) = PA \cdot BC + PD \cdot AC + PC \cdot AB$
$\geq PB \cdot AC + PD \cdot AC$ by Ptolemy's Theorem   (1)
$= (PB + PD) \cdot AC \geq BD \cdot AC$ by Triangle Inequality.   (2)
Hence, the minimal possible value of $f(P)$ is $BD \cdot AC$. This is only attainable if the equality holds in both (1) and (2), i.e., we must have that P lies on the circumcircle of $\triangle ABC$ and P lies on BD (i.e., B, P, D are collinear). This complete the proof.   □

Note: Refer to the diagram on the right. It is easy to see that D must be outside the circumcircle of $\triangle ABC$ since $\angle B + \angle D < 180°$. Hence, P must lie between B and D. Indeed, P is the intersection of BD and the circumcircle of $\triangle ABC$.

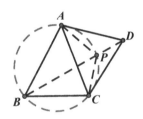

Example 6.2.2 (RUS 04)   Let $\odot O$ be inscribed inside a quadrilateral $ABCD$. It is given that the external angle bisectors of $\angle A, \angle B$ intersect at K, the external angle bisectors of $\angle B, \angle C$ intersect at L, the external angle bisectors of $\angle C, \angle D$ intersect at M, and the external angle bisectors of $\angle D, \angle A$ intersect at N. If the orthocenters of $\triangle ABK, \triangle BCL, \triangle CDM, \triangle DAN$ are P, Q, R, S respectively, show that $PQRS$ is a parallelogram.

Insight. Refer to the following left diagram. How are the external angle bisectors related to the orthocenter (right angles)? Recall that the angle bisectors of neighboring supplementary angles are perpendicular. Can you see that $OA \perp AK$? Can you see that $OA \text{//} BP$ and similarly, $OB \text{//} AP$? We have a parallelogram $AOBP$.

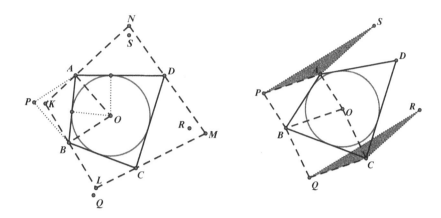

Proof.   It is easy to see that OA bisects $\angle BAD$. Hence, $OA \perp AK$ because they bisect neighboring angles which are supplementary. We also have $BP \perp AK$ because P is the orthocenter of $\triangle ABK$.

Hence, $OA \, // \, BP$ and similarly, $OB \, // \, AP$. It follows that $AOBP$ is a parallelogram. Similarly, $BOCQ$, $CODR$ and $DOAS$ are parallelograms. Now $AP = OB = CQ$ and $AP \, // \, OB \, // \, CQ$. Similarly, $AS = CR$ and $AS \, // \, CQ$. It follows that $\triangle APS \cong \triangle CQR$ (S.A.S.). Refer to the right diagram above. We conclude that $PS = QR$, $PS \, // \, QR$ and hence, $PQRS$ is a parallelogram.   □

Note: One may recall that $AB + CD = BC + AD$ since $\odot O$ is inscribed inside $ABCD$. However, this is not related to the conclusion.

Example 6.2.3 (IMO 12)   Let J be the ex-center of $\triangle ABC$ opposite the vertex A. This ex-circle (i.e., the circle centered at J and tangent to BC, AB extended and AC extended) is tangent to BC at M, and is tangent to the lines AB, AC at K, L respectively. Let the lines LM, BJ meet at F and the lines KM, CJ meet at G. If AF extended and AG extended meet the line BC at S, T respectively, show that M is the midpoint of ST.

Insight.   Refer to the following diagram. We are to show $SM = TM$. Notice that there are many lines intersecting each other and the ex-circle gives us many equal line segments. Hence, we may apply Menelaus' Theorem involving SM and TM.

Apply Menelaus' Theorem to $\triangle ASC$ intersected by the line FL and we obtain

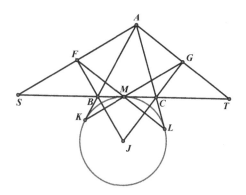

$$\frac{SM}{CM} \cdot \frac{CL}{AL} \cdot \frac{AF}{SF} = 1.$$

Since $CM = CL$, we have

$$SM = \frac{SF \cdot AL}{AF}.$$

Similarly, apply Menelaus' Theorem to $\triangle ATB$ intersected by the line GK and we have $TM = \dfrac{TG \cdot AK}{AG}$.

We are to show $SM = TM$, i.e., $\dfrac{SF \cdot AL}{AF} = \dfrac{TG \cdot AK}{AG}$.

Clearly, $AK = AL$. Hence, it suffices to show that $\dfrac{SF}{AF} = \dfrac{TG}{AG}$, i.e., we **should** have $FG /\!/ BC$.

Notice that we can simplify the diagram significantly because A, S, T can be neglected. Refer to the diagram on the right. It suffices to show that $\angle 1 = \angle 2$. Since $BJ \perp MK$ and $CJ \perp ML$, one sees a number of concyclicity and hence, many equal angles.

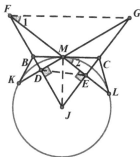

Indeed, we have $\angle 1 = \angle MDE$ (since D, E, G, F are concyclic)
$= \angle MJE$ (since D, J, E, M are concyclic)
$= 90° - \angle EMJ = \angle 2$ (since $MJ \perp BC$). This completes the proof.

Example 6.2.4 (IND 11)   Refer to the diagram on the right. A quadrilateral $ABCD$ is inscribed inside a circle. Let E, F, G, H be the midpoints of arcs $\overparen{AB}, \overparen{BC}, \overparen{CD}, \overparen{DA}$ respectively.

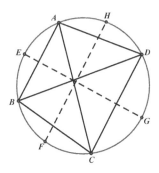

It is known that $AC \cdot BD = EG \cdot FH$. Show that AC, BD, EG and FH are concurrent.

Insight.   Apparently there are very few clues. In particular, we do not know how $AC \cdot BD = EG \cdot FH$ can be applied. How are these line segments related?

While AC may not be related to EG, it is not difficult to see that AC is related to EF, because $\angle EDF = \dfrac{1}{2} \angle ADC$! (Can you see it?) Refer to the left diagram below.

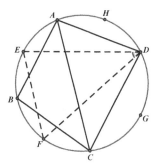 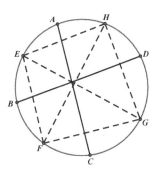

Similarly, AC is related to HG and BD is related to EH, FG. (Can you see that $\angle GBH = 90° - \angle EDF$?) Refer to the right diagram above. If we can replace AC and BD by EF, FG, GH and EH, the condition given becomes a relationship between the sides of a cyclic quadrilateral and its diagonals. Is it reminiscent of Ptolemy's Theorem?

Proof. It is easy to see that $\angle EDB + \angle BDF = \dfrac{1}{2}\angle ADB + \dfrac{1}{2}\angle BDC$, i.e.,

$\angle EDF = \dfrac{1}{2}\angle ADC$. By Sine Rule, $\dfrac{AC}{\sin \angle ADC} = 2R = \dfrac{EF}{\sin \angle EDF}$, where R is the radius of the circle. Let $\angle EDF = \alpha$.

We have $AC = \dfrac{EF}{\sin \alpha}\sin 2\alpha = \dfrac{EF}{\sin \alpha}2\sin \alpha \cos \alpha = 2EF\cos \alpha.$ (1)

Similarly, $AC = 2HG\cos \angle GBH = 2HG\cos(90° - \alpha) = 2HG\sin \alpha.$ (2)
Let $\angle FAG = \beta$. We also have $BD = 2FG\cos \beta = 2EH\sin \beta$.
Ptolemy's Theorem states $EF \cdot HG + FG \cdot EH = EG \cdot FH.$ (3)
By (1) and (2),

$$EF \cdot HG = \frac{AC^2}{4\sin \alpha \cos \alpha} = \frac{AC}{2}\cdot\frac{AC}{\sin 2\alpha} = \frac{AC}{2}\cdot 2R = AC \cdot R.$$

Similarly, $FG \cdot EH = BD \cdot R$.
Now (3) gives $EF \cdot HG + FG \cdot EH = (AC + BD)\cdot R = EG \cdot FH$.

Since $EG \cdot FH = AC \cdot BD$, we have $(AC + BD)\cdot 2R = 2AC \cdot BD.$ (4)
Notice that $2R$ is the diameter, i.e., $AC, BD \le 2R$.

It follows that $(AC + BD)\cdot 2R = AC \cdot 2R + BD \cdot 2R \ge AC \cdot BD + BD \cdot AC = 2AC \cdot BD$, where the equality holds by (4). This is only possible if $AC = BD = 2R$, i.e., AC, BD are both diameters of the circle.
Since $AC \cdot BD = EG \cdot FH$, EG, FH are also diameters. In conclusion, AC, BD, EG, EH are concurrent at the center of the circle. $\qquad \square$

Note: In (1), we applied the double angle formula $\sin 2\alpha = 2\sin \alpha \cos \alpha$, which could be found in most pre-calculus textbooks.

Example 6.2.5 (UKR 11) Given a trapezium $ABCD$, $AD // BC$ and F is a point on CD. AF and BD intersect at E. Draw $EG // AD$, intersecting AB at G. BD and CG intersect at H. AB and FH extended intersect at I. Show that the lines AD, CI, FG are concurrent.

Insight. Refer to the diagram on the right. It seems not easy to show the lines AD, CI, FG are concurrent. Notice that the intersection of these lines is far from the trapezium $ABCD$.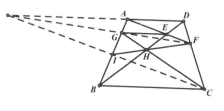

Since $AD \,//\, EG \,//\, BC$, the Intercept Theorem gives many equations of line segments in the trapezium $ABCD$. Is it possible for us to derive the conclusion from these line segments instead of the extensions of AD, CI, FG? Recall Desargues' Theorem. Can we find two triangles whose vertices are A, D, C, I, F, G, while the lines connecting corresponding vertices are AD, CI, FG respectively?

It is not a difficult task. In fact, since A, G, I and C, D, F are collinear, we do not have many choices left, one of which is $\triangle AFI$ and $\triangle DGC$. Can we show that these two triangles satisfy the condition for Desargues' Theorem, i.e., say the lines AB, CD intersect at P and AF, DG intersect at Q, can we show that P, Q, H are collinear? Refer to the diagram below.

Proof. Let the lines AB, CD intersect at P and AF, DG intersect at Q. We claim that P, Q, H are collinear. By Menelaus' Theorem, it suffices to show that $\dfrac{AQ}{EQ} \cdot \dfrac{EH}{BH} \cdot \dfrac{BP}{AP} = 1$.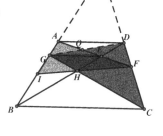

Since $EG \,//\, AD \,//\, BC$, we have

$$\frac{AQ}{EQ} = \frac{AD}{EG}, \quad \frac{EH}{BH} = \frac{EG}{BC} \quad \text{and} \quad \frac{BP}{AP} = \frac{BC}{AD}.$$

It follows that $\dfrac{AQ}{EQ} \cdot \dfrac{EH}{BH} \cdot \dfrac{BP}{AP} = \dfrac{AD}{EG} \cdot \dfrac{EG}{BC} \cdot \dfrac{BC}{AD} = 1$.

Since P, Q, H are collinear, the conclusion follows by applying Desargues' Theorem to $\triangle AFI$ and $\triangle DGC$.   □

6.3 Unusual and Unused Conditions

A typical geometry problem in competitions comes with a few given conditions. Besides those more standard conditions (parallel and perpendicular lines, midpoints, angle bisectors, circles and tangency), there may be unusual conditions which easily catch the attention of the contestants. For example:

- Angles or line segments which are far apart but equal
- $30°$, $45°$ or $60°$ angles
- Points constructed in an unusual manner
- Equations of line segments or angles, the geometric meanings of which are apparently not clear
- Points, lines or circles which coincide unexpectedly

One naturally expects such conditions to play a critical role when solving the geometry problem. Hence, it is worthwhile to spend time and effort focusing on these conditions, which may lead to an important intermediate step.

On the other hand, it is also common that one cannot find any clue after exploring the unusual condition, or even cannot see any *sense* about it. Do not be frustrated! It could be a wise strategy to leave it aside and focus on other conditions, writing down intermediate steps which could be derived. We shall attempt to link those steps to the conclusion and expect to be stuck during the process (because we still have unused conditions). Now you may find the unused condition handy: it may be exactly the missing link needed!

Geometry problems in competitions are generally well constructed and the conditions given are exactly sufficient (because unnecessary extra conditions may cause inconsistency). Hence, if all the conditions given have been applied and a chain of derivations is constructed, most probably you are very close to the complete proof.

Example 6.3.1 (CZE-SVK 09)   Given a rectangle $ABCD$ inscribed inside $\odot O$, P is a point on the minor arc $\overset{\frown}{CD}$. Let K, L, M be the feet of the perpendiculars from P to AB, AC, BD respectively. Show that $\angle LKM = 45°$ if and only if $ABCD$ is a square.

Insight.   One immediately notices that $\angle LKM = 45°$ is an unusual condition. How is it related to $ABCD$, in which case a square? Refer to the diagram on the right. If $ABCD$ is a square, we must have $\angle CAD = \angle CBD = 45°$.

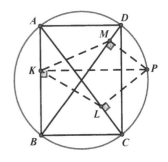

It seems we shall apply the angle properties around the circles. (Notice that the right angles give a number of concyclicity.)

Proof.   Since $\angle BKP = \angle BMP = 90°$, B, K, M, P are concyclic and $\angle PKM = \angle PBM$. Similarly, A, K, L, P are concyclic, which implies $\angle PKL = \angle PAL = \angle PBC$ (angles in the same arc).
Now $\angle LKM = \angle PKM + \angle PKL = \angle PBM + \angle PBC = \angle CBD$.   Hence, $\angle LKM = 45°$ if and only if $\angle CBD = 45°$, i.e., $ABCD$ is a square.   □

Note: One who attempts to show $\angle LKM = \angle CBD$ by concyclicity directly may find it difficult because we do not know much about the intersection of the lines KL and BC.

Example 6.3.2 (IRN 11)   Given $\triangle ABC$ where $\angle A = 60°$, D, E are on AB, AC extended respectively such that $BD = CE = BC$. If the circumcircle of $\triangle ACD$ intersects DE at D and P, show that P lies on the angle bisector of $\angle BAC$.

Insight.   Refer to the diagram on the right. One immediately notice that $\angle A = 60°$ is an unusual condition. Moreover, it is not easy to draw $BD = BC = CE$ when given $\triangle ADE$. How could we apply these conditions?

Notice that $BD = BC = CE$ gives two isosceles triangles $\triangle BCD$ and $\triangle CBE$, where $\angle 1$ and $\angle 2$ are related to $\angle A$.
In fact, $\angle ABC = 2\angle 2$ and $\angle ACB = 2\angle 1$. Since $\angle A = 60°$, we must have $\angle ABC + \angle ACB = 120°$ and hence, $\angle 1 + \angle 2 = 60°$.
Let BE and CD intersect at F. One sees that $\angle BFD = \angle 1 + \angle 2 = 60°$ $= \angle A$ and hence, A, B, F, C are concyclic.

Notice that the diagram should be *symmetric*: since P **should** lie on the angle bisector of $\angle BAC$, if A, C, P, D are concyclic, then A, B, P, E **should** be concyclic as well. Can you show it?

Refer to the left diagram below. Since A, B, F, C are concyclic, we have $\angle ABF = \angle ECF$. Since A, C, P, D are concyclic, we must have $\angle ECF = 180° - \angle ACD = 180° - \angle APD = \angle APE$. It follows that $\angle ABF = \angle APE$ and hence, A, B, P, E are concyclic.

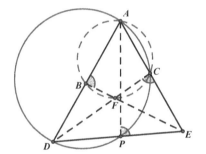 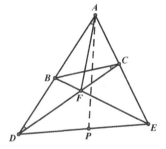

We are to show AP bisects $\angle A$, i.e., $\angle BAP = \angle CAP$. Since $\angle BAP = \angle BEP$ and $\angle CAP = \angle CDP$ by concyclicity, it suffices to show $DF = EF$. Refer to the right diagram above.
Can you see that $DF = EF = AF$, i.e., F is the circumcenter of $\triangle ADE$? (**Hint**: Can you see $\angle BAF = \angle BCF = \angle BDF$?) We leave it to the reader to complete the proof.

Note: Upon showing the concyclicity of A, B, F, C and A, B, P, E, there are many ways to show that AP is the angle bisector. For example, can you see that P is of the same distance from the lines AB and AC? Refer to the following left diagram. Can you see $\triangle BDP \cong \triangle ECP$

(A.A.S.)? Since BD and CE are corresponding sides, the heights from P to BD, CE respectively must be the same.

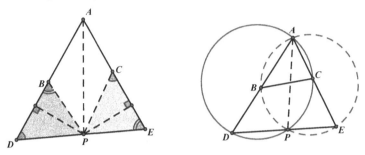

Alternatively, one may also show the conclusion by the Angle Bisector Theorem, i.e., we are to show $\dfrac{AD}{AE} = \dfrac{PD}{PE}$.

Refer to the right diagram above. By the Tangent Secant Theorem, $AD \cdot BD = PD \cdot DE$ and $AE \cdot CE = PE \cdot DE$.

Since $BD = CE$, we have $\dfrac{AD}{AE} = \dfrac{AD \cdot BD}{AE \cdot CE} = \dfrac{PD \cdot DE}{PE \cdot DE} = \dfrac{PD}{PE}$, which completes the proof.

Example 6.3.3 (RUS 08)   Given $\odot I$ inscribed inside $\triangle ABC$, AB, AC touch $\odot I$ at X, Y respectively. Let CI extended intersect the circumcircle of $\triangle ABC$ at D. If the line XY passes through the midpoint of AD, find $\angle BAC$.

Insight.   One immediately notices that the line XY passing through the midpoint of AD is an unusual condition, without which $\triangle ABC$, its incenter I and CD give a standard diagram (Example 3.4.2). Refer to the diagram below.

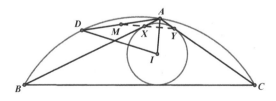

We have $AD = DI$, i.e., $\triangle AID$ is an isosceles triangle. It is easy to see that $XY \perp AI$. Let M be the midpoint of AD. How could you apply the condition that X, Y, M are collinear?

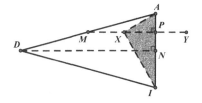

Can you see where AI and XY intersect? Refer to the diagram above.

Let N be the midpoint of AI. Clearly, $DN \perp AI$ and hence, $DN \,/\!/\, XY$. Let AI intersect XY at P. Since X, Y, M are collinear, P is the midpoint of AN. It follows that $AP = \dfrac{1}{4} AI$.

What can you say about the right angled triangle $\triangle AXI$? Can you see that $\dfrac{AP}{PX} = \dfrac{PX}{PI}$ and hence, $\left(\dfrac{AP}{PX}\right)^2 = \dfrac{AP}{PX} \cdot \dfrac{PX}{PI} = \dfrac{AP}{PI} = \dfrac{1}{3}$?

Now it is easy to see that $\angle XAI = 60°$ and hence, $\angle BAC = 120°$. We leave the details to the reader.

Note: If one draws an acute angled triangle $\triangle ABC$, the line XY will not even intersect the line segment AD. By constructing the diagram carefully, one should realize that $\angle BAC$ is obtuse.

Example 6.3.4 (CGMO 11)   Let $ABCD$ be a quadrilateral where AC, BD intersect at E. Let M, N be the midpoints of AB, CD respectively and the perpendicular bisectors of AB, CD intersect at F. If the line EF intersects BC, AD at P, Q respectively and it is given that $FM \cdot CD = FN \cdot AB$ and $BP \cdot DQ = CP \cdot AQ$, show that $PQ \perp BC$.

Insight.   One immediately notices the unusual conditions $FM \cdot CD = FN \cdot AB$ and $BP \cdot DQ = CP \cdot AQ$, but apparently, they refer to different properties.

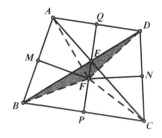

$FM \cdot CD = FN \cdot AB$ implies $\dfrac{FM}{FN} = \dfrac{AB}{CD}$. Refer to the previous diagram. What can you conclude about the (isosceles) triangles $\triangle ABF$ and $\triangle CDF$?

Can you see that $\triangle BDF$ and $\triangle ACF$ are congruent? What can you conclude upon obtaining the equal angles, say $\angle CAF = \angle DBF$?

$BP \cdot DQ = CP \cdot AQ$ gives $\dfrac{BP}{CP} = \dfrac{AQ}{DQ}$. Is it reminiscent of the Angle Bisector Theorem? Where is the angle bisector?

Indeed, the diagram is symmetric. If one sees that F is the center of the circle where $ABCD$ is inscribed, the proof is almost complete.

Proof.   Notice that $FM \cdot CD = FN \cdot AB$ implies $\dfrac{FM}{FN} = \dfrac{AB}{CD}$. Since M, N are the midpoints of AB, CD respectively, it is easy to see that the isosceles triangles $\triangle ABF$ and $\triangle CDF$ are similar.

In particular, we have $\angle AFB = \angle CFD$, which implies $\angle AFC = \angle BFD$. Since $AF = BF$ and $CF = DF$, we have $\triangle BDF \cong \triangle ACF$ (S.A.S.). It follows that $\angle CAF = \angle DBF$ and hence, A, B, F, E are concyclic. Similarly, $\angle BDF = \angle ACF$ and C, D, E, F are concyclic.

We have $\angle BEF = \angle BAF$ and $\angle CEF = \angle CDF$ by concyclicity. Notice that $\angle BAF = \angle CDF$ (because $\triangle ABF \sim \triangle DCF$). Now $\angle BEF = \angle CEF$, i.e., EP bisects $\angle BEC$. By the Angle Bisector Theorem, $\dfrac{BE}{CE} = \dfrac{BP}{CP}$. (1)

Similarly, EQ bisects $\angle AED$ and $\dfrac{AE}{DE} = \dfrac{AQ}{DQ}$. (2)

We are given $BP \cdot DQ = CP \cdot AQ$, i.e., $\dfrac{BP}{CP} = \dfrac{AQ}{DQ}$. (3)

(1), (2) and (3) imply that $\dfrac{BE}{CE} = \dfrac{AE}{DE}$, i.e., $BE \cdot DE = AE \cdot CE$.

By the Intersecting Chords Theorem, $ABCD$ is cyclic. Clearly, it is inscribed in a circle centered at F.

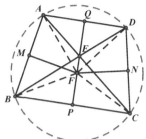

Now $AF = BF = CF = DF$ and hence, $AB = CD$. Refer to the diagram on the right. It is easy to see that $\triangle ABF \cong \triangle DCF$ (S.A.S.). Hence, $AB = CD$ and $ABCD$ is an isosceles trapezium where $AD \parallel BC$. It follows that $PQ \perp BC$. □

Note: Since the diagram is symmetric, we **should** have $PQ \perp AD$ as well, i.e., $AD \parallel BC$. Upon showing that EF bisects $\angle BEC$, one naturally expects that $ABCD$ is an isosceles trapezium.

Example 6.3.5 (BGR 11) Let P be a point inside the acute angled triangle $\triangle ABC$. D, E, F are the feet of the perpendiculars from P to BC, AC, AB respectively. Q is a point inside $\triangle ABC$ such that $AQ \perp EF$ and $BQ \perp DF$. Draw $QH \perp AB$ at H. Show that D, E, F, H are concyclic.

Insight. We are given a lot of right angles. In particular, one notices that the construction of Q is unusual. What can we obtain from Q?
Refer to the left diagram below. By applying the properties of right angles, can you see that $\angle 1 = \angle 2 = \angle 3$?

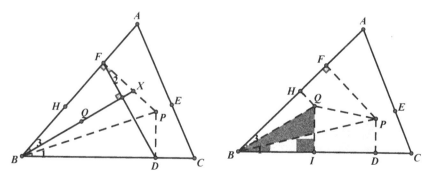

If we draw $QI \perp BC$ at I, can you see that the right angled triangles $\triangle PBD, \triangle PBF, \triangle QBH, \triangle QBI$ are closely related? (**Hint:** Can you see similar triangles?) Refer to the right diagram above. How are BI, BH, BD, BF related to BP and BQ, say via similar triangles? How are HF and DI related to PQ?

Proof. Since $PD \perp BC$ and $PF \perp AB$, B, D, P, F are concyclic and hence, $\angle 1 = \angle 2$. Let BQ extended intersect BF at X. We have $\angle 2 = \angle 3$ in the right angled triangle $\triangle BFX$. Hence, $\angle 1 = \angle 3$ and it follows that $\angle CBQ = \angle ABP$.
Draw $QI \perp BC$ at I. We have $BI \cdot BD = BQ \cos \angle CBQ \cdot BP \cos \angle 1$ and $BH \cdot BF = BQ \cos \angle 3 \cdot BP \cos \angle ABP$. It follows that $BI \cdot BD = BH \cdot BF$. Now H, F, D, I are concyclic by the Tangent Secant Theorem.

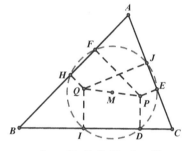

Let M be the midpoint of PQ. It is easy to see that the perpendicular bisectors of both HF and DI pass through M. Let $\odot M$ denote the circle centered at M with radius DM. Clearly, H, F, D, I lie on $\odot M$. Refer to the diagram on the right.
Similarly, if we draw $QJ \perp AC$ at J, one sees that E, J, I, D also lie on $\odot M$, i.e., D, E, F, H, I, J are concyclic. This completes the proof. □

Example 6.3.6 (IMO 13) Given $\triangle ABC$ with $\angle B > \angle C$, Q is on AC and P is on CA extended such that $\angle ABP = \angle ABQ = \angle C$. D is a point on BQ such that $PB = PD$. AD extended intersect the circumcircle of $\triangle ABC$ at R. Show that $QB = QR$.

Insight. Refer to the diagram on the right. One immediately notices the condition $\angle 1 = \angle 2 = \angle ACB$. We also have $\angle ACB = \angle 3$ (angles in the same arc).

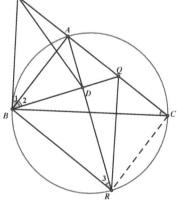

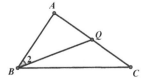

Recall a basic result of similar triangles as shown in the diagram above. Since $\angle 2 = \angle C$, we must have $\triangle ABQ \sim \triangle ACB$. Since $\angle 2 = \angle 3$, we also have $\triangle ABD \sim \triangle ARB$. It follows that $\angle ABR = \angle ADB = \angle QDR$. Since $\angle ABR + \angle ACR = 180°$, we must have $\angle QDR + \angle ACR = 180°$, which implies C, Q, D, R are concyclic.

Refer to the diagram on the right. We are to show $QB = QR$. Of course, the most straightforward method is to show that $\angle QBR = \angle QRB$. Since we have two circles and a few pairs of similar triangles, perhaps we shall seek more equal angles.

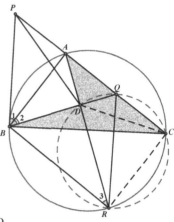

We can write:

$$\angle QRB = \angle 3 + \angle QRD = \angle ACB + \angle QCD$$

$$\angle QBR = \angle CBR + \angle CBQ = \angle CAR + \angle CBQ.$$

Now it suffices to show that $\angle ACB + \angle QCD = \angle CAR + \angle CBQ$. (1)

Notice that all these angles are related to the shaded region in the diagram. In particular, $\angle ACB + \angle CAR + \angle CBQ = \angle ADB$ (exterior angles of $\triangle ACD$ and $\triangle BCD$). How is this related to (1)? If one cannot see the clue, substitute $\angle CAR + \angle CBQ = \angle ADB - \angle ACB$ into (1)! Now it suffices to show that $2\angle ACB + \angle QCD = \angle ADB$. (2)

How can we show (2)? Notice that this is not true for an arbitrary (concave) quadrilateral $ABCD$. Which are the conditions given we have **not** used yet? We have not used:

- $PB = PD$
- PB is a tangent (i.e., $\angle 1 = \angle C$).

Could these two conditions help us?

Since $PB = PD$, we immediately have $\angle PDB = \angle PBD = 2\angle ACB$. This is awesome! Now (2) becomes $\angle PDB + \angle QCD = \angle ADB$ and it suffices to show that $\angle QCD = \angle ADB - \angle PDB = \angle PDA$.

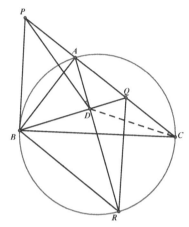

Refer to the diagram on the right. We could reach the conclusion by showing $\triangle PAD \sim \triangle PDC$. In fact, these two triangles **should** be similar.

Can we show $\dfrac{PA}{PD} = \dfrac{PD}{PC}$, or equivalently, $PA \cdot PC = PD^2$?

Notice that we have $PB = PD$ and $PB^2 = PA \cdot PC$ since PB is a tangent. Now both unused conditions have made their contributions, which complete the proof.

Proof. Refer to the diagram on the right. It is given that $\angle 1 = \angle 2 = \angle ACB = \angle 3$. Hence, $\triangle ABD \sim \triangle ARB$ and we have $\angle ADB = \angle ABR = 180° - \angle ACR$. It follows that $\angle BDR = \angle ACR$, which implies that C, Q, D, R are concyclic. Since $\angle 1 = \angle ACB$, PB is tangent to the circumcircle of $\triangle ABC$. Given $PB = PD$, we have $PD^2 = PB^2 = PA \cdot PC$.

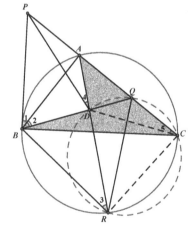

This implies $\dfrac{PA}{PD} = \dfrac{PD}{PC}$ and hence, $\triangle PAD \sim \triangle PDC$. We have $\angle 4 = \angle 5$.

Now $\angle QBR = \angle CBR + \angle CBQ = \angle CAR + \angle CBQ$

$= \angle ADB - \angle ACB$ (exterior angles of $\triangle ACD$ and $\triangle BCD$)

$= \angle PDB + \angle 4 - \angle ACB = \angle PBD + \angle 4 - \angle ACB$ (since $PB = PD$)

$= 2\angle ACB + \angle 4 - \angle ACB = \angle ACB + \angle 4 = \angle ACB + \angle 5$

$= \angle 3 + \angle ARQ = \angle QRB$, which implies $PD = PB$. □

Note:

(1) The last section of angle-chasing is a concise argument and one needs to be very familiar with basic properties of angles, especially in circles. In fact, such angle-chasing is commonly seen in geometry problems and is considered a fundamental technique. Nevertheless, we should point out that such a compact argument presented is only for mathematical elegance. In fact, it is not inspiring as the reader following the argument may not see **how** to search for clues and reach the conclusion. This is exactly why we spend a few more pages in explaining the insight.

(2) One may find an alternative solution starting from the Angle Bisector Theorem: Since BA bisects of $\angle PBD$, we have $\dfrac{QB}{QA} = \dfrac{PB}{PA}$. Since we are to show $QB = QR$, it suffices to show that $\dfrac{QR}{QA} = \dfrac{PB}{PA} = \dfrac{PD}{PA}$ because $PB = PD$. Upon showing C, Q, D, R concyclic, it is easy to see that $\triangle ACD \sim \triangle ARQ$ and hence, $\dfrac{QR}{QA} = \dfrac{CD}{AD}$. Now it suffices to show $\dfrac{PD}{PA} = \dfrac{CD}{AD}$, but this is because $\triangle PCD \sim \triangle PDA$.

6.4 Seeking Clues from the Diagram

A well-constructed diagram could be very helpful in problem-solving, especially for those more challenging problems in competition where the insight is not clear. Although referring to the diagram is **not** a valid proof, it may give us hints on what **could** be correct.

One should always construct a diagram according to the description in the problem without any loss of generality. For example, given a triangle $\triangle ABC$ where P is an arbitrary point on BC, one should avoid drawing an isosceles or right angled triangle, and choose P to be distinct from the midpoint of BC and the feet of the perpendicular from A. If one constructs a **general** diagram and observes any geometric fact from the diagram, for example, a right angle, collinearity or concyclicity, it **may** be true! One may attempt to show it, or assume it is true and seek intermediate steps which could be deduced.

Drawing a (reasonably) **accurate** diagram may help us substantially in seeking clues. Note that if circumcircles, incircles, tangent lines or equal angles are given, one should **not** construct these geometric objects casually. For example, when given a circle inscribed in a triangle, it is recommended that one draws the triangle and its angle bisectors to locate the incenter, constructs the incircle with a compass, and introduces heights to find the points of tangency. A poorly constructed or distorted diagram may be misleading and heavily distract one from acquiring the insight.

One should also learn to **simplify** the diagram, erasing lines, points and circles during problem-solving when necessary. Indeed, when the diagram is complicated, one may fail to recognize even the most elementary geometric facts (for example, radii of a circle which are the same, equal tangent segments, perpendicular bisectors which give isosceles triangles, etc.). In particular, if circumcenters or orthocenters are given, one should only draw explicitly the circles and altitudes which are necessary. Otherwise, the diagram may become unreadable!

When exploring a part of the diagram which demonstrates a specific geometric structure, one may consider drawing a **separate** diagram focused on that part. In a much simpler setting, one may find it easier to seek clues or recognize a well-known result. Refer to Example 6.3.3 for an illustration on this strategy.

Example 6.4.1 (APMO 91) Let G be the centroid of $\triangle ABC$. Draw a line $XY /\!/ BC$ passing through G, intersecting AB, AC at X, Y respectively. BG and CX intersect at P. CG and BY intersect at Q. If M is the midpoint of BC, show that $\triangle ABC \sim \triangle MQP$.

Insight. Refer to the diagram on the right. It *seems* that the corresponding sides of $\triangle ABC$ and $\triangle MQP$ are parallel. Can we show it, say $PQ /\!/ BC$? Since we are given the centroid and a parallel line, we can find the ratio of the line segments easily.

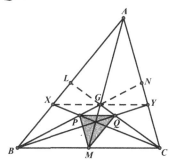

Proof. Let L, N be the midpoints of AB, AC respectively. Since $XY /\!/ BC$ and G is the centroid, it is easy to see that $GX = GY$.

Hence, $\dfrac{PX}{CP} = \dfrac{GX}{BC} = \dfrac{GY}{BC} = \dfrac{QY}{BQ}$ and by the Intercept Theorem, $PQ /\!/ BC$. ·

One also sees that $\dfrac{GY}{CM} = \dfrac{AG}{AM} = \dfrac{2}{3}$. Hence, $\dfrac{GQ}{CQ} = \dfrac{GY}{BC} = \dfrac{1}{3}$.

Since $\dfrac{CG}{CL} = \dfrac{2}{3}$, we have $\dfrac{CQ}{CL} = \dfrac{CQ}{CG} \cdot \dfrac{CG}{CL} = \dfrac{3}{4} \cdot \dfrac{2}{3} = \dfrac{1}{2}$, i.e., Q is the midpoint of CL. Hence, $MQ /\!/ AB$ by the Midpoint Theorem.

Similarly, $MP /\!/ AC$ and the conclusion follows. $\qquad\square$

Example 6.4.2 (TUR 10) Given a circle Γ_1 where AB is a diameter, C, D lie on Γ_1 and are on different sides of the line AB. Draw a circle Γ_2 passing through A, B, intersecting AC at E and AD extended at F. Let P be a point on DA extended such that PE is tangent to Γ_2 at E. Let Q be a point (different from E) on the circumcircle of $\triangle AEP$ such that $PE = PQ$. Let M be the midpoint of EQ. If CD and EF intersect at N, show that $PM /\!/ BN$.

Insight. One may notice that constructing such a diagram following the instructions given is not a simple task. However, it could be rewarding. Refer to the diagram on the right. It *seems* that E, F, Q are collinear. Is it true?

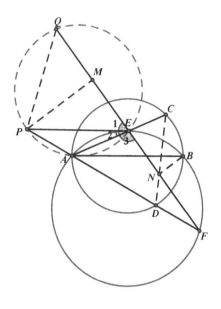

Since $PE = PQ$ and M is the midpoint of EQ, we immediately have $PM \perp EQ$. If E, F, Q are indeed collinear, we should have $BN \perp EF$.

What can we say about BN and EF? Can you see $BC \perp AC$ and $BD \perp AD$?

Can you see $BN \perp EF$ while CD is a Simson's Line of $\triangle AEF$?

We are to show $PM \parallel BN$. Hence, we **should** have E, F, Q collinear. Can we show that $\angle 1 + \angle 2 + \angle 3 = 180°$?

Proof. Since $PE = PQ$, we have $PM \perp EQ$. Since AB is a diameter of Γ_1, C, D are the feet of the perpendiculars from B to AE, AF respectively. Hence, CD is the Simson's Line of $\triangle AEF$ with respect to B. It follows that $BN \perp EF$. Now it suffices to show that E, F, Q are collinear.

Notice that $\angle 1 = \angle Q = \angle EAF$ (Corollary 3.1.5) and $\angle 2 = \angle F$ (Theorem 3.2.10). Now $\angle 1 + \angle 2 + \angle 3 = \angle EAF + \angle F + \angle 3 = 180°$. This completes the proof. □

Example 6.4.3 (VNM 09) Let Γ be the circumcircle of an acute angled triangle $\triangle ABC$, where D, E, F are the feet of the altitudes from A, B, C respectively. Let D', E', F' be the points of reflection of D, E, F about the midpoints of BC, AC, AB respectively. The circumcircles of $\triangle AE'F'$, $\triangle BD'F'$, $\triangle CD'E'$ meet Γ again at A', B', C' respectively. Show that $A'D, B'E$ and $C'F$ are concurrent.

Insight. Apparently, the construction of the diagram is complicated. For example, we draw the diagram on the right to locate A'. (Notice that we have already omitted the midpoints of AB, BC, CA.) If we continue to construct B' and C', the diagram might be unreadable! Perhaps we should examine the property of A' first.

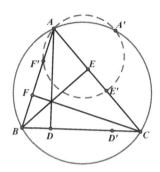

It *seems* from the diagram that A' is symmetric to A with respect to the perpendicular bisector of BC, i.e., $A'D' \perp BC$. If we can show that A' is indeed symmetric to A, the diagram could be significantly simplified.

Proof. We first show the following lemma.

Let $ABCD$ be an isosceles trapezium where $AD // BC$ and $AB = CD$. Draw $BE \perp AC$ at E and $CF \perp AB$ at F. Let E', F' be on AC, AB respectively such that $AE = CE'$ and $AF = BF'$. We have A, D, E', F' concyclic.

Refer to the diagram on the right. Let P be the reflection of F' about the perpendicular bisector of BC. Since $ABCD$ is an isosceles trapezium, we have $CP = BF' = AF$.

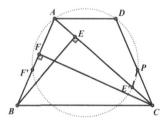

Notice that $\dfrac{AF}{AC} = \cos \angle BAC = \dfrac{AE}{AB}$.

Since $AE = CE'$ and $AB = CD$, we must have $\dfrac{CP}{CA} = \dfrac{AF}{CA} = \dfrac{AE}{AB} = \dfrac{CE'}{CD}$, i.e., $\dfrac{CP}{CA} = \dfrac{CE'}{CD}$. Now $CP \cdot CD = CA \cdot CE'$, which implies A, D, P, E' are concyclic.

Since $ADPF'$ is an isosceles trapezium, we conclude that A, D, P, E', F' are concyclic.

We apply this lemma to the original problem. Let X be the point symmetric to A with respect to the perpendicular bisector of BC. Now $ABCX$ is an isosceles trapezium and by the lemma, A, X, E', F' are concyclic. This implies A' and X coincide (since the circumcircle of $\triangle AE'F'$ intersect Γ only at A and A'). In particular, $ADD'A'$ is a rectangle.

We are to show that $A'D, B'E, C'F$ are concurrent. Let us examine the property of $A'D$. Refer to the diagram on the right where M is the midpoint of BC.

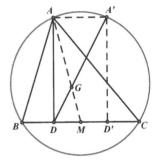

Since $ADD'A'$ is a rectangle, we have

$$DM = \frac{1}{2}AA'.$$

It follows that $\dfrac{MG}{AG} = \dfrac{DM}{AA'} = \dfrac{1}{2}$ and hence, G is the centroid of $\triangle ABC$.

We conclude that $A'D$ passes through the centroid of $\triangle ABC$. Similarly, $B'E$ and $C'F$ must pass through the centroid of $\triangle ABC$ as well. This completes the proof.   □

Note:

(1) One may also use the power of a point to show that A' is symmetric to A with respect to the perpendicular bisector of BC. In particular, one may show that $BF' \cdot BA = CE' \cdot CA$ (because $\dfrac{BF'}{CE'} = \dfrac{AF}{AE} = \dfrac{AC}{AB}$) and hence, B and C have the same power with respect to the circumcircle of $\triangle AE'F'$.

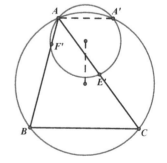

Refer to the diagram on the right. It follows that the circumcenter of $\triangle AE'F'$ is equidistant to B and C and hence, lies on the perpendicular bisector of BC. Now the line passing through the circumcenters of $\triangle AE'F'$ and $\triangle ABC$ is perpendicular to BC.

This line must be perpendicular to AA' as well (Theorem 3.1.20). We conclude that $AA' /\!/ BC$.

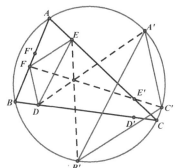

(2) One may also observe the diagram and attempt to show $DE /\!/ A'B'$. Given the reflections A' and B', this is not difficult (Exercise 3.13). Similarly, we have $EF /\!/ B'C'$ and $DF /\!/ A'C'$. Refer to the diagram on the right. Now $A'D, B'E, C'F$ are concurrent by Theorem 2.5.11.

Example 6.4.4 (BLR 11) Given an acute angled triangle $\triangle ABC$, M is the midpoint of AB. Let P, Q be the feet of the perpendiculars from A to BC and from B to AC respectively. If the circumcircle of $\triangle BMP$ is tangent to the line segment AC, show that the circumcircle of $\triangle AMQ$ is tangent to the line BC.

Insight. Refer to the diagram on the right. It is not easy to show a circle tangent to a line. However, notice that the circle passing through A, M and tangent to the line BC is unique. Hence, we may draw this circle and show that it intersects AC exactly at Q.

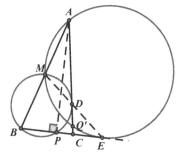

Let AC touch the circumcircle of $\triangle BMP$ at D. If BC extended touches the circumcircle of $\triangle AMQ$ at E, we would have $BE^2 = BM \cdot BA$. Since M is the midpoint, we **should** have $BE^2 = BM \cdot BA = AM \cdot AB = AD^2$, i.e., $BE = AD$.

It *seems* from the diagram that M, D, E are collinear. Can we show it? If M, D, E are indeed collinear, we **should** have, by Menelaus' Theorem, that $\dfrac{AM}{BM} \cdot \dfrac{BE}{CE} \cdot \dfrac{CD}{AD} = 1$, which implies $CD = CE$. (Notice that we have utilized the condition $AM = BM$ once more, even though it is not clear at first glance how this condition could be applied.)

Can we show $CD = CE$, say by showing $\angle CED = \angle CDE$? Notice that $\angle CDE = \angle ADM = \angle ABD$.

Proof. Refer to the diagram on the right. Let the circumcircle of $\triangle BMP$ touch AC at D and MD extended intersect BC extended at E. We claim that $\angle CED = \angle CDE$. Notice that:

$$\angle CDE = \angle ADM = \angle ABD. \quad (1)$$

$$\angle CED = 180° - \angle BME - \angle ABE. \quad (2)$$

Since PM is the median on the hypotenuse of the right angled triangle $\triangle ABP$, we must have $\angle ABE = \angle BPM = \angle BDM$.

By (2), $\angle CED = 180° - \angle BME - \angle BDM = \angle ABD.$ (3)

(1) and (3) imply $\angle CED = \angle CDE$ and hence, $CD = CE$.

By Menelaus' Theorem, $\dfrac{AM}{BM} \cdot \dfrac{BE}{CE} \cdot \dfrac{CD}{AD} = 1$. Since $AM = BM$ and $CD = CE$, we must have $BE = AD$. It follows that $BE^2 = AD^2 = AB \cdot AM = AB \cdot BM$. By the Tangent Secant Theorem, BE touches the circumcircle of $\triangle AME$ at E.

Let the circumcircle of $\triangle AME$ intersect AC at Q'. We claim that $BQ' \perp AC$. Since $AM = BM$, it suffices to show $AM = MQ'$ (Example 1.1.8), or equivalently, $\angle AQ'M = \angle MAQ'$.

Refer to the diagram on the right. Notice that

$$\angle AQ'M = \angle ADM - \angle Q'ME.$$

Since $\angle ADM = \angle CDE = \angle CED = \angle MAE$ and $\angle Q'ME = \angle Q'AE$, we have

$$\angle AQ'M = \angle MAE - \angle Q'AE = \angle MAQ'.$$

This completes the proof. □

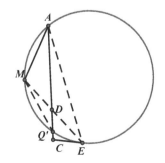

Example 6.4.5 (USA 10) Given $\triangle ABC$, M, N are on AC, BC respectively such that $MN \, /\!/ \, BC$, and P, Q are on AB, BC respectively such that $PQ \, /\!/ \, AC$. Given that the incircle of $\triangle CMN$ touches AC at E and the incircle of $\triangle BPQ$ touches AB at F, the lines EN, AB intersect at R and the lines FQ, AC intersect at S. Show that if $AE = AF$, then the incenter of $\triangle AEF$ lies on the incircle of $\triangle ARS$.

Insight. First, we draw the diagram according to the description. Refer to the left diagram below. There are many circles and lines and it becomes difficult to seek clues. Since the incircles of $\triangle BPQ$ and $\triangle CMN$ are constructed similarly, we may focus on one of them.

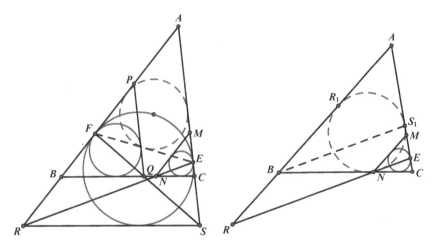

Refer to the right diagram above. It is easy to see that $\triangle ABC \sim \triangle MNC$. Hence, if we draw the incircle of $\triangle ABC$, which touches AB, AC at R_1, S_1 respectively, then S_1 and E are corresponding points in $\triangle ABC$ and $\triangle MNC$ respectively. It follows that $BS_1 \, /\!/ \, EN$. Similarly, we have $CR_1 \, /\!/ \, FQ$. It *seems* from the left diagram above that $BC \, /\!/ \, RS$. Can you prove it by the Intercept Theorem? (Notice that $AE = AF$ and $AR_1 = AS_1$.)

Now $\triangle ABC \sim \triangle ARS$ and hence, the incircle of $\triangle ABC$ corresponds to the incircle of $\triangle ARS$. Since R_1 and F (and similarly S_1 and E) are corresponding points of the similar triangles $\triangle ABC$ and $\triangle ARS$, the incircle of $\triangle ARS$ touches AR, AS at E, F respectively! Refer to the left diagram below. Notice that we have removed the unnecessary lines and points.

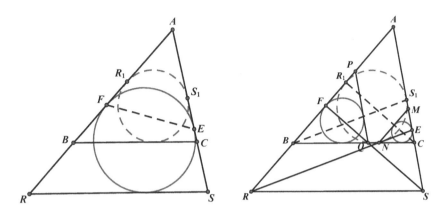

Proof. Let the incircle of $\triangle ABC$ touch AB, AC at R_1, S_1 respectively. Since $MN /\!/ AB$, $\triangle ABC \sim \triangle MNC$. Notice that S_1 and E are corresponding points in the similar triangles $\triangle ABC$ and $\triangle MNC$. We conclude that $BS_1 /\!/ ER$.

It follows that $\dfrac{AB}{AR} = \dfrac{AS_1}{AE}$. Similarly, we must have $CR_1 /\!/ FS$ and $\dfrac{AC}{AS} = \dfrac{AR_1}{AF}$. Since $AR_1 = AS_1$ and $AE = AF$, we must have $\dfrac{AB}{AR} = \dfrac{AC}{AS}$.

By the Intercept Theorem, $BC /\!/ RS$. Refer to the right diagram above.

Now we have $\triangle ARS \sim \triangle ABC$. We are to show the incenter of $\triangle AEF$ lies on the incircle of $\triangle ARS$. Notice that R_1 and F are corresponding points in the similar triangles $\triangle ABC$ and $\triangle ARS$, because $\dfrac{AR_1}{AF} = \dfrac{AC}{AS} = \dfrac{AB}{AC}$. A similar argument applies for S_1 and E as well. Now it suffices to show that the incenter of $\triangle AR_1S_1$ lies on the incircle of $\triangle ABC$.

Since AR_1, AS_1 are tangent to the incircle of $\triangle ABC$, called $\odot I$, the incenter of $\triangle AR_1S_1$ is exactly the intersection of AI and $\odot I$, i.e., the midpoint of the arc $\overarc{R_1S_1}$ (Exercise 3.5). This completes the proof. $\square$

Note: We used correspondence between similar triangles extensively in the proof above. One not familiar with these properties could always use similar triangles to argue instead, although it will make the proof unnecessarily lengthy.

Example 6.4.6 (CHN 12)   Refer to the diagram below. I is the incenter of $\triangle ABC$, whose incircle $\odot I$ touches AB, BC, CA at D, E, F respectively. If the line EF intersects the lines AI, BI, DI at M, N, K respectively, show that $DM \cdot KE = DN \cdot KF$.

Insight.   Since there is a circle in the diagram, the conclusion reminds us of the Tangent Secant Theorem. However, it seems DM, KE are not part of a secant line of $\odot I$.

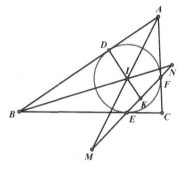

Can we show $\dfrac{DM}{DN} = \dfrac{KF}{KE}$ instead?

It seems not easy either because we do not see similar triangles immediately which relate $DM, DN,$ KE and KF.

Where does the difficulty come from? We do not know the properties of the line MN (including E, F and K). Perhaps we should first study the properties of this line and the points on it. Let us focus on one side of the triangle and its incircle. Refer to the diagram on the right. We have erased the unnecessary lines and points.

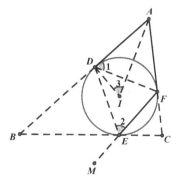

Now it is clear that $\angle 1 = \angle 2$ because AD is tangent to $\odot I$. We also have $\angle 1 = \angle 3$ because $AD \perp DI$ and $AI \perp DF$. Hence, $\angle 2 = \angle 3$, which implies $\angle DEM = \angle DIM$ (since A, I, M are collinear). It follows that D, I, E, M are concyclic.

Similarly, we also have D, I, F, N concyclic. Refer to the diagram on the right. Notice that the three circles give $KE \cdot MK = KI \cdot DK = KF \cdot NK$ (Tangent Secant Theorem).

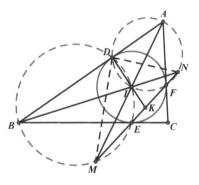

Now we have $\dfrac{KE}{KF} = \dfrac{NK}{MK}$ and hence, it suffices to show $\dfrac{DM}{DN} = \dfrac{MK}{NK}$.

Notice that this is equivalent to DK bisecting $\angle MDN$ (Angle Bisector Theorem).

It seems from the diagram that B, D, I, E are concyclic. One may easily see this because $\angle BDI = \angle BEI = 90°$. Now B, D, I, E, M are concyclic (where BI is a diameter). Hence, $\angle BMI = 90°$ and $AM \perp BM$.

Can you see that I is the orthocenter of a larger triangle? How is it related to the angle bisector of $\angle MDN$?

Proof. Refer to the diagram on the right. We have

$$\angle DEF = \frac{1}{2}\angle DIF = \angle AID.$$

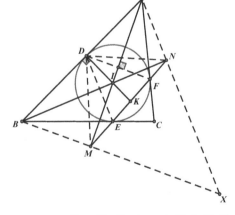

Hence, $\angle DIM = \angle DEM$ and we must have D, I, E, M concyclic. It is easy to see that B, D, I, E are also concyclic.

We conclude that B, D, I, E, M are concyclic. Similarly, A, D, I, F, N are concyclic.

Now $KE \cdot MK = KI \cdot DK = KF \cdot NK$, which implies $\dfrac{KE}{KF} = \dfrac{NK}{MK}$. (1)

Notice that $\angle BMA = \angle BEI = 90°$, i.e., $AM \perp BM$. Similarly, we have $AN \perp BN$. Let the lines AM, BM intersect at X.
Now I is the orthocenter of $\triangle ABX$ and hence, the incenter of $\triangle DMN$ (Example 3.1.6).

By the Angle Bisector Theorem, $\dfrac{MK}{NK} = \dfrac{DM}{DN}$. (2)

(1) and (2) give $\dfrac{DM}{DN} = \dfrac{KF}{KE}$, or equivalently, $DM \cdot KE = DN \cdot KF$. $\square$

Example 6.4.7 (CHN 06) Let $ABCD$ be a cyclic quadrilateral inscribed in $\odot O$, where O does not lie on any side of the quadrilateral. The diagonals AC, BD intersect at P. Let O_1, O_2, O_3, O_4 denote the circumcenters of $\triangle OAB, \triangle OBC, \triangle OCD, \triangle ODA$ respectively. Show that the lines O_1O_3, O_2O_4 and OP are concurrent.

Insight. Refer to the left diagram below. We draw the circumcenters only, but hide other related details like the perpendicular bisectors and the circumcircles. It seems not clear how the lines O_1O_3, O_2O_4 and OP are related. (Notice that applying Ceva's Theorem is not feasible.)

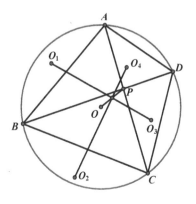 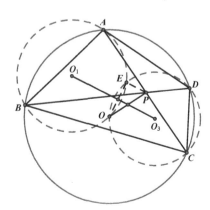

However, it *seems* that both O_1O_3 and O_2O_4 pass through the midpoint of OP. Is it true?

We focus on the line O_1O_3. Let $\odot O_1$ and $\odot O_2$ denote the circumcircles of $\triangle OAB$ and $\triangle OCD$ respectively and the circles intersect at O and E. Refer to the right diagram above. We know that O_1O_3 is the perpendicular bisector of OE. If O_1O_3 indeed passes through the midpoint of OP, we **should** have $PE \perp OE$ (Midpoint Theorem).

Can we show $PE \perp OE$? One may consider calculating the angles, as there are many circles (and circumcenters) in the diagram. Refer to the diagram on the right. It suffices to show that $\angle 1 + \angle DEP = 90°$.

We do not know much about $\angle DEP$, but we know

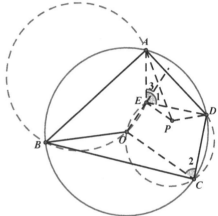

$$\angle 1 = \angle 2 = 90° - \frac{1}{2}\angle COD \text{ (because } OC = OD \text{)}.$$

Similarly, $\angle 3 = \angle ABO = 90° - \dfrac{1}{2}\angle AOB$.

Now $\angle 1 + \angle 3 = 180° - \dfrac{1}{2}(\angle AOB + \angle COD) = 180° - (\angle ADB + \angle CAD)$

$= \angle APD$, which implies A, D, P, E are concyclic.

Notice that we have used the properties of the circumcenters extensively. Indeed, we are not given many conditions other than the circumcenters.

We have obtained one more circle. One should be able to show the conclusion easily using the properties of angles.

Proof.

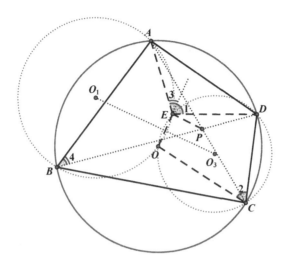

Let the circumcircles of $\triangle ABO$ and $\triangle CDO$ intersect at O and E. Refer to the diagram above. Consider the isosceles triangle $\triangle OCD$.

We have $\angle 1 = \angle 2 = 90° - \dfrac{1}{2}\angle COD = 90° - \angle CAD.$ (1)

Similarly, $\angle 3 = \angle 4 = 90° - \angle ADB.$

Now $\angle AED = \angle 1 + \angle 3 = 180° - \angle ADB - \angle CAD = \angle APD.$ We conclude that A, D, P, E are concyclic.

Hence, $\angle DEP = \angle DAP = 90° - \angle 1$ by (1), which implies $PE \perp OE.$

Since $O_1 O_3$ is the perpendicular bisector of OE, we must have $PE \;//\; O_1 O_3$ and hence, $O_1 O_3$ passes through the midpoint of OP.

Similarly, $O_2 O_4$ also passes through the midpoint of OP. It follows that $O_1 O_3$, $O_2 O_4$ and OP are concurrent (at the midpoint of OP). □

6.5   Exercises

1. (CZE-SVK 89)   Let O be the circumcenter of $\triangle ABC$. D, E are points on AB, AC respectively. Show that B, C, E, D are concyclic if and only if $DE \perp OA$.

2. (IWYMIC 14)   In $\triangle ABC$, $\angle A = \angle C = 45°$. M is the midpoint of BC. P is a point on AC such that $BP \perp AM$. If $PC = \sqrt{2}$, find AB.

3. (JPN 14)   Let $ABCDEF$ be a cyclic hexagon where the diagonals AD, BE, CF are concurrent. If $AB = 1$, $BC = 2$, $CD = 3$, $DE = 4$ and $EF = 5$, find AF.

4. (IND 11) $\triangle ABC$ is an acute angled triangle where D is the midpoint of BC. BE bisects $\angle B$, intersecting AC at E. $CF \perp AB$ at F. Show that if $\triangle DEF$ is an equilateral triangle, then $\triangle ABC$ is also an equilateral triangle.

5. (USA 90) $\triangle ABC$ is an acute angled triangle where AD, BE are heights. Let the circle with diameter BC intersect AD and its extension at M, N respectively. Let the circle with diameter AC intersect BE and its extension at P, Q respectively. Show that M, P, N, Q are concyclic.

6. (CAN 11) $ABCD$ is a cyclic quadrilateral. BA extended and CD extended intersect at X. AD extended and BC extended intersect at Y. If the angle bisector of $\angle X$ intersects AD, BC at E, F respectively, and the angle bisector of $\angle Y$ intersects AB, CD at G, H respectively, show that $EGFH$ is a parallelogram.

7. (ROU 08) Given $\triangle ABC$, D, E, F are on BC, AC, AB respectively such that $\dfrac{BD}{CD} = \dfrac{CE}{AE} = \dfrac{AF}{BF}$. Show that if the circumcenters of $\triangle ABC$ and $\triangle DEF$ coincide, then $\triangle ABC$ is an equilateral triangle.

8. (IMO 04) Given a non-isosceles acute angled triangle $\triangle ABC$ where O is the midpoint of BC, draw $\odot O$ with diameter BC, intersecting AB, AC at D, E respectively. Let the angle bisectors of $\angle A$ and $\angle DOE$ intersect at P. If the circumcircles of $\triangle BPD$ and $\triangle CPE$ intersect at P and Q, show that Q lies on BC.

9. (CHN 04) Given $\triangle ABC$, D is a point on BC and P is on AD. A line ℓ passing through D intersects AB, PB at M, E respectively, and intersects AC extended and PC extended at F, N respectively. Show that if $DE = DF$, then $DM = DN$.

10. (IMO 08) Given an acute angled triangle $\triangle ABC$ where O_1, O_2, O_3 are the midpoints of BC, AC, AB respectively, H is the orthocenter of $\triangle ABC$. Draw $\odot O_1$, $\odot O_2$, $\odot O_3$ whose radii are O_1H, O_2H, O_3H respectively. If $\odot O_1$ intersects BC at A_1, A_2, $\odot O_2$ intersects AC at B_1, B_2 and $\odot O_3$ intersects AB at C_1, C_2, show that A_1, A_2, B_1, B_2, C_1, C_2 are concyclic.

11. (IMO 14) Given an acute angled triangle $\triangle ABC$, P, Q are on BC such that $\angle PAB = \angle C$ and $\angle CAQ = \angle B$. M, N are on the lines AP, AQ respectively such that $AP = PM$ and $AQ = QN$. Show that the intersection of the lines BM and CN lies on the circumcircle of $\triangle ABC$.

12. (CHN 13) Given $\triangle ABC$ where $AB < AC$, M is the midpoint of BC. $\odot O$ passes through A and is tangent to BC at B, intersecting the lines AM, AC at D, E respectively. Draw $CF // BE$, intersecting BD

extended at F. Let the lines BC and EF intersect at G. Show that $AG = DG$.

13. (RUS 13) Let $\odot I$ denote the incircle of $\triangle ABC$, which touches BC, AC, AB at D, E, F respectively. Let J_1, J_2, J_3 be the ex-centers opposite A, B, C respectively. If $J_2 F$ and $J_3 E$ intersect at P, $J_3 D$ and $J_1 F$ intersect at Q, $J_1 E$ and $J_2 D$ intersect at R, show that I is the circumcenter of $\triangle PQR$.

14. (IMO 10) Refer to the diagram below. $ABCDE$ is a pentagon such that $BC \,/\!/\, AE$, $AB = BC + AE$ and $\angle B = \angle D$. Let M be the midpoint of CE and O be the circumcenter of $\triangle BCD$. Show that if $OM \perp DM$, then $\angle CDE = 2\angle ADB$.

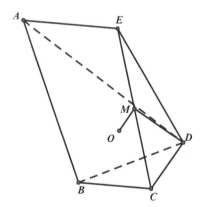

Insights into Exercises

Chapter 1

1.1 Notice that $\angle B + \angle C = \angle A$. If $\angle PAB = \angle C$, what can you say about $\angle PAC$?

1.2 We are to show $AC = AB + BD$. If we choose E on AC such that $AB = AE$, it suffices to show $CE = BD$. Since AD bisects $\angle A$, can you see that E is the reflection of B about AD, i.e., $\triangle ABD \cong \triangle AED$? How can we use the condition $\angle B = 2\angle C$? Can you see $\angle B = \angle AED$?

1.3 Can you see congruent triangles? It is similar to Example 1.2.6.

1.4 Notice that the ex-center is still about properties of angle bisectors. How did we show the existence of the incenter?

1.5 Notice that AI, AJ_1 are angle bisectors of neighboring supplementary angles. Can you see $AI \perp AJ_1$? Refer to Example 1.1.9.

1.6 We have $\angle EAF = \dfrac{1}{2}\angle BAD$. However, the remaining portions of $\angle BAD$ are far apart. How can we put them together? Moreover, BE and DF are far apart as well. Cut and paste! It is similar to Example 1.2.9.

1.7 Can you see congruent triangles? Given that $BP = AC$ and $CQ = AB$, which two triangles are probably congruent?

1.8 We are given the angle bisector of $\angle CBE$ and $BE = AB$. Notice that $\triangle ABC$ is an equilateral triangle. Can you see congruent triangles (say by the reflection about the angle bisector BD)? Can you see D is on the perpendicular bisector of AB?

1.9 Since I is the incenter of $\triangle ABC$, can you express both $\angle BID$ and $\angle CIH$ in terms of $\angle A, \angle B$ and $\angle C$? Alternatively, you may apply Theorem 1.3.3.

1.10 One may immediately see that $\triangle ABC \cong \triangle ADC$. Even though this is not related to PE and PF, we have more equal angles and line segments now. Can you find more congruent triangles which lead to $PE = PF$?

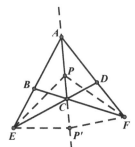

Note:

(1) P could be on the line segment AC or its extension.

(2) One may see many pairs of congruent triangles in the diagram, but careful justification is needed for each pair and the argument must not depend on the diagram.

1.11 By definition, O lies on the perpendicular bisector of BC and I lies on the angle bisector of $\angle A$. What can you conclude if $AB = AC$?

1.12 Can you see D is an ex-center of $\triangle ABP$, i.e., AD bisects the exterior angle of $\angle BAC$? Now can you express both $\angle PAD$ and $\angle BDP$ in terms of $\angle ABP$ and $\angle APC$?

1.13 If $ABCD$ is a parallelogram, one sees that $BC - AB = AD - CD$ holds. Since $AD \, / \! / \, BC$, we may draw a parallelogram $ABCD'$ such that D' lies on the line AD. Now $AD - CD = BC - AB = AD' - CD'$. This is only possible when D and D' coincide. (You may show it using triangle inequality. Notice that you need to discuss both cases when $AD > AD'$ and $AD < AD'$.)

1.14 Notice that the condition AB is equal to the distance between ℓ_1, ℓ_2 is important. If we move ℓ_2 downwards, $\angle GIH$ will be smaller, i.e., it is not a fixed value.

If we draw a perpendicular from E to ℓ_2, say $EP \perp \ell_2$ at P, we have $AB = EP$. Does it help us to find congruent triangles? Are there any other equal angles or sides? If $EH \perp FG$, then we have $EH = FG$ (Example 1.4.12). However, it seems from the diagram that EH and FG are **not** perpendicular. Moreover, EH and FG are apparently not equal. What should we do? It is difficult to calculate EH and FG because we do not know the positions of E, F, G, H on the sides of the square. Perhaps we can use the same technique as in Example 1.4.12, say to push EH upwards.

If we draw $AQ /\!/ EH$, intersecting CD at Q, it is easy to see that $AQ = EH$. Now we have $\triangle EPH \cong \triangle ADQ$ (H.L.) and hence, $\angle 1 = \angle 2$. This implies that EH bisects the exterior angle of $\angle CHG$. A similar argument applies for FG as well. Can you see I is an ex-center of $\triangle CGH$ (Exercise 1.4)? Now we can calculate $\angle GIH$ using the properties of angle bisectors.

Chapter 2

2.1 Can you express $[BCXD]$, $[ACEY]$ and $[ABZF]$ in terms of $[\triangle ABC]$?

2.2 It is easy to show $BG = CE$ (Example 1.2.6). How are BG, CE related to (the midpoints) O_1, O_2, M, N?

2.3 Can you see right angled isosceles triangles in the diagram (for example, $CD = CF + AF$)? Since we are to show $AE < \frac{1}{2}CD$, what do we know about $CD - 2AE$?

2.4 M, N are midpoints, but we cannot apply the Midpoint Theorem directly on MN. What if we consider more midpoints (Example 2.2.8)?

2.5 Can you see *EFGH* is a parallelogram? Now we can focus on the parallelogram *EFGH*, which is a simpler problem. Can we use the techniques of congruent triangles to solve it?

2.6 Notice that every point in the diagram is uniquely determined once the square is drawn. Let $AB = a$. We can calculate AP, for example, by drawing $PQ \perp AD$ at Q and applying Pythagoras' Theorem. Can you find AQ and PQ? Can you find $\dfrac{PF}{PD}$?

2.7 Given $BG \perp CG$, can you see AB, BC, AC can all be expressed in terms of the medians BD, CE (by the Midpoint Theorem and Pythagoras' Theorem)?

2.8 Given $\triangle ABC$, we can calculate $[\triangle DEF]$ by subtracting $[\triangle ADF]$, $[\triangle BDE]$ and $[\triangle CEF]$ from $[\triangle ABC]$, while the areas of the small triangles are determined once the positions of D, E, F are known.

$[\triangle D'E'F']$ can be calculated in a similar manner, while the positions of D', E', F' are determined by D, E, F.

Since D, E, F are arbitrarily chosen, the conclusion **should** hold if we let $\dfrac{AD}{AB} = a$, $\dfrac{BE}{BC} = b$, $\dfrac{CF}{CA} = c$ and express both areas in terms of a, b, c and $[\triangle ABC]$.

2.9 We are to show $BD \cdot CD = BE \cdot CF$, or equivalently, $\dfrac{BD}{BE} = \dfrac{CF}{CD}$. Since $\angle B = \angle C = 60°$, we **should** have $\triangle BDE \sim \triangle CFD$. Can we prove it, say by equal angles? Notice that A and D are symmetric about MN, i.e., $\angle EDF = \angle A = 60°$.

2.10 Example 1.2.7 is a special case of this problem, where $\angle A = 45°$ and $AH = BC$. We solved Example 1.2.7 using congruent triangles. Can you see a pair of similar triangles in this problem?

2.11 We know how to calculate a median, but what about trisection points? Can you see AD is a median of $\triangle ABE$? Similarly, AE is a median of $\triangle ACD$.

2.12 Notice that the parallel line is almost the only condition. If we apply Ceva's Theorem, the conclusion would be concurrency instead of collinearity. Nevertheless, we can show GM passes through D, which is equivalent to the conclusion.

Applying Menelaus' Theorem directly to D, G, M will probably not show the collinearity because it is not related to the condition $AB /\!/ CE$. How about applying Menelaus' Theorem more than once?

2.13 This is similar to Example 2.5.3.

2.14 It seems natural to apply Menelaus' Theorem. Even though the line where D, E, F should lie does **not** intersect any triangle, Menelaus' Theorem still holds when the points of division are on the extension of the sides of the triangle.

One may also consider applying the Angle Bisector Theorem to the exterior angle bisectors.

2.15 Refer to the diagram on the right. If we apply Menelaus' Theorem when the line DE intersects $\triangle ABC$, we have $\dfrac{AD}{BD} \cdot \dfrac{BM}{CM} \cdot \dfrac{CE}{AE} = 1$. Alternatively, if we consider the line BC intersecting $\triangle ADE$, we have

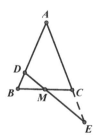

$\dfrac{AB}{DB} \cdot \dfrac{DM}{EM} \cdot \dfrac{EC}{AC} = 1$. However, neither gives us a clue for $\dfrac{1}{AD} + \dfrac{1}{AE}$ or $\dfrac{2}{AB}$. Perhaps we shall apply Menelaus' Theorem to another triangle, but which triangle (and the line intersecting it) should we choose?

We are to show $\dfrac{AB}{AD} + \dfrac{AB}{AE} = 2$, where $AB = AC$. How could we obtain say $\dfrac{AB}{AD}$, If we apply Menelaus' Theorem, BD should be a side of the triangle and the line should pass through A. It seems we should choose the line AE intersecting $\triangle BDM$. Even though AE intersects BD, DM and BM only at the extension, we could still apply Menelaus' Theorem.

Can you give a similar argument for $\dfrac{AC}{AE}$?

Chapter 3

3.1 Apply Corollary 3.1.4.

3.2 This is similar to Example 3.1.7. Connect EF and one could see concyclicity.

3.3 Can you see $\angle AIJ = \dfrac{1}{2}(\angle A + \angle B)$? How does this relate to the exterior angle of $\angle C$?

3.4 Since AB is the diameter, $AC \perp BC$ and $AD \perp BD$. Can you construct a triangle whose orthocenter is P? Example 3.1.6 relates the orthocenter of a triangle to the incenter of another triangle.

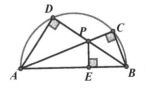

3.5 Let M be the midpoint of $\overset{\frown}{AB}$. Clearly the angle bisector of $\angle PAB$ passes through M. Can you find another angle bisector which passes through M? You may apply Theorem 3.2.10 for angles related to tangent lines.

3.6 Notice that $\angle BHC = 180° - \angle A$ because H is the orthocenter.

3.7 It is easy to see that OD is the perpendicular bisector of BC. How can we show $OM \perp PM$? Draw a diagram and one may see many equal angles and right angles. It should not be difficult to find concyclicity.

3.8 This is similar to Example 3.1.17. Besides, one may also recall the property of $\triangle ACD$, i.e., an isosceles triangle with $120°$ at the vertex (Example 2.3.4).

3.9 There are many right angles in this diagram due to the orthocenter and diameters. (Draw a diameter of $\odot O$.)

3.10 Since we are to show $CDEF$ is a rectangle, it suffices to show CF and DE bisect each other and are equal. We know $CM = DM = EM$. Hence, it suffices to show $CF = DE$.

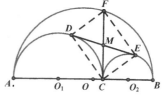

Notice that both CF and DE are uniquely determined by AC and BC. In particular, CF and DE can be calculated by Pythagoras' Theorem.

3.11 How can we apply the condition $\angle B = 2\angle C$? Since AD is the angle bisector, it is natural to reflect $\triangle ABD$ about AD, i.e., choose E on AC such that $AB = AE$.

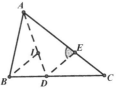

Now $\angle AED = 2\angle C$, which implies $\angle C = \angle CDE$, i.e., $DE = CD$. It *seems* that $BDEI$ is a rhombus. Can you show it? (Notice that if $BDEI$ is indeed a rhombus, then E is the circumcenter of $\triangle CDI$.)

3.12 Let the circumcircle of $\triangle ABP$ intersect AQ at M'. What do you know about M'? Can you see M' is the midpoint of AQ?

3.13 Can you see $AA'CB$ and $ABB'C$ are isosceles trapeziums? Notice that there are many equal angles in the diagram due to concyclicity, heights, parallel lines and equal arcs.

3.14 This follows immediately from Example 3.4.2.

3.15 Can you see I is the orthocenter of $\triangle J_1 J_2 J_3$?

3.16 One may see many right angles from the diagram. (Notice that the diameter also gives right angles.) Moreover, P, Q, R, S are the feet of the perpendiculars from Y, a point on the circumference. Is it reminiscent of Simson's Line? What if you draw $YC' \perp AB$ at C'?

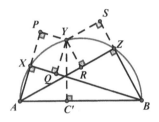

We are to show $\angle PCS = \dfrac{1}{2} \angle XOZ$. Can we replace $\dfrac{1}{2} \angle XOZ$ by an angle on the circumference? Those right angles should give plenty of concyclicity. It seems we are not far from the conclusion.

Alternatively, one may also notice that $PXQY$ and $SYRZ$ are rectangles. What can we say about these rectangles?

3.17 Since $AD \,/\!/\, BC$ and we are to show $\ell_1 \,/\!/\, \ell_2$, we **should** have a parallelogram enclosed by AD, BC, ℓ_1 and ℓ_2. Can we show it?

By extending the sides of $ABCD$ and ℓ_1, ℓ_2, we will have many equal tangent segments. Hence, we may be able to find an equation of various line segments. (Refer to Example 3.2.7. You may need to draw a large diagram.)

Now we may identify the parallelogram by applying Exercise 1.13. Even though this is not a commonly used result, it is most closely related to the parallelogram given the sum or difference of neighboring sides. (If you are not familiar with this result, you may prove it first as a lemma.)

Chapter 4

4.1 Draw a common tangent at C. Can you see $AB // DE$? What other equal angles can you obtain if A, C, D, E are concyclic?

4.2 Draw $DE \perp AP$ at E. By definition, $\sin \angle PAD = \dfrac{DE}{AD}$.

Can you see C is the midpoint of AP? Can you see a number of right angled isosceles triangles?

4.3 It seems not easy to see the geometrical sense of AB^3 and AD^3. However, there are many right angles and we know AB^2 and AD^2 (by Example 2.3.1). In particular, if G, H are the feet of the perpendiculars from D, B to AC respectively, one can show that $\dfrac{AB^2}{AD^2} = \dfrac{AG}{AH}$.

Now it suffices to show $\dfrac{AB}{AD} \cdot \dfrac{AG}{AH} = \dfrac{AF}{AE}$. Since $DE // BF$, applying the Intercept Theorem will probably solve the problem. Are you fluent and skillful in manipulating ratios?

4.4 Since P is an arbitrary point and $\angle OPF = \angle OEP$ **should** always hold, can we replace P by a special point on the circumference? Unfortunately, we cannot use M because M lies on the line OE.

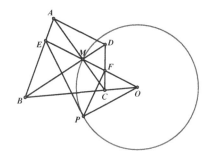

What can we say about P? Notice that we **should** have $\triangle OPE \sim \triangle OFP$, or equivalently, $OP^2 = OE \cdot OF$. Since $OP = OM$, can we show that $OM^2 = OE \cdot OF$? Notice that we do not need the circle anymore! Since $EF \,/\!/\, AD$, we may probably show $\dfrac{OE}{OM} = \dfrac{OM}{OF}$ using the Intercept Theorem. (Are you skillful in applying the Intercept Theorem? Refer to the remarks after Corollary 2.2.2.)

4.5 The only equal lengths we have are $PA = PB$. Apparently, it is not easy to place QE, QF in congruent triangles. Notice that there are many equal angles in the diagram due to the circle, tangents and parallel lines. Can you identify similar triangles involving QE and QF? For example, can you see $\triangle AEQ \sim \triangle ABC$? If we express QE, QF as ratios of line segments, perhaps we can show that the ratios are the same.

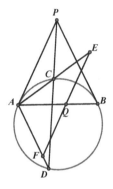

Note that it is not easy to solve the problem by applying the Intercept Theorem even though we have $AP \,/\!/\, EF$: we do not know $\dfrac{QE}{AP}$ or $\dfrac{QF}{AP}$.

4.6 Apply the Tangent Secant Theorem. (You may need Example 2.3.1.)

4.7 Can you see AD is both an angle bisector and a height? Can you construct the isosceles triangle? Can you find BC using similar triangles or the Tangent Secant Theorem? (You are given CE and BD. How are they related to BC?)

4.8 We are to show D, E, F are collinear where D, E, F are closely related to $\triangle ABC$: shall we apply Menelaus' Theorem? Can you show that $\dfrac{AF}{BF}\cdot\dfrac{BD}{CD}\cdot\dfrac{CE}{AE}=1$? What do we know about $\dfrac{AF}{BF}$, $\dfrac{BD}{CD}$ and $\dfrac{CE}{AE}$?

We know $AF\cdot BF=CF^2$ by the Tangent Secant Theorem, i.e., $\dfrac{AF}{CF}=\dfrac{CF}{BF}$. Can you see that $\dfrac{AF}{BF}=\dfrac{AF}{CF}\cdot\dfrac{CF}{BF}$?

Notice that the circumcircle of $\triangle ABC$ and the tangent lines give similar triangles. For example, can you see that $\triangle BCF\sim\triangle CAF$?

Now $\dfrac{AF}{CF}=\dfrac{CF}{BF}=\dfrac{AC}{BC}$ and hence, $\dfrac{AF}{BF}=\dfrac{AF}{CF}\cdot\dfrac{CF}{BF}=\left(\dfrac{AC}{BC}\right)^2$. This implies $\dfrac{AF}{BF}$ is uniquely determined by $\triangle ABC$. Can you express $\dfrac{BD}{CD}$ and $\dfrac{CE}{AE}$ similarly?

4.9 We see that AJ, AK are not related to the choice of P. How are CE, BF related to $\triangle ABC$? One easily sees that $CJ = AJ$ and $BK = AK$ (because of the perpendicular bisectors). Now CE, BF are in $\triangle CEJ$ and $\triangle BFK$ respectively.

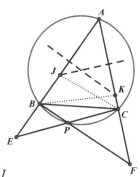

If we have $\triangle CEJ\sim\triangle FBK$, then $\dfrac{CE}{BF}=\dfrac{JE}{BK}=\dfrac{CJ}{KF}$.

Hence, $\dfrac{CE^2}{BF^2}=\dfrac{JE}{BK}\cdot\dfrac{CJ}{KF}$. The conclusion follows because $\dfrac{CJ}{BK}=\dfrac{AJ}{AK}$.

Can we show $\triangle CEJ\sim\triangle BFK$? There are many equal angles in the diagram due to the circle and the perpendicular bisectors.

4.10 Since $AP = AQ$, one immediately sees that $\angle 2=\angle 1=\angle 3$ (angles in the same arc). Can you see similar triangles?

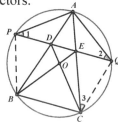

Since we are to show $DE \,//\, BC$, $BCDE$ **should** be a trapezium. Can you see that $BCDE$ **should** be an isosceles trapezium? How is O related to $BCDE$? (**Hint:** $OB = OC$.) Now it suffices to show that B, C, D, E are concyclic. What can you conclude from $AP = AQ$ and the similar triangles?

4.11 Naturally, we suppose two common tangents intersect at P and show that P lies on the third common tangent. One may see this as a special case of Theorem 4.3.6, while the radical axes are the common tangent. We still apply the Tangent Secant Theorem and construct a proof by contradiction.

4.12 Since three circles intersect (or touch) each other, one may consider applying Theorem 4.3.6. Can you see which lines are the radical axes? What can you obtain by applying the Tangent Secant Theorem?

4.13 Since $\angle B = 2\angle C$, drawing the angle bisector of $\angle B$ gives an isosceles triangle. One may attempt a few techniques with the angle bisector, but notice that applying the Angle Bisector Theorem or reflecting the diagram about the angle bisector would not give AC^2. Since we have an isosceles triangle, how about reflecting the diagram about the perpendicular bisector of BC?

4.14 Refer to the left diagram below. What property do we know about the circumcenter of $\triangle ACE$? By Example 4.3.3, the circumcircle of $\triangle ACE$, say I, is the incircle of $\triangle O_1O_2O_3$ and moreover, A, C, E are the feet of the perpendiculars from I to O_1O_2, O_2O_3, O_1O_3 respectively.

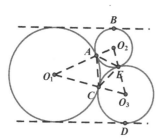 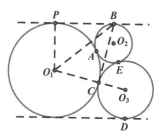

We are to show B, I, C are collinear. It suffices to show $BC \perp O_1O_2$. Let $O_1P \perp \ell_1$ at P. Refer to the right diagram above. Since $O_1C = O_1P$, we **should** have $\triangle BPO_1 \cong \triangle BCO_1$. However, it may not be easy to find equal angles since we do not know how the line segments, say O_1B or BC, intersect the circles given. Can we show $BC = BP$?

Would it be easier to show $O_1B^2 - O_1C^2 = O_3B^2 - O_3C^2$? Notice that all these line segments are uniquely determined by the radii of the three circles (by Pythagoras' Theorem).

Observe that those radii are not independent. Let the radii of $\odot O_1$, $\odot O_2$ and $\odot O_3$ be r_1, r_2, r_3 respectively. For example, if we draw $O_3X \perp O_2B$ at X, we have $O_2X^2 + O_3X^2 = O_2O_3^2$, where $O_2X = 2r_1 - r_2 - r_3$, $O_2O_3 = r_2 + r_3$ and $O_3X = DQ - BP$.

Refer to the diagram on the right. One may find BP via the right angled trapezium BPO_1O_2 and similarly DQ as well. Applying Pythagoras' Theorem repeatedly should lead to the conclusion.

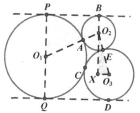

4.15 How can we use the condition that PQ is tangent to the circumcircle of $\triangle MNL$? Notice that PQ only touches the circumcircle of $\triangle MNL$ once, i.e., at L. We are to show $OP = OQ$. Hence, it suffices to show $OL \perp PQ$. Regrettably, this seems not clear because O is **not** the circumcenter of $\triangle MNL$.

Refer to the diagram on the right. Once we draw $\triangle MNL$, it is easy to see $AB // ML$ and $AC // NL$ because M, N, L are midpoints. Are there any similar triangles?

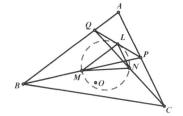

Clearly, $\angle BAC = \angle MLN$. We also have $\angle LMN = \angle PLN = \angle APQ$ because of Theorem 3.2.10 and $AC \, // \, NL$. Similarly, $\angle LNM = \angle AQP$. We must have $\triangle LMN \sim \triangle APQ$.

Notice that L is the midpoint, i.e., $\dfrac{LM}{BQ} = \dfrac{PL}{PQ} = \dfrac{QL}{PQ} = \dfrac{LN}{CP}$. Can you see this implies $AP \cdot CP = AQ \cdot BQ$? How does this remind you of OP and OQ? Consider the power of points P, Q with respect to $\odot O$, the circumcircle of $\triangle ABC$!

Chapter 5

5.1 Recall Example 3.4.1.

5.2 There are many right angles in the diagram. One immediately sees that $PH^2 = MH \cdot BH$. Hence, it suffices to show $MH \cdot BH = AH \cdot OH$, or $\dfrac{MH}{AH} = \dfrac{OH}{BH}$. Can we show it by similar triangles?

Notice that M and O are midpoints. If we cannot find many angle properties related to them, perhaps we can calculate more lengths.

On a side note, all the points are uniquely determined in the circle because $\triangle PAB$ is a right angled isosceles triangle. One may calculate PH, AH, OH explicitly, say by Pythagoras' Theorem and Cosine Rule. Of course, this would not lead to an elegant solution, but is still a valid proof.

5.3 One may solve it by either similar triangles or angle properties in a circle. Can you see any pair of angles which **should** be equal? Can you see that A, I, E, P **should** be concyclic?

5.4 Can you see $DR = DQ$? Can you see that DX is the perpendicular bisector of QR? What can you say about EY and CZ?
Hint: This is an easy question if you construct the diagram wisely. Do **not** draw all the points explicitly as it only complicates the diagram unnecessarily and distracts you from seeking the clues.

5.5 Given the orthocenter H and the midpoint M, one immediately sees that $A'BHC$ is a parallelogram, where AA' is a diameter of $\odot O$ (Example 3.4.4).
In particular, A', H, Q are collinear and N is the midpoint of $A'H$.
We are to show M, N, P, Q are concyclic.

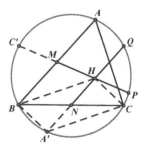

It seems we may consider the Intersecting Chords Theorem. Refer to the diagram above. Can you see that $A'H \cdot QH = C'H \cdot PH$, where C' is obtained by PH extended intersecting $\odot O$? (CC' is also a diameter!)

5.6 Recall that $r = \dfrac{1}{2}(AB + AC - BC)$ since $\angle A = 90°$. Notice that a similar argument applies for r_1, r_2 as well.

5.7 Consider the reflection of C about BD, called C'. Can you see that $CP + PQ = C'P + PQ \geq C'Q$? What is the smallest possible value of $C'Q$? (Notice that C' does not depend on the choice of P and Q.)

5.8 We **should** have $PM = QM$. However, it is not easy to show because BQ, CP are **not** the altitudes. How are $\triangle APQ$ and $\triangle BCH$ related? Notice that $\angle BHC = 180° - \angle A$. Does it remind you of any technique? Double the median HM!

5.9 It suffices to show $\angle ACQ = \angle BAD$. Notice that $\angle CAQ = \angle ADB$ (because AP is tangent to $\odot O$). Hence, we **should** have $\triangle ACQ \sim \triangle DAB$.

Can we show $\dfrac{AQ}{AC} = \dfrac{BD}{AD}$?

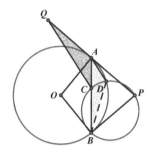

Notice that $AQ = AP = BP$. Can we show $\dfrac{BP}{AC} = \dfrac{BD}{AD}$? Is there another pair of similar triangles which imply this? We have two circles and hence, plenty of equal angles.

5.10 If X lies on BE extended, then P, H coincide and Q, X coincide, where H is the orthocenter of $\triangle ABC$. It is easy to see that E is the midpoint of PQ (Example 3.4.3). Refer to the left diagram below.

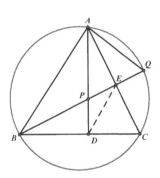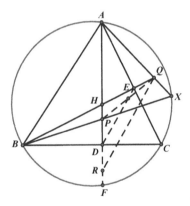

Let X be an arbitrary point. Now it is not easy to show DE passes through the midpoint of PQ since P, H do not coincide and Q does not lie on the circumcircle of $\triangle ABC$. Nevertheless, since P still lies on the line AD, perhaps we can draw $QR /\!/ DE$, intersecting the line AD at R. Refer to the right diagram above.

We **should** have $PD = DR$. Since $\triangle BFH$ is an isosceles triangle (Example 3.4.3), P and R **should** be symmetric about the line BC.

On the other hand, what properties do we know about Q? It is easy to see that $\angle BFH = \angle BHF = \angle AHQ$ and $\angle QAH = \angle PBF$.

Hence, we have $\triangle AHQ \sim \triangle BFP$, where $\triangle BFP$ **should** be the reflection of $\triangle BHR$. Refer to the diagram on the right. Note that $\triangle BHR$ and $\triangle AHQ$ are related by the parallel lines DE and QR.

If we equate the ratios of the line segments via the similar triangles and the parallel lines, we will probably see the conclusion.

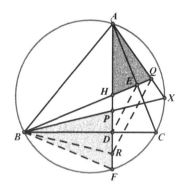

5.11 One immediately notices that the point A could be neglected. Let DE intersect PH at G. We are to show that G is the midpoint of PH. In fact, we have a midpoint H if we extend PH, intersecting Γ at Q. How can we apply the condition $PD = PE$? Can you see that $\triangle PEG \sim \triangle PQE$?

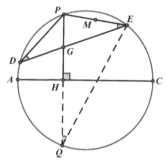

If we choose M as the midpoint of PE, M and H are corresponding points in the similar triangles. Is it reminiscent of Example 5.2.8?

Alternatively, one recognizes that P is the circumcenter of $\triangle DEH$. Notice that the circumcircle of $\triangle DEH$ intersects Γ exactly at D and E. If DE intersects PH at G, one may probably show $PG = HG$ by considering the power of point G (or by the Intersecting Chords Theorem).

Note: One may refer to Example 3.5.1, the diagram of which apparently shows a similar structure.

In fact, if PP' is a diameter of Γ, one may draw $\odot P'$ with radius $P'D$ (where PP' is the perpendicular bisector of DE). Notice that $PD \perp P'D$ and $PE \perp P'E$, i.e., PD, PE are tangent to $\odot P'$. Refer to the diagram on the right.

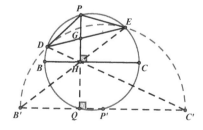

Let $B'C'$ be the diameter passing through Q. It is easy to see that $B'C' \, // \, BC$. By Example 3.5.1, $B'E$, $C'D$ and AQ are concurrent at H. Unfortunately, knowing this fact is not helpful when showing $PG = GH$.

5.12 Can you see $\angle CPD = 90° - \angle CAD$ and $\angle CQD = 180° - 2\angle CAD$? How are P and Q related?

Since B, C, E **should** be related, can you see that B should be the orthocenter of $\triangle APE$? Can you show that $AB \perp PE$?

5.13 Can you see that DE is the perpendicular bisector of CI? Can you show F lies on the perpendicular bisector of CI? It may not be easy because we do not know much about the line segments CF and FI. We are given a parallel line ℓ_1 and a tangent line ℓ_2. If ℓ_1 and the line DE intersect at F', can we show that $F'C$ is tangent to $\odot O$ (i.e., F and F' coincide) by angle properties?

5.14 We are to show $AB - AC = BP - CP$, where $\angle A = 90°$ and angle bisectors are given. It is natural to consider reflecting A about the angle bisectors. In particular, if we draw $DF \perp BC$ at F and $EG \perp BC$ at G, it is easy to see that $AB - AC = BG - CF$.

Hence, P **should** be the midpoint of FG. Can we show it? (Notice that there are many right angles in the diagram.)

5.15 One may notice that the condition and the conclusion are probably related to similar triangles sharing a common vertex. In particular, we are to show $\angle AED = \angle BEP$ and we know that $\angle DAE = \angle DBE$. Hence, we **should** have $\triangle ADE \sim \triangle BPE$.

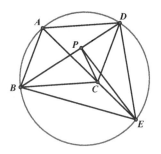

However, showing $\triangle ADE \sim \triangle BPE$ may not be easy because we know neither BP nor $\angle BPE$. Can we show $\triangle ABE \sim \triangle DPE$ instead? It seems the difficulties remain: what do we know about P?

Perhaps we should seek more clues from the condition. We are given a circle and a parallelogram, the properties of which should give us many pairs of equal angles. For example, $\angle BDC = \angle ABD = \angle AED$. Notice that we also have $\angle PCD = \angle ACB = \angle CAD$ from the given condition.

It follows that $\triangle PCD \sim \triangle DAE$, which gives us $\dfrac{PD}{CD} = \dfrac{DE}{AE}$. Now we know more properties of P. Can you see $\triangle ABE \sim \triangle DPE$?

5.16 Upon constructing the diagram, one may notice that this problem is very similar to Example 5.2.8. Can we still apply the technique by introducing a perpendicular from O to the chord AB?
We are to show $AC \perp CE$, i.e., if CE intersects $\odot O$ at A', then AA' must be a diameter of $\odot O$.

5.17 Given A, B, C, D are concyclic and A, B, F, E are concyclic, can you see that $\angle DAE = \angle CBF$? Is this useful? (Notice that AD and BF should **not** be parallel because F could be arbitrarily chosen on CD).
Given the circumcenters G and H, can you see that $\angle DGE = \angle 2DAE$? What can you conclude about the (isosceles) triangles $\triangle DEG$ and $\triangle CFH$? Can you see that $DG /\!/ FH$?

Since P, G, H **should** be collinear, can you see similar triangles from $DG /\!/ FH$? How are $\triangle APE$ and $\triangle BPC$ related? Clearly they are not similar, but how are $\dfrac{PD}{DE}$ and $\dfrac{PF}{CF}$ related?

Chapter 6

6.1 Recall Example 3.4.1.

6.2 Let AM and BP intersect at D. It is easy to find $\dfrac{AD}{DM}$ in the right angled triangle $\triangle ABM$. Can you find $\dfrac{AP}{CP}$ by Menelaus' Theorem?

Alternatively, one may draw $PE \perp BC$ at E. Can you see that $\triangle PEC$ is also a right angled isosceles triangle?

Indeed, there are many ways to calculate $\dfrac{AP}{CP}$. One may also draw the square $ABCX$. Can you see that BP extended pass through the midpoint of CX, called F? Can you see $\triangle ABM \cong \triangle BCF$? Can you see $\dfrac{AP}{CP} = \dfrac{AB}{CF}$?

6.3 Since AD, BE, CF are concurrent, can you see many pairs of similar triangles?

6.4 We are given a median, an angle bisector and an altitude. Can you show that BE is an altitude as well (by considering the median on the hypotenuse BC)? Can you see $EF = \dfrac{1}{2} AC$?

6.5 How will the circles drawn (with diameters BC and AC) intersect $\triangle ABC$? If you draw a circle with a diameter AC, can you see that it must intersect BC, AC at D, E respectively?
Can you see that MN, PQ intersect at the orthocenter of $\triangle ABC$, called H? Can you show that $MH \cdot NH = PH \cdot QH$ by the Intersecting Chords Theorem?

Alternatively, one easily sees that $CM = CN$ and $CP = CQ$. Since M, N, P, Q **should** be cyclic, this circle **should** be centered at C. Can you show $CM = CP$? (Notice that they are in right angled triangles!)

6.6 What can you say about $\dfrac{AX}{DX}$, by the Angle Bisector Theorem or similar triangles? How are $\triangle ADX$ and $\triangle CDY$ (**not** similar) related?

6.7 Since $\triangle ABC$ **should** be an equilateral triangle, one should draw an *almost* equilateral triangle. Suppose the circumcircle of $\triangle DEF$ intersects BC at D, D'. Can you see that D and D' are symmetric about the midpoint of BC? (Notice that the perpendicular bisector from O to BC is also the perpendicular bisector of DD'.)

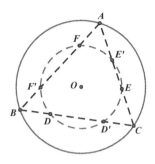

How are BD and BD' related? How are BD and BF' related?

6.8 How to show Q lies on BC? One strategy is to show that if the circumcircle of $\triangle BPD$ intersects BC at Q', then C, E, P, Q' are concyclic (say by angle properties). However, this may not be easy because we do not know how $\triangle BDP$ and $\triangle CEP$ are related.

Where should $\triangle BDP$ intersect BC? Refer to the diagram on the right. It *seems* that A, P, Q' are collinear. If this is true, we **should** have $\angle B = \angle APD$.

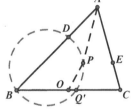

Notice that $\angle B = \angle AED$ (because BE, CD are heights). Hence, we **should** have $\angle AED = \angle APD$, i.e., A, D, P, E **should** be concyclic. It is easy to see that $PD = PE$ because OP is the perpendicular bisector of DE. Can you see why A, D, P, E are concyclic?

6.9 How can we construct such a diagram? If we choose D and P casually, it is difficult to introduce ℓ which gives $DE = DF$.

Let us construct the diagram in the reverse manner. Refer to the diagram on the right. First we draw a line segment EF with its midpoint D, and N is on EF extended. Now if P and C are chosen, A and B will be uniquely determined (illustrated by the broken lines).

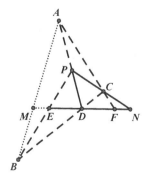

Hence, M is uniquely determined (by the dotted lines), where we **should** have $DM = DN$.

It seems that DM can be calculated via other line segments. Is it reminiscent of Menelaus' Theorem?

Which triangle should we apply Menelaus' Theorem to? We should have line segments DM, DN (or equivalently, EM, FN) in the equation, and probably DE, DF as well. Apparently, more than one triangle will be involved.

6.10   Notice that drawing all the circles given, $\odot O_1$, $\odot O_2$ and $\odot O_3$, only makes the diagram unnecessarily complicated. Instead, we may study the properties of two circles, say $\odot O_2$ and $\odot O_3$. Similar properties should apply to $\odot O_1$ as well.

Let $\odot O_2$ and $\odot O_3$ intersect at P and H.
One immediately sees that $PH \perp O_2 O_3$.

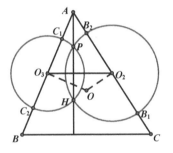

Since O_2, O_3 are the midpoints of AC, AB respectively, we have $O_2 O_3 \,/\!/\, BC$ and hence, $PH \perp BC$. This implies A, P, H are collinear.

Now a simple application of the Tangent Secant Theorem shows that B_1, B_2, C_1, C_2 are concyclic. Similarly, we should have A_1, A_2, B_1, B_2 concyclic as well. How can we show that $A_1, A_2, B_1, B_2, C_1, C_2$ all lie on the same circle?

Which circle does B_1, B_2, C_1, C_2 lie on? Do you know the center and the radius of that circle? (You may identify the center by drawing the perpendicular bisectors of $B_1 B_2$ and $C_1 C_2$.) How about the circle which A_1, A_2, B_1, B_2 lie on?

6.11   Suppose BM and CN intersect at X. Since we are to show X lies on the circumcircle of $\triangle ABC$, the most straightforward method might be showing that $\angle BXC = 180° - \angle BAC$.

One notices that $\angle PAB = \angle C$ is a useful condition, with which one easily sees that $\triangle ABC \sim \triangle PBA$.
Similarly, $\triangle ABC \sim \triangle QAC$. (*)
Refer to the diagram on the right. Can you see that $\angle 1 = \angle 2 = \angle BAC$?
Hence, we **should** have $\angle 4 = \angle BAC = \angle 1 = \angle BQN$, which implies B, N, X, Q are concyclic.

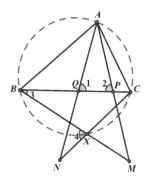

Now we **should** have $\angle 3 = \angle N$ and similarly, $\angle BCN = \angle M$. This implies that $\triangle BPM \sim \triangle NQC$. Can we show it? Since $\angle 1 = \angle 2$, it suffices to show $\dfrac{BP}{PM} = \dfrac{NQ}{CQ}$. Notice that we have **not** used the condition that P, Q are midpoints of AM, AN respectively. Now it suffices to show $\dfrac{BP}{AP} = \dfrac{AQ}{CQ}$. Can you see it from (*)?

6.12 We are given a circle and a triangle, but the condition $CF \, // \, BE$ seems not closely related to circle geometry. Perhaps we can find equal angles through the parallel lines and the property of $\odot O$.

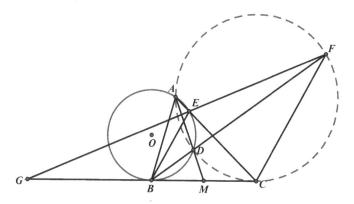

Refer to the diagram above. Since $BE \, // \, CF$, we have $\angle BFC = \angle EBF = \angle CAD$, which implies A, D, C, F are concyclic. Alternatively, one may obtain this result by $\angle ACF = \angle BEC = \angle ADF$. Suppose A, D, C, F lie on $\odot O_1$.

We are to show $AG = DG$, which implies G **should** lie on the perpendicular bisector of AD. Since AD is the common chord of $\odot O$ and $\odot O_1$, its perpendicular bisector is the line OO_1. Can we show that G, O, O_1 are collinear? Refer to the diagram below. Can we show $\angle BGO = \angle BGO_1$? Notice that $BE \mathbin{/\mkern-5mu/} CF$ gives us similar triangles $\triangle BEG \sim \triangle CFG$. Hence, it suffices to show O and O_1 are corresponding points in $\triangle BEG$ and $\triangle CFG$.

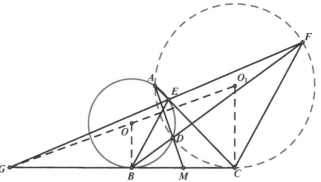

O is obtained by intersecting the perpendicular bisector of BE and the line passing through B perpendicular to BC. Hence, it suffices to show BC is tangent to $\odot O_1$. Notice that we have not used the condition $BM = CM$. Observe the position of M and the two circles. Does it remind you of the Tangent Secant Theorem?

6.13 Refer to the diagram on the right. We are to show I is the circumcenter of $\triangle PQR$, which is equivalent to $PI = QI = RI$. How is I related to P, Q, R? We know $EI = FI$ and indeed, AI is the perpendicular bisector of EF. Notice that I **should** lie on the perpendicular bisector of QR.

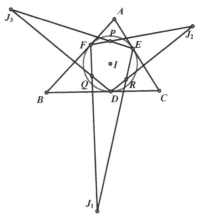

It *seems* from the diagram that $EF // QR$. Is it true?

If we can show $\dfrac{J_1Q}{FQ} = \dfrac{J_1R}{ER}$, then $QR // EF$, which implies $QI = RI$ (because A, I, J_1 are collinear and $J_1E = J_1F$). Similarly, $PI = QI$ and the conclusion follows. This is probably the critical step we need!

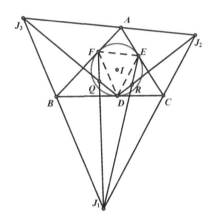

On the other hand, we know $EF // J_2J_3$ because $AI \perp J_2J_3$ (Exercise 1.5). Similarly, we have $DF // J_1J_3$ and $DE // J_1J_2$. Refer to the diagram on the right.

Notice that the parallel lines give $\Delta DEF \sim \Delta J_1J_2J_3$. Now can you see $\dfrac{J_1Q}{FQ} = \dfrac{J_1R}{ER}$?

6.14 We are given many conditions. It is easy to seek clues from some of the conditions. Refer to the left diagram below. Since $AB = BC + AE$ and $AE // BC$, it is natural to move BC up (i.e., extend AE to G such that $BC = EG$). (*)

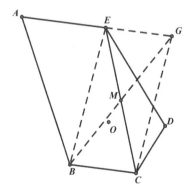 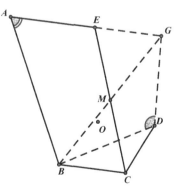

We obtain a parallelogram $BCGE$ where M is the center, as well as an isosceles triangle ΔABG. Given $AE // BC$, can you see that BG bisects

$\angle ABC$? Now it suffices to show that $\angle ADB = \dfrac{1}{2}\angle CDE = \dfrac{1}{2}\angle ABC$ $= \angle AGB$, i.e., we **should** have A, B, D, G concyclic.

Apparently, we do not know much about the line segments, but only about the angles. Refer to the right diagram above. Can we show that $\angle BDG = 180° - \angle A$? Notice that $180° - \angle A = \angle ABC = \angle CDE$. Hence, we **should** have $\angle BDG = \angle CDE$, or equivalently, $\angle BDC = \angle EDG$.

Notice that we have not used the following conditions:
- M is the midpoint of CE (and hence the center of the parallelogram $BCGE$.
- O is the circumcenter of $\triangle BCD$.
- $OM \perp DM$

It seems that these properties are related to *symmetry*. Refer to the diagram on the right. Let D' be the reflection of D about OM. What can you say about D'? Can you see congruent triangles related to D'? How is the parallelogram $BCGE$ related to D'? How is D' related to O, the circumcenter of $\triangle BCD$?

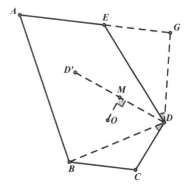

Note: If one extends BC instead of AE at (*) to G such that $CG = AE$, an isosceles triangle $\triangle ABG$ will be obtained where AG bisects $\angle A$. Unfortunately, this is not useful because we need angles related to half of $\angle ABC$ or $\angle CDE$.

Solutions to Exercises

Chapter 1

1.1 Since $AP = BP$, we have $\angle 1 = \angle B$. Now $\angle 2 = 90° - \angle 1 = 90° - \angle B = \angle C$, which implies $AP = CP$. The conclusion follows.

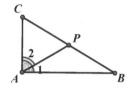

1.2 Choose E on AC such that $AB = AE$. Since AD bisects $\angle BAC$, one sees that $\triangle ABD \cong \triangle AED$ (S.A.S.). Hence, $BD = DE$ and $\angle AED = \angle ABD = 2\angle C$.
Since $\angle AED = \angle C + \angle CDE$, we conclude that $\angle C = \angle CDE$, i.e., $CE = DE$.
Now $CE = DE = BD$. We have $AC = AE + CE = AB + BD$.

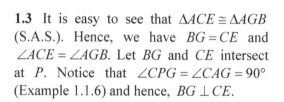

1.3 It is easy to see that $\triangle ACE \cong \triangle AGB$ (S.A.S.). Hence, we have $BG = CE$ and $\angle ACE = \angle AGB$. Let BG and CE intersect at P. Notice that $\angle CPG = \angle CAG = 90°$ (Example 1.1.6) and hence, $BG \perp CE$.

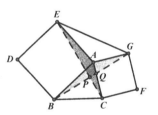

1.4 Refer to the left diagram below. Let BP, CP bisect the exterior angles of $\angle B, \angle C$ respectively. We are to show AP bisects $\angle A$. Draw $PD \perp BC$ at D, $PE \perp AB$ at E and $PF \perp AC$ at F. It is easy to see that $\triangle BPE \cong \triangle BPD$ (A.A.S.) and hence, $PD = PE$. Similarly, $PD = PF$.

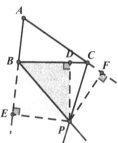

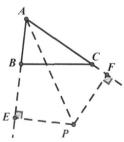

Now we have $PE = PF$. Refer to the right diagram above. One sees that $\triangle APE \cong \triangle APF$ (H.L.) and hence, AP bisects $\angle A$.

1.5 Connect AJ_1. Since AI and AJ_1 are the angle bisectors of neighboring supplementary angles, we have $AI \perp AJ_1$ (Example 1.1.9, or one may simply see that

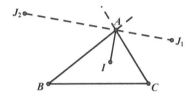

$$\angle IAJ_1 = \angle CAI + \angle CAJ_1 = \frac{1}{2} \angle BAC + \frac{1}{2}\left(180° - \angle BAC\right) = 90°.)$$

Similarly, $AI \perp AJ_2$. Now $J_1AJ_2 = 90° + 90° = 180°$, which implies A, J_1, J_2 are collinear and hence, $AI \perp J_1J_2$.

1.6 Choose E' on CD extended such that $DE' = BE$. Connect AE'. It is easy to see that $\triangle ABE \cong \triangle ADE'$ (S.A.S.). Hence, $AE = AE'$ and $\angle BAE = \angle DAE'$. Now we see that $\angle EAF = \angle E'AF = 45°$ and $\triangle AEF \cong \triangle AE'F$ (S.A.S.). Hence, $EF = E'F = DF + BE$.

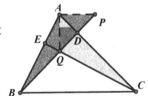

1.7 We have $\angle ABD = \angle ACE = 90° - \angle BAC$. Hence, $\triangle ABP \cong \triangle QCA$ (S.A.S.). It follows that $AQ = AP$ and $\angle QAD = \angle APD = 90° - \angle PAC$, i.e., $\angle QAD + \angle PAC = \angle PAQ = 90°$. Thus, $\angle AQP = 45°$.

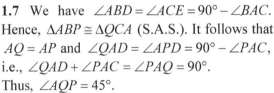

1.8 Connect CD. Since $BE = AB = BC$ and BD bisects $\angle CBE$, we have $\triangle BCD \cong \triangle BED$ (S.A.S.). Hence, $\angle BED = \angle BCD$.
Since $AD = BD$, D (and similarly C) lie on the perpendicular bisector of AB, which is indeed the line CD. It follows that CD bisects $\angle ACB$.

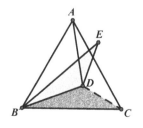

Now $\angle BED = \angle BCD = \frac{1}{2} \angle ACB = 30°$.

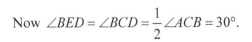

1.9 Since I is the incenter, CI bisects $\angle C$.

Theorem 1.3.3 gives $\angle AIB = 90° + \dfrac{1}{2}\angle C$.

Hence, $\angle BID = 180° - \angle AIB = 90° - \dfrac{1}{2}\angle C$

$= 90° - \angle BCI = \angle CIH$.

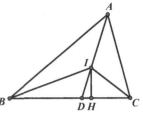

1.10 Since $\angle 1 = \angle 2$ and $\angle 3 = \angle 4$, we have $\triangle ABC \cong \triangle ADC$ (A.A.S.). Hence, $AB = AD$ and $\angle ABF = \angle ADE$. Now $\triangle ABF \cong \triangle ADE$ (A.A.S.), which implies $AE = AF$. It follows that $\triangle AEP \cong \triangle AFP$ (S.A.S.) and $PE = PF$. Note that the proof holds regardless of the position of P.

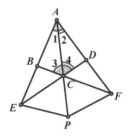

1.11 Let M be the midpoint of BC. Since O is the circumcenter of $\triangle BCD$, OM is the perpendicular bisector of BC. On the other hand, since I is the incenter of $\triangle ACD$, AI is the angle bisector $\angle A$, which passes through M since $AB = AC$. Thus, A, I, O lie on the perpendicular bisector of BC. The conclusion follows.

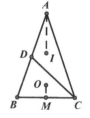

1.12 Let $\angle ABC = 2\alpha$ and $\angle APC = 2\beta$. We have $\angle BAP = \angle APC - \angle ABC = 2(\alpha - \beta)$. Since BD, PD are angle bisectors, we have $\angle CBD = \alpha$ and $\angle CPD = \beta$. It follows that $\angle BDP = \angle CPD - \angle CBD = \alpha - \beta$.

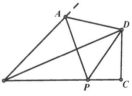

Notice that D is the ex-center of $\triangle ABP$ opposite B (Exercise 1.4), which implies that AD bisects the exterior angle of $\angle BAP$.

Now $\angle PAD = \dfrac{1}{2}(180° - \angle BAP) = 90° - \dfrac{1}{2} \cdot 2(\alpha - \beta) = 90° - \angle BDP$. This completes the proof.

1.13 Suppose otherwise. Draw $CD' /\!/ AB$, intersecting the line AD at D'. Now $ABCD'$ is a parallelogram and $AB = CD'$, $BC = AD'$. We have $AD' - CD' = BC - AB = AD - CD$.

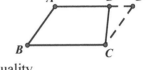

Case I: $AD < AD'$
Refer to the diagram on the right.
We have $DD' = AD' - AD = CD' - CD$, i.e.,
$DD' + CD = CD'$. This contradicts triangle inequality.

Case II: $AD > AD'$
Similarly, we have $DD' = AD - AD' = CD - CD'$, i.e., $DD' + CD' = CD$. This contradicts triangle inequality.

It follows that D and D' coincide, i.e., $ABCD$ is a parallelogram.

1.14 Draw $EP \perp \ell_2$ at P and $AQ /\!/ EH$, intersecting CD at Q. It is easy to see that $AEHQ$ is a parallelogram and hence, $EH = AQ$. Given that $EP = AD$, we must have $\triangle EPH \cong \triangle ADQ$ (H.L.). It follows that $\angle 1 = \angle AQD = \angle 2$.
Similarly, we have $\angle BGF = \angle HGF$.

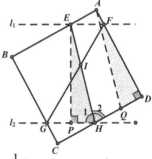

Now $\angle GIH = 180° - \angle HGF - \angle 1$, where $\angle 1 = \dfrac{1}{2}\left(180° - \angle CHG\right)$

$= 90° - \dfrac{1}{2}\angle CHG$ and similarly, $\angle HGF = 90° - \dfrac{1}{2}\angle CGH$.

Hence, $\angle GIH = 180° - \left(90° - \dfrac{1}{2}\angle CHG\right) - \left(90° - \dfrac{1}{2}\angle CGH\right)$

$= \dfrac{1}{2}\left(\angle CGH + \angle CHG\right) = 45°$, because $\triangle CGH$ is a right angled triangle where $\angle C = 90°$.

Note: One may observe that I is the ex-center of $\triangle CGH$ opposite C (Exercise 1.4). Indeed, one may show, following a similar argument as above, that if J is the ex-center of $\triangle ABC$ opposite A, then we always have $\angle BJC = 90° - \frac{1}{2}\angle A$. (You may compare this result with Theorem 1.3.3.)

Chapter 2

2.1 Since $AB = BD$, $[\triangle ABC] = [\triangle BCD] = \frac{1}{2}[BCXD]$, i.e., we have $[BCXD] = 2[\triangle ABC]$.

Similarly, $[ACEY] = [\triangle ABC]$ and $[ABZF] = 4[\triangle ABC]$.

Now the total area of parallelograms is $175 = 7[\triangle ABC]$. It follows that $[\triangle ABC] = 25$ cm^2.

2.2 It is easy to see that $\triangle ACE \cong \triangle AGB$, which implies $CE = BG$ and $BG \perp CE$ (Exercise 1.3). Since O_1M is a midline of $\triangle BEC$, we have $O_1M = \frac{1}{2}CE$ and $O_1M \,/\!/\, CE$.

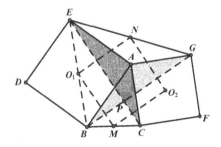

Similarly, $O_2N = \frac{1}{2}CE$ and $O_2N \,/\!/\, CE$. Now $O_1M = O_2N$ and $O_1M \,/\!/\, O_2N$ imply MO_1NO_2 is a parallelogram.

A similar argument gives $O_1N = O_2M = \dfrac{1}{2}BG$ and $O_1N \,//\, O_2M \,//\, BG$.

Now $BG = CE$ implies $O_1M = O_1N$ while $BG \perp CE$ implies $O_1M \perp O_2N$. It follows that MO_1NO_2 is a square.

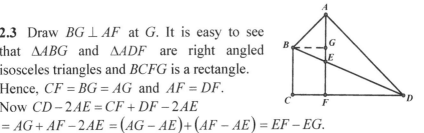

2.3 Draw $BG \perp AF$ at G. It is easy to see that $\triangle ABG$ and $\triangle ADF$ are right angled isosceles triangles and $BCFG$ is a rectangle. Hence, $CF = BG = AG$ and $AF = DF$.

Now $CD - 2AE = CF + DF - 2AE$

$= AG + AF - 2AE = (AG - AE) + (AF - AE) = EF - EG$.

Since $DF = AF > AG = CF$, $\dfrac{EF}{BC} = \dfrac{DF}{DF + CF} > \dfrac{1}{2}$, i.e., $EF > EG$.

In conclusion, $CD - 2AE > 0$ and $AE < \dfrac{1}{2}CD$.

2.4 Notice that PM is a midline of $\triangle BDE$. Hence, $PM = \dfrac{1}{2}BD$ and $PM \,//\, BD$. Similarly,

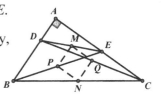

$QN = \dfrac{1}{2}BD = PM$ and $QN \,//\, BD \,//\, PM$.

We also have $QM = PN = \dfrac{1}{2}CE$ and $QM \,//\, PN \,//\, CE$. It follows that $MPNQ$ is a parallelogram. Since $PM \,//\, AB$, $QM \,//\, AC$ and $AB \perp AC$, we must have $PM \perp QM$. Hence, $MPNQ$ is a rectangle and $MN = PQ$.

2.5 It is easy to see that $EFGH$ is a parallelogram (Example 2.2.6). We focus on $EFGH$. Refer to the diagram on the right. Let EM extended and FG extended intersect at Q.

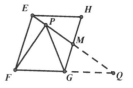

Since $EH \parallel FQ$ and $GM = HM$, $\triangle EHM \cong \triangle QGM$ (A.A.S.). Hence, $QG = EH = FG$. It is given that $FG = PG$. We have $PG = FG = QG = \frac{1}{2}FQ$. It follows that $FP \perp PQ$ (Example 1.1.8).

2.6 Let $AB = a$. Since $ABCD$ is a square, it is easy to see that $BE = CF$ and $\triangle BCE \cong \triangle CDF$. Now $\angle BCE = \angle CDF = 90° - \angle CFD$, which implies $CE \perp DF$.

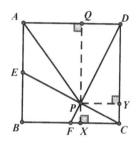

Notice that $\dfrac{PF}{PD} = \left(\dfrac{CF}{CD}\right)^2 = \dfrac{1}{4}$ (Example 2.3.1).

Draw $PX \perp BC$ at X, $PY \perp CD$ at Y and $PQ \perp AD$ at Q. We have $\dfrac{CY}{DY} = \dfrac{PF}{PD} = \dfrac{1}{4}$ and $\dfrac{FX}{CX} = \dfrac{PF}{PD} = \dfrac{1}{4}$. Hence, $DY = \dfrac{4}{5}a = PQ$ and $CX = \dfrac{4}{5}CF = \dfrac{2}{5}BC$, which implies $AQ = BX = \dfrac{3}{5}a$.

By Pythagoras' Theorem, $AP = \sqrt{AQ^2 + PQ^2} = a\sqrt{\left(\dfrac{3}{5}\right)^2 + \left(\dfrac{4}{5}\right)^2} = a$, i.e., $AP = AB$.

Note: There is an alternative solution based on the median CE doubled. Refer to the diagram on the right. Extend CE to X such that $CE = EX$. It is easy to see that $\triangle BCE \cong \triangle AXE$.

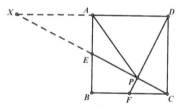

Hence, X lies on the line AD and $AD = AX$. Notice that $CE \perp DF$ as shown in the proof above. It follows that AP is the median on the hypotenuse DX of the right angled triangle $\triangle PXD$. Hence, $AP = \dfrac{1}{2}DX = AD = AB$ (Theorem 1.4.6).

This is an elegant solution, even though the previous solution using Pythagoras' Theorem is more straightforward.

2.7 Let BD, CE be the medians. By the Midpoint Theorem, $BG = 2DG$ and $CG = 2EG$.

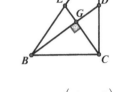

Let $DG = a$ and $EG = b$. Since $BG \perp CG$, we have $BE^2 = (2a)^2 + b^2 = 4a^2 + b^2$.

Hence, $AB^2 = (2BE)^2 = 4 \cdot (4a^2 + b^2) = 16a^2 + 4b^2$.

Similarly, $AC^2 = 4a^2 + 16b^2$. It follows that $AB^2 + AC^2 = 20(a^2 + b^2)$, while $BC^2 = (2a)^2 + (2b)^2 = 4(a^2 + b^2)$. The conclusion follows.

2.8 Refer to the following diagrams. Since $DD' /\!/ BC$, by the Intercept Theorem, we have $\dfrac{AD}{BD} = \dfrac{AD'}{CD'}$. Let $\dfrac{AD}{AB} = \dfrac{AD'}{AC} = a$.

Similarly, let $\dfrac{BE}{BC} = \dfrac{BE'}{AB} = b$ and $\dfrac{CF}{AC} = \dfrac{CF'}{BC} = c$.

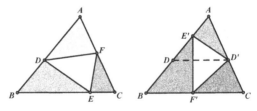

Let $S = [\triangle ABC]$. Note that $[\triangle DEF] = S - ([\triangle ADF] + [\triangle BDE] + [\triangle CEF])$ and $[\triangle D'E'F'] = S - ([\triangle AD'E'] + [\triangle BE'F'] + [\triangle CD'F'])$.

One sees that $\dfrac{[\triangle ADF]}{[\triangle ABC]} = \dfrac{AD}{AB} \cdot \dfrac{AF}{AC} = a(1 - c)$, i.e., $[\triangle ADF] = a(1 - c)S$.

Similarly, $[\triangle BDE] = b(1 - a)S$ and $[\triangle CEF] = c(1 - b)S$.

We also have $[\triangle AD'E'] = \dfrac{AD'}{AC} \cdot \dfrac{AE'}{AB} = a(1-b)S, \; [\triangle BE'F'] = b(1-c)S$

and $[\triangle CD'F'] = c(1-a)S.$

Hence, $[\triangle ADF] + [\triangle BDE] + [\triangle CEF] = [a(1-c) + b(1-a) + c(1-b)] \cdot S$

$= [a+b+c-ac-ab-bc] \cdot S$

$= [a(1-b) + b(1-c) + c(1-a)] \cdot S = [\triangle AD'E'] + [\triangle BE'F'] + [\triangle CD'F'].$

The conclusion follows.

2.9 Connect DE, DF. We claim that $\triangle BDE \sim \triangle CFD$. Notice that $\angle B = \angle C = 60°$. It suffices to show that $\angle 1 = \angle 2$. Since EF is the perpendicular bisector of AD, we must have $AE = DE$ and $AF = DF$. Hence, $\triangle AEF \cong \triangle DEF$ (S.S.S.). Now $\angle EDF = \angle EAF = 60°$ and hence, $\angle 2 = 180° - \angle EDF - \angle 3$ $= 180° - 60° - \angle 3 = 180° - \angle B - \angle 3 = \angle 1$.

We conclude that $\triangle BDE \sim \triangle CFD$. It follows that $\dfrac{BD}{BE} = \dfrac{CF}{CD}$, or equivalently, $BD \cdot CD = BE \cdot CF$.

2.10 It is easy to see that $\angle EAH = \angle DCH$. Hence, $\triangle BCE \sim \triangle HAE$ and we have

$$\frac{BC}{AH} = \frac{CE}{AE} = \tan \angle A.$$

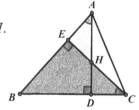

2.11 Let $BC = a, \; AC = b, \; AB = c, \; AD = x$ and $AE = y$. Clearly, $BE = CD = \dfrac{2}{3}a.$

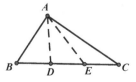

Since AD is a median of $\triangle ABE$, we have $x^2 = \dfrac{1}{2}c^2 + \dfrac{1}{2}y^2 - \dfrac{1}{4}\left(\dfrac{2}{3}a\right)^2$

by Theorem 2.4.3. Similarly, $y^2 = \dfrac{1}{2}b^2 + \dfrac{1}{2}x^2 - \dfrac{1}{4}\left(\dfrac{2}{3}a\right)^2$ because AE is

a median of $\triangle ACD$. Hence, we have

$$x^2 + y^2 = \frac{1}{2}b^2 + \frac{1}{2}c^2 + \frac{1}{2}x^2 + \frac{1}{2}y^2 - \frac{1}{4}\left(\frac{2}{3}a\right)^2 - \frac{1}{4}\left(\frac{2}{3}a\right)^2 ,$$ which could

be simplified to $x^2 + y^2 = \left(b^2 + c^2\right) - \dfrac{4}{9}a^2$.

Pythagoras' Theorem gives $b^2 + c^2 = a^2$ and the conclusion follows.

2.12 Let AE and BC intersect at G. Suppose GD extended intersects AB at M'. By Ceva's

Theorem, $\dfrac{AM'}{BM'} \cdot \dfrac{BC}{GC} \cdot \dfrac{GE}{AE} = 1$. Since $CE /\!/ AB$,

we have $\dfrac{BC}{GC} = \dfrac{AE}{GE}$.

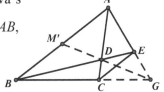

Hence, $\dfrac{AM'}{BM'} = 1$, i.e., M' coincides with M. We conclude that the line

MD passes through G, i.e., the lines AE, BC, MD are concurrent at G.

Note: One may solve this problem by Menelaus' Theorem as well.

Consider the line BE intersecting $\triangle ACG$: $\dfrac{AE}{GE} \cdot \dfrac{GB}{CB} \cdot \dfrac{CD}{AD} = 1$. (*)

Since $AB /\!/ CE$, $\dfrac{AE}{GE} = \dfrac{BC}{CG}$. We obtain $\dfrac{GB}{GC} \cdot \dfrac{CD}{AD} = 1$ from (*).

Now $\dfrac{AM}{BM} = 1$ implies $\dfrac{AM}{BM} \cdot \dfrac{BG}{CG} \cdot \dfrac{CD}{AD} = 1$. It follows from Menelaus'

Theorem that D, G, M are collinear.

2.13 Refer to the diagram on the right. Let AQ intersect BC at Q', BR intersect AC at R' and CP intersect AB at P'. We claim that $\dfrac{BQ'}{CQ'} \cdot \dfrac{CR'}{AR'} \cdot \dfrac{AP'}{BP'} = 1$.

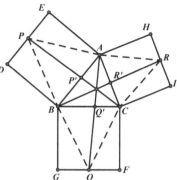

Notice that $\dfrac{BQ'}{CQ'} = \dfrac{[\triangle ABQ]}{[\triangle ACQ]}$

$$= \frac{\dfrac{1}{2} AB \cdot BQ \sin \angle ABQ}{\dfrac{1}{2} AC \cdot CQ \sin \angle ACQ} = \frac{AB \sin(\angle ABC + \alpha)}{AC \sin(\angle ACB + \alpha)} \quad \text{where} \quad BQ = CQ \quad \text{and}$$

$\alpha = \angle BCQ = \angle CBQ$. It is easy to see that $\alpha = \angle ACR = \angle ABP$.

Similarly, $\dfrac{CR'}{AR'} = \dfrac{BC \sin(\angle ACB + \alpha)}{AB \sin(\angle BAC + \alpha)}$ and $\dfrac{AP'}{BP'} = \dfrac{AC \sin(\angle BAC + \alpha)}{BC \sin(\angle ABC + \alpha)}$.

It follows that $\dfrac{BQ'}{CQ'} \cdot \dfrac{CR'}{AR'} \cdot \dfrac{AP'}{BP'} = 1$ and by Ceva's Theorem, AQ, BR, CP are concurrent.

.

2.14 By Menelaus' Theorem, it suffices to show $\dfrac{AF}{BF} \cdot \dfrac{BD}{CD} \cdot \dfrac{CE}{AE} = 1$.

By the Angle Bisector Theorem, $\dfrac{AF}{BF} = \dfrac{AC}{BC}$, $\dfrac{BD}{CD} = \dfrac{AB}{AC}$ and $\dfrac{CE}{AE} = \dfrac{BC}{AB}$. The conclusion follows.

Note:
(1) One may find it easier to solve this problem by applying Menelaus' Theorem and the Angle Bisector Theorem *mechanically* instead of referring to the diagram.
(2) One may also solve this problem using Desargues' Theorem.

Refer to the diagram on the right, where P, Q, R are the ex-centers of $\triangle ABC$ opposite A, B, C respectively. Apply Desargues' Theorem to $\triangle ABC$ and $\triangle PQR$.

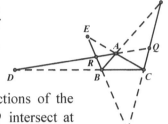

One sees that D, E, F are the intersections of the corresponding sides extended: AB, PQ intersect at F, BC, QR intersect at D, AC, PR intersect at E.

Now D, E, F are collinear if the lines AP, BQ, CR are concurrent. This is clear because they all pass through the incenter of $\triangle ABC$.

2.15 Refer to the diagram on the right.

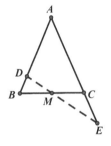

Apply Menelaus' Theorem when the line AE intersects $\triangle BDM$: $\dfrac{BA}{DA} \cdot \dfrac{DE}{ME} \cdot \dfrac{MC}{BC} = 1$.

Since $\dfrac{MC}{BC} = \dfrac{1}{2}$, we have $\dfrac{AB}{AD} = 2\dfrac{ME}{DE}$.

Apply Menelaus' Theorem when the line AB intersects $\triangle CEM$:

$\dfrac{CA}{EA} \cdot \dfrac{ED}{MD} \cdot \dfrac{MB}{CB} = 1$. Since $\dfrac{MB}{CB} = \dfrac{1}{2}$, we have $\dfrac{AC}{AE} = 2\dfrac{MD}{DE}$.

Since $AB = AC$, we have $\dfrac{AB}{AD} + \dfrac{AB}{AE} = \dfrac{AB}{AD} + \dfrac{AC}{AE} = 2\left(\dfrac{DM}{DE} + \dfrac{EM}{DE}\right)$

$= 2 \cdot \dfrac{DM + EM}{DE} = 2$, i.e., $\dfrac{AB}{AD} + \dfrac{AB}{AE} = 2$. This completes the proof.

Note: One may find an alternative solution using the area method. We are to show $\dfrac{1}{AD} + \dfrac{1}{AE} = \dfrac{2}{AB}$, i.e., $\dfrac{AD + AE}{AD \cdot AE} = \dfrac{2}{AB}$. We claim that $AD \cdot AB + AE \cdot AB = 2AD \cdot AE$.

Notice that $[\triangle ACD] = \dfrac{1}{2} AD \cdot AB \sin \angle A$ (since $AB = AC$),

$$[\triangle ABE] = \frac{1}{2} AB \cdot AE \sin \angle A \text{ and } [\triangle ADE] = \frac{1}{2} AD \cdot AE \sin \angle A.$$

Hence, it suffices to show that $[\triangle ACD] + [\triangle ABE] = 2[\triangle ADE]$.

Refer to the diagram on the right. Since $BM = CM$, we have $[\triangle BDE] = [\triangle CDE]$. (Can you see it?)

Hence, $[\triangle ADE] - [\triangle ACD] = [\triangle ABE] - [\triangle ADE]$, which completes the proof.

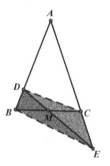

Chapter 3

3.1 (a) We always have $\angle A = \angle C$ in the parallelogram $ABCD$. Now $ABCD$ is cyclic if and only if $\angle A + \angle C = 180°$, which implies $\angle A = \angle C = 90°$. Hence, $ABCD$ is cyclic if and only if $ABCD$ is a rectangle.

(b) In a trapezium $ABCD$, say $AD /\!/ BC$, we always have $\angle A + \angle B = 180°$. Now $ABCD$ is cyclic if and only if $\angle A + \angle C = 180°$, which implies $\angle B = \angle C$, i.e., $ABCD$ is cyclic if and only if it is an isosceles trapezium.

3.2 Since $\angle BAF = \angle CDE$, A, D, F, E are concyclic. Hence, $\angle BAD = \angle CFE$ (Corollary 3.1.5). Since $\angle BAD + \angle ABC = 180°$, we have $\angle ABC + \angle CFE = 180°$, i.e., B, C, F, E are concyclic.

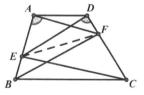

Since $\angle AFE = \angle ADE$ and $\angle BFE = \angle BCE$ (Corollary 3.1.3), we have $\angle AFB = \angle AFE + \angle BFE = \angle ADE + \angle BCE$. One can easily see that $\angle ADE + \angle BCE = \angle CED$ (Example 1.4.15). The conclusion follows.

3.3 Since $\angle 1 = \angle BAI + \angle ABI = \dfrac{1}{2}(\angle A + \angle B)$

and $\angle 2 = \dfrac{1}{2}(180° - \angle C) = \dfrac{1}{2}(\angle A + \angle B)$,   we

have $\angle 1 = \angle 2$. Hence, A, I, C, J are concyclic.

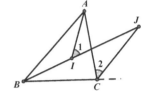

Note: One may also show that $\angle CAI = \angle CJI$.

3.4 Let the lines AD, BC intersect at X. Since AB is the diameter of the semicircle, we must have $AC \perp BC$, $AD \perp BD$ (Corollary 3.1.13). Hence, P is the orthocenter of $\triangle ABX$. It follows that $XP \perp AB$, i.e., X, P, E are collinear. Example 3.1.6 states that P is the incenter of $\triangle CDE$.

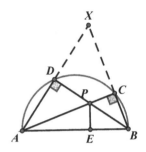

3.5 Let OP intersect $\odot O$ at M. It is easy to see $\triangle PAO \cong \triangle PBO$ (H.L.). Hence, $\angle AOM = \angle BOM$ and they must correspond to equal arcs. It follows that M is the midpoint of $\overparen{AB}$.

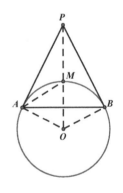

Since $\angle PAM = \dfrac{1}{2}\angle AOM$ (Theorem 3.2.10), we

have $\angle BAM = \dfrac{1}{2}\angle BOM = \dfrac{1}{2}\angle AOM = \angle PAM$.

Now AM bisects $\angle PAB$ and clearly, PM bisects $\angle APB$. It follows that M is the incenter of $\triangle PAB$.

3.6 (a) Since H is the orthocenter, $\angle BHC = 180° - \angle A$ (Example 2.5.5). Since B, C, O, H are cyclic, we have $\angle BOC = 2\angle A = \angle BHC$. It follows that $2\angle A = 180° - \angle A$, or $\angle A = 60°$.

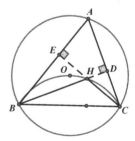

(b) Let the circumradii of $\triangle ABC$ and $\triangle BHC$ be R_1, R_2 respectively.

By Sine Rule, $\dfrac{BC}{\sin\angle A} = 2R_1$ and $\dfrac{BC}{\sin\angle BHC} = 2R_2$.

Since $\angle BHC = 180° - \angle A$, we have $\sin\angle A = \sin(180° - \angle A)$. It follows that $R_1 = R_2$.

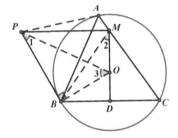

3.7 It is easy to see that OD is the perpendicular bisector of BC. Hence, $BM = CM$ and we have $\angle 2 = \angle CMD = 90° - \angle C$. On the other hand, consider the right angled triangles $\triangle AOP$ and $\triangle BOP$.

We have $\angle 1 = 90° - \angle 3$ and $\angle 3 = \dfrac{1}{2}\angle AOB = \angle C$ (Theorem 3.1.1).

It follows that $\angle 1 = 90° - \angle C = \angle 2$ and hence, B, O, M, P are concyclic. Now $\angle OMP = \angle OBP = 90°$. This completes the proof.

3.8 Extend CD to P such that $CD = PD$. We have $BP = 2DE$ and $\triangle ADP$ is an equilateral triangle (because $\triangle DAC$ is an isosceles triangle and $\angle ADC = 120°$). It follows that $\angle APD = 60° = \angle ABD$, i.e., A, P, B, D are concyclic. Refer to the left diagram below.

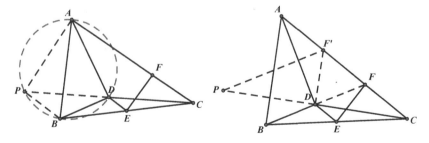

Let F and F' be the trisection points of AC. Notice that $\triangle DFF'$ is an equilateral triangle (Example 2.3.4). Clealy, $PF'\,/\!/\,DF$. We must have $\angle AF'P = \angle AFD = 60° = \angle ADP$. It follows that A, F', D, P are concyclic. Refer to the right diagram above.

Now A, P, B, D, F' lie on the same circle where PF' is a diameter (since $\angle PAC = 90°$). We have $DE /\!/ BP$, $EF /\!/ BF'$ and $BP \perp BF'$ (since PF' is the diameter). It follows that $DE \perp EF$.

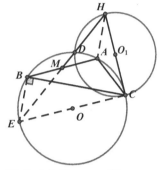

3.9 Let CE be a diameter of $\odot O$. Now $BE \perp BC$ and $AH \perp BC$, which implies $BE /\!/ AH$. Similarly, $AE /\!/ BH$ since both are perpendicular to AC. It follows that $AEBH$ is a parallelogram.
It suffices to show that H, M, E are collinear, in which case the diagonals of $AEBH$ bisect each other.
Notice that $\angle CDH = 90° = \angle CDE$. Hence, H, D, M, E are collinear.

3.10 Let the midpoints of AB, AC, BC be O, O_1, O_2 respectively. We have $CM = DM = EM$ (equal tangent segments). Draw $O_2 D' \perp O_1 D$ at D'. Notice that $DEO_2 D'$ is a rectangle and hence, $DE = D'O_2$.

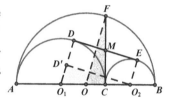

We denote $O_1 C = r_1$ and $O_2 C = r_2$. Notice that $AB = 2(r_1 + r_2)$ and hence, $OF = OB = r_1 + r_2$. It follows that $O_1 D' = O_1 D - O_2 E = r_1 - r_2$ and $OC = OB - BC = (r_1 + r_2) - 2r_2 = r_1 - r_2$, i.e., $O_1 D' = OC$. We also notice that $O_1 O_2 = O_1 C + O_2 C = \dfrac{1}{2} AC + \dfrac{1}{2} BC = \dfrac{1}{2} AB = r_1 + r_2 = OF$.

Now $\triangle OCF \cong \triangle O_1 D'O_2$ (H.L.), which implies $CF = D'O_2$. Hence, $CF = DE$. Since $CM = DM = EM$, we must have $FM = DM$.

Now $CDFE$ is a parallelogram since CF and DE bisect each other. Moreover, $CDFE$ is a rectangle since $CF = DE$.

Note: One may also show $CF = DE$ using Pythagoras' Theorem, i.e., $CF = \sqrt{OF^2 - OC^2}$ and $DE = D'O_2 = \sqrt{O_1 O_2^2 - O_1 D'^2}$.

3.11 Choose E on AC such that $AB = AE$. It is easy to see that $\triangle ABD \cong \triangle AED$, $BD = DE$ and AD is the perpendicular bisector of BE.

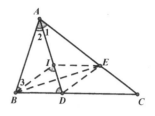

Now $\angle AED = \angle ABD = 2\angle C$, which implies $\angle CDE = \angle AED - \angle C = \angle C$.

Hence, $CE = DE = BD$. We claim that E is the circumcenter of $\triangle CDI$ and it suffices to show that $EI = DE$, or equivalently, $BI = BD$.

Notice that $\angle BDI = \angle 1 + \angle C = \dfrac{1}{2}\angle A + \angle C$ and $\angle BID = \angle 2 + \angle 3$ $= \dfrac{1}{2}\angle A + \angle 3$. Since $\angle 3 = \dfrac{1}{2}\angle ABC = \angle C$, we have $\angle BDI = \angle BID$, i.e., $BI = BD$. This completes the proof.

3.12 Let M' denote the midpoint of AQ. Since $\angle A = 90°$ and Q is the midpoint of BC, we have $AQ = BQ$ $= CQ$.

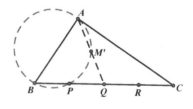

Hence, $AM' = \dfrac{1}{2}AQ = \dfrac{1}{2}BQ = BP$.

Since $QA = QB$, we have $PM' /\!/ AB$ by the Intercept Theorem. It follows that $ABPM'$ is an isosceles trapezium. Hence, A, B, P, M' are concyclic (Exercise 3.1), i.e., M' lies on the circumcircle of $\triangle ABP$. Similarly, M' also lies on the circumcircle of $\triangle ACQ$. We conclude that M and M' coincide and hence, A, M, Q are collinear.

3.13 It is easy to see that $AA'CB$ and $ABB'C$ are isosceles trapeziums. Hence, $\overset{\frown}{BC} = \overset{\frown}{AB'}$, which extend equal angles on the circumference, i.e., $\angle 1 = \angle 2$.

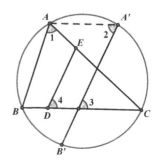

We also have $\angle 2 = \angle 3$ since $AA' /\!/ BC$.

Notice that A, B, D, E are concyclic (because AD, BE are heights) and hence, $\angle 1 = \angle 4$. It follows that $\angle 3 = \angle 4$ and hence, $A'B' \parallel DE$.

3.14 Refer to the left diagram below. Let AI extended intersect the circumcircle of $\triangle ABC$ at D. Example 3.4.2 gives $BD = CD = DI$, which implies that D is the circumcenter of $\triangle BIC$.

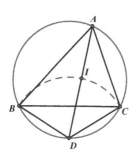
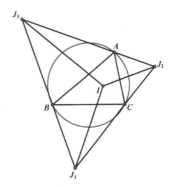

3.15 Refer to the right diagram above. By definition, A, I, J_1 are collinear. Since $AI \perp J_2 J_3$ (Exercise 1.5), we have $AJ_1 \perp J_2 J_3$. Similarly, $BJ_2 \perp J_1 J_3$ and $CJ_3 \perp J_1 J_2$.

Now A, B, C are the feet of altitudes of $\triangle J_1 J_2 J_3$ whose orthocenter is I. It follows that the midpoints of $IJ_1, IJ_2, IJ_3, JJ_1, JJ_2, JJ_3$ lie on the nine-point circle of $\triangle J_1 J_2 J_3$.

3.16 Draw $YC' \perp AB$ at C'. Since P, Q, C' are the feet of the perpendiculars from Y to the sides of $\triangle ABX$, we must have P, Q, C' collinear (Simson's Line). Similarly, S, R, C' are also collinear. It follows that PQ and SR intersect at C', i.e., C and C' coincide.

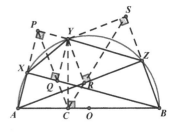

Since $\angle APY = \angle ARY = \angle ACY = 90°$, we have A, P, Y, R are concyclic and A, P, Y, C are concyclic. It follows that A, P, R, Y, C are concyclic.

Now $\angle PCS = \angle PAR = \frac{1}{2}\angle XOZ$, which completes the proof.

Note: One may also find the following alternative solution, which does not requires the fact that C lies on AB. Refer to the diagram on the right. It is easy to see that

$$\frac{1}{2}\angle XOZ = \angle XAZ = 180° - \angle XYZ.$$

Hence, it suffices to show that $\angle PCS + \angle XYZ = 180°$.

Consider the shaded quadrilateral $PCSY$, where the sum of the interior angles is $360°$, i.e.,
$$\angle PCS + \angle XYZ + \angle PYX + \angle CPY + \angle SYZ + \angle CSY = 360°. \quad (*)$$
Since AB is the diameter, we have $\angle AXB = 90° = \angle XPY = \angle XQY$.
Hence, $PXQY$ must be a rectangle. Now $\angle PYX = \angle CPY = 90° - \angle 1$.
Similarly, $SYRZ$ is also a rectangle and we have $\angle SYZ = \angle CSY = \angle 2$.
Now (*) gives $\angle PCS + \angle XYZ + 2 \times (90° - \angle 1 + \angle 2) = 360°$.
This leads to the conclusion $\angle PCS + \angle XYZ = 180°$ as one observes that $\angle 1 = \angle 2$ (Corollary 3.1.5).

3.17 Refer to the diagram below.

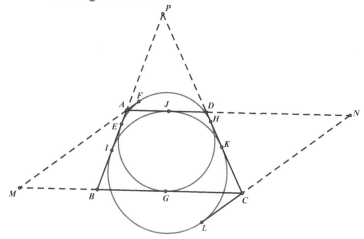

Let the lines AB, CD intersect at P. Let ℓ_1 touch Γ_2 at F and intersect the line BC at M. Let ℓ_2 touch Γ_1 at L and intersect the line AD at N. It suffices to show that $AMCN$ is a parallelogram.

We claim that $CM - AM = AN - CN$. (*)

By applying equal tangent segments repeatedly, we have

$CM - AM = (CG + MG) - (MF - AF) = CH + AE$, because $MG = MF$.

Similarly, $AN - CN = (AJ + NJ) - (NL - CL) = AI + CK$

Now $(CH + AE) - (AI + CK) = HK - EI$

$= (PK - PH) - (PI - PE) = 0$ since $PI = PK$ and $PE = PH$.

This completes the proof of (*).

Now it is easy to see that $AMCN$ is a parallelogram (Exercise 1.13).

Chapter 4

4.1 Draw the common tangent of $\odot O$ and $\odot P$ at C. By applying Theorem 3.2.10 repeatedly, we have $\angle A = \angle 1$ $= \angle 2 = \angle D$. Hence, $AB \, /\!/ \, DE$. Since A, B, D, E are concyclic, we have $\angle A = \angle E$ (angles in the same arc).

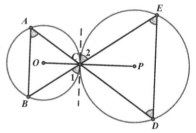

Now $\angle D = \angle E$ and since $AB \, /\!/ \, DE$, $\angle B = \angle E = \angle A$. It is easy to see that $ABDE$ is an isosceles trapezium and $\triangle ABC \sim \triangle EDC$.

We have $\dfrac{AC}{OC} = \dfrac{EC}{PC}$ since they are corresponding line segments. The conclusion follows.

Note: One may also see that $\dfrac{AC}{OC} = 2\sin \angle B$ and $\dfrac{EC}{PC} = 2\sin \angle D$ by Sine Rule. Since $\angle B = \angle D$, we must have $\dfrac{AC}{OC} = \dfrac{EC}{PC}$.

4.2 We are given that $ACDO$ is a parallelogram, i.e., $CD \, /\!/ \, AB$. Since D is the midpoint of BP, we must have $AC = CP$.

Connect BC. Since AB is the diameter, we have $BC \perp AP$. Since $AC = CP$, one sees that $\triangle ABP$ is a right angled isosceles triangle where $AB = BP$ (because $\triangle ABC \cong \triangle PBC$).

Draw $DE \perp AP$ at E. We have $DE /\!/ BC$. Since D is the midpoint of BP, we must have $CE = PE = \dfrac{1}{4} AP$ by the Intercept Theorem.

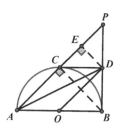

Let $OA = 1$. It is easy to see that $AD = \sqrt{AB^2 + BD^2} = \sqrt{2^2 + 1^2} = \sqrt{5}$ and $DE = \dfrac{1}{2} BC = \dfrac{\sqrt{2}}{2}$. Now $\sin \angle PAD = \dfrac{DE}{AD} = \dfrac{\sqrt{10}}{10}$.

4.3 Let AC intersect DE, BF at G, H respectively. Since $\angle B = 90°$ and $ABCD$ is cyclic, we must have $\angle ADC = 90°$. Since $DE \perp AC$, we have $AB^2 = AH \cdot AC$ (Example 2.3.1). Similarly, $AD^2 = AG \cdot AC$.

It follows that $\dfrac{AB^2}{AD^2} = \dfrac{AH}{AG}$.   (1)

Since $DE /\!/ EF$, we have $\dfrac{AE}{AB} = \dfrac{AG}{AH} = \dfrac{AD}{AF}$.

By (1), $\dfrac{AE}{AB} = \dfrac{AD^2}{AB^2} = \dfrac{AD}{AF}$.   (2)

Now $\dfrac{AE}{AF} = \dfrac{AE}{AB} \cdot \dfrac{AB}{AD} \cdot \dfrac{AD}{AF} = \left(\dfrac{AD}{AB}\right)^2 \cdot \dfrac{AB}{AD} \cdot \left(\dfrac{AD}{AB}\right)^2 = \dfrac{AD^3}{AB^3}$.

Note: There are many ways to derive the conclusion from (2). For example, one may write $AE = \dfrac{AD^2}{AB}$, $AF = \dfrac{AB^2}{AD}$ and hence obtain $\dfrac{AE}{AF}$.

4.4 Refer to the left diagram below. We claim that $\dfrac{OF}{OP} = \dfrac{OP}{OE}$. Since $OP = OM$, it suffices to show $\dfrac{OF}{OM} = \dfrac{OM}{OE}$.

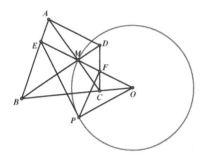 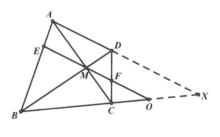

Refer to the right diagram above. Let AD extended intersect BC extended at X. We have $\dfrac{OF}{DX} = \dfrac{CO}{CX} = \dfrac{OM}{AX}$, i.e., $\dfrac{OF}{OM} = \dfrac{DX}{AX}$.

Similarly, $\dfrac{OM}{OE} = \dfrac{DX}{AX}$. We conclude that $\dfrac{OF}{OM} = \dfrac{OM}{OE}$, or equivalently, $\dfrac{OF}{OP} = \dfrac{OP}{OE}$. It follows that $\triangle OFP \sim \triangle OPE$ and hence, $\angle OPF = \angle OEP$.

4.5 Notice that $\angle ABC = \angle PAE = \angle E$, which implies $\triangle ABC \sim \triangle AEQ$. Hence, $\dfrac{QE}{AQ} = \dfrac{BC}{AC}$.

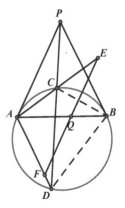

It follows that $QE = \dfrac{BC \cdot AQ}{AC}$.

Similarly, $\triangle ABD \sim \triangle AFQ$ and $QF = \dfrac{BD \cdot AQ}{AD}$.

Now it suffices to show that $\dfrac{BC}{AC} = \dfrac{BD}{AD}$, but this is by Example 4.1.1.

4.6 Connect OA. In the right angled triangle $\triangle AOP$, $PA^2 = PO \cdot PM$ (Example 2.3.1). We also have $PA^2 = PC \cdot PD$ by the Tangent Secant Theorem. Hence, $PC \cdot PD = PO \cdot PM$ and the conclusion follows.

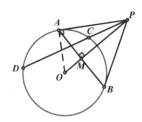

4.7 Let AC extended and BD extended intersect at P. One sees that AD bisects $\angle BAC$ (Corollary 3.3.3). Since AB is the diameter, we have $AD \perp BP$ and hence, $\triangle ABP$ is an isosceles triangle where $AB = AP$ (because $\triangle ABD \cong \triangle APD$). Now $BP = 2BD = 4\sqrt{5}$.

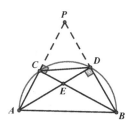

It is also easy to see that $\triangle BDE \sim \triangle BCP$ since both are right angled triangles. Hence, we have $BE \cdot BC = BD \cdot BP$. (One may also see this by the Tangent Secant Theorem because C, E, D, P are concyclic.) It follows that $BE \cdot (BE + 3) = 2\sqrt{5} \cdot 4\sqrt{5}$, solving which gives $BE = 5$. Hence, $BC = 8$ and by Pythagoras' Theorem,

$$CP = \sqrt{BP^2 - BC^2} = \sqrt{\left(4\sqrt{5}\right)^2 - 8^2} = 4.$$

Since $PA \cdot PC = PB \cdot PD$ by the Tangent Secant Theorem, we must have $PA \cdot 4 = 2\sqrt{5} \cdot 4\sqrt{5}$. We conclude that $AB = PA = 10$.

4.8 Since $\angle BCF = \angle BAC$ (Theorem 3.2.10), we have $\triangle BCF \sim \triangle CAF$. Hence, $\dfrac{AF}{CF} = \dfrac{CF}{BF} = \dfrac{AC}{BC}$ and we have $\dfrac{AF}{BF} = \dfrac{AF}{CF} \cdot \dfrac{CF}{BF} = \left(\dfrac{AC}{BC}\right)^2$.

Similarly, $\dfrac{BD}{CD} = \left(\dfrac{AB}{AC}\right)^2$ and $\dfrac{CE}{AE} = \left(\dfrac{BC}{AB}\right)^2$.

It follows that $\dfrac{AF}{BF}\cdot\dfrac{BD}{CD}\cdot\dfrac{CE}{AE}=\left(\dfrac{AC}{BC}\right)^2\cdot\left(\dfrac{AB}{AC}\right)^2\cdot\left(\dfrac{BC}{AB}\right)^2=1$ and by Menelaus' Theorem, D,E,F are collinear.

Note: This is an example of Menelaus' Theorem where the line does not intersect the triangle, but the division points are on the extension of the sides instead. In this case, writing down the equation mechanically could be easier than referring to the diagram, especially for beginners.

4.9 Refer to the diagram below. Connect CJ, BK. It is easy to see that $CJ = AJ$ and $BK = AK$. (*)
Notice that $\angle E = \angle ABF - \angle BPE$, while $\angle BPE = \angle A$ (Corollary 3.1.5) $= \angle ABK$ (since $AK = BK$).
Hence, $\angle E = \angle ABF - \angle ABK = \angle FBK$.
Similarly, $\angle F = \angle ECJ$.
It follows that $\triangle CEJ \sim \triangle FBK$.

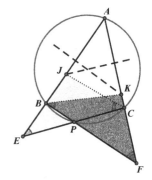

Now $\dfrac{CE}{BF}=\dfrac{JE}{BK}=\dfrac{CJ}{KF}$.

Hence, $\dfrac{CE^2}{BF^2}=\dfrac{JE}{BK}\cdot\dfrac{CJ}{KF}=\dfrac{AJ\cdot JE}{AK\cdot KF}$ by (*).

4.10 Connect BP, CP. Since $AP = AQ$, we have $\angle 2 = \angle 1 = \angle 3$ (angles in the same arc). Now $\triangle ADQ \sim \triangle AQC$, which implies $\dfrac{AQ}{AC}=\dfrac{AD}{AQ}$, or $AQ^2 = AC\cdot AD$.

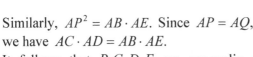

Similarly, $AP^2 = AB\cdot AE$. Since $AP = AQ$, we have $AC\cdot AD = AB\cdot AE$.
It follows that B, C, D, E are concyclic and hence, $\angle ABO = \angle ACO$. Notice that $\triangle OBE \sim \triangle OCD$. Since $OB = OC$, we have $\triangle OBE \cong \triangle OCD$ (A.A.S.) and hence, $OD = OE$. Now $\triangle OBC \sim \triangle ODE$ since both are isosceles triangles. Hence, $\angle OBC = \angle ODE$, which implies $DE \parallel BC$.

Note: Since B, C, D, E concyclic, one sees that $\triangle ABC$ and $\triangle ADE$ are isosceles triangles where $AB = AC$ and $AD = AE$.

4.11 Let the common tangents passing through A and B intersect at P, i.e., $PA \perp O_1O_2$ and $PB \perp O_2O_3$. Notice that $PA = PB$ (equal tangent segments). Refer to the diagram on the right. We claim that PC must be a common tangent of $\odot O_1$ and $\odot O_3$.

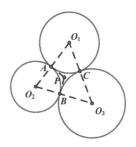

Suppose otherwise, say the line PC intersects $\odot O_1$ at C and D. By the Tangent Secant Theorem, $PB^2 = PA^2 = PC \cdot PD$. If the line PC touches $\odot O_3$ at C, we have $PB = PC$. This is only possible if PC is a common tangent of $\odot O_1$ and $\odot O_3$, i.e., C, D coincide. If the line PC intersects $\odot O_3$ at C and E, we have $PB^2 = PC \cdot PE$ and hence, D and E coincide. Since $\odot O_1$ and $\odot O_3$ are tangent to each other at C, this implies C, D, E coincide and hence, PC is a common tangent of $\odot O_1$ and $\odot O_3$.

In conclusion, PC is the radical axis of $\odot O_1$ and $\odot O_3$. Hence, ℓ_1, ℓ_2, ℓ_3 are concurrent at P.

4.12 Draw PA, PB tangent to $\odot O$ at A, B respectively. Notice that AP is the common tangent of $\odot O$ and $\odot O_1$ and hence, the powers of P with respect to $\odot O$ and $\odot O_1$ are the same. Similarly, the power of P with respect to $\odot O$ and $\odot O_2$ are the same. It follows that P lies on the radical axis of $\odot O_1$ and $\odot O_2$ (Theorem 4.3.6), i.e., P lies on the line CD.

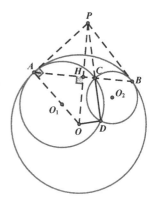

Let OP intersect AB at H. Clearly, OP is the perpendicular bisector of AB. Hence, we have $PA^2 = PH \cdot OP$ (Example 2.3.1).

Since $PA^2 = PC \cdot PD$, we must have $PC \cdot PD = PH \cdot PO$. This implies C, D, O, H are concyclic. It follows that $\angle ODC = \angle OHC = 90°$.

4.13 Let D be the point on the angle bisector of $\angle B$ such that $AD /\!/ BC$. Since $\angle B = 2\angle C$, we have $\angle 1 = \angle 2 = \angle C$.
Since $AD /\!/ BC$, we have $\angle 1 = \angle 3 = \angle C$. It follows that $ABCD$ is an isosceles trapezium which is obviously cyclic.

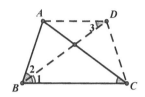

By Ptolemy's Theorem, $AC \cdot BD = AD \cdot BC + AB \cdot CD$.
Since $AC = BD$ and $CD = AB = AD$ (because $\angle 2 = \angle 3$), we have $AC^2 = AB \cdot BC + AB^2 = AB \cdot (AB + BC)$.

Note: Once we have $AB = AD = CD$, one may also show the conclusion by the area method. Refer to the diagram on the right.
Extend CB to E such that $BE = AB$.
It is easy to see that $AE = BD = AC$.

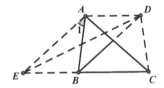

Hence, $[\triangle ACE] = \dfrac{1}{2} AC \cdot AE \sin \angle CAE = \dfrac{1}{2} AC^2 \sin \angle CAE$ and

$$[\triangle CDE] = \frac{1}{2} CD \cdot CE \sin \angle BCD = \frac{1}{2} AB(AB + BC) \sin \angle BCD.$$

Since $\angle CAE = \angle BAC + \angle 1$ and $\angle 1 = \angle AEB = \angle ACB$ by isosceles triangles, we have $\angle CAE = \angle BAC + \angle ACB = 180° - \angle ABC$. It follows that $\sin \angle CAE = \sin \angle ABC = \sin \angle BCD$.

Since $[\triangle ACE] = [\triangle CDE]$, we must have $AC^2 = AB \cdot (AB + BC)$.

One may notice that applying Ptolemy's Theorem is much faster.

4.14 Refer to the left diagram below. Let $\odot O_1$ touch ℓ_1, ℓ_2 at P, Q respectively. It is easy to see that PQ is a diameter of $\odot O_1$ and $O_2 B \,/\!/\, O_3 D \,/\!/\, PQ$. Let I be the circumcenter of $\triangle ACE$.

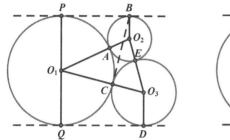 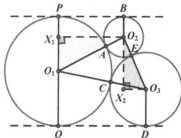

Example 4.3.3 states that the circumcircle of $\triangle ACE$ is the incircle of $\triangle O_1 O_2 O_3$. In particular, $IC \perp O_1 O_3$. We claim that $BC \perp O_1 O_3$.

Let the radii of $\odot O_1$, $\odot O_2$ and $\odot O_3$ be r_1, r_2, r_3 respectively. Refer to the right diagram above. Draw $O_2 X_1 \perp PQ$ at X_1 and $O_3 X_2 \perp BO_2$ at X_2. Pythagoras' Theorem gives us $BP = O_2 X_1 = \sqrt{O_1 O_2{}^2 - O_1 X_1{}^2}$
$= \sqrt{(r_1 + r_2)^2 - (r_1 - r_2)^2} = 2\sqrt{r_1 r_2}$. Similarly, $DQ = 2\sqrt{r_1 r_3}$.

In the right angled triangle $\triangle O_2 O_3 X_2$, $O_2 O_3{}^2 = O_2 X_2{}^2 + O_3 X_2{}^2$. Observe that $O_2 O_3 = r_2 + r_3$, $O_2 X_2 = PQ - BO_2 - DO_3 = 2r_1 - r_2 - r_3$ and $O_3 X_2 = DQ - BP = 2\left(\sqrt{r_1 r_3} - \sqrt{r_1 r_2}\right)$. Hence, $(r_2 + r_3)^2 = (2r_1 - r_2 - r_3)^2$
$+ 4\left(\sqrt{r_1 r_3} - \sqrt{r_1 r_2}\right)^2$, the simplification of which gives $r_1 = 2\sqrt{r_2 r_3}$. (*)

Now $BO_3{}^2 - CO_3{}^2 = BX_2{}^2 + O_3 X_2{}^2 - r_3{}^2$

$= (2r_1 - r_3)^2 + 4\left(\sqrt{r_1 r_3} - \sqrt{r_1 r_2}\right)^2 - r_3{}^2 = 4r_1{}^2 + 4r_1 r_2 - 8r_1\sqrt{r_2 r_3}$

$= 4r_1 r_2$ by (*).

This implies $BO_3{}^2 - CO_3{}^2 = BP^2 = BO_1{}^2 - O_1 P^2 = BO_1{}^2 - O_1 C^2$.

It follows that $BC \perp O_1O_3$ by Theorem 2.1.9. Since $IC \perp O_1O_3$, B, I, C are collinear. Similarly, A, I, D are collinear. The conclusion follows.

4.15 Since PQ is tangent to the circumcircle of $\triangle MNL$, i.e., PQ touches the circle exactly once, the point of tangency must be L. It is easy to see $AB \,/\!/\, ML$ and $AC \,/\!/\, NL$ because M, N, L are midpoints.

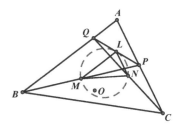

Clearly, $\angle BAC = \angle MLN$. Notice that $\angle APQ = \angle PLN = \angle LMN$ and similarly, $\angle LNM = \angle AQP$. It follows that $\triangle LMN \sim \triangle APQ$ and hence, $\dfrac{LM}{LN} = \dfrac{AP}{AQ}$. Since $PL = QL$, we must have $\dfrac{LM}{BQ} = \dfrac{PL}{PQ} = \dfrac{QL}{PQ} = \dfrac{LN}{CP}$, i.e., $\dfrac{LM}{LN} = \dfrac{BQ}{CP}$.

Now we have $\dfrac{AP}{AQ} = \dfrac{BQ}{CP}$, or equivalently, $AP \cdot CP = AQ \cdot BQ$.

Let $\odot O$ denote the circumcircle of $\triangle ABC$. Consider the power of point P with respect to $\odot O$. We see that $OP^2 - r^2 = -AP \cdot CP$ where r is the circumradius of $\triangle ABC$. (Refer to Definition 4.3.5.) Similarly, we have $OQ^2 - r^2 = -AQ \cdot BQ$.

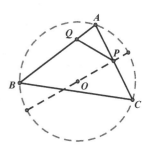

Since $AP \cdot CP = AQ \cdot BQ$, we obtain $OP^2 = OQ^2$, i.e., $OP = OQ$.

Note: It is easier to write down the expression for the power of a point *without* referring to the diagram. Indeed, those irrelevant lines in the diagram could be very confusing.

Chapter 5

5.1 Let BD, CE intersect at H, the orthocenter of $\triangle ABC$. By Example 3.4.1, $\angle BAO = \angle CAH = 90° - \angle C$. It is easy to see that B, C, D, E are concyclic. Hence, $\angle C = \angle AED$. It follows that $\angle BAO + \angle AED = 90°$, i.e., $AO \perp DE$.

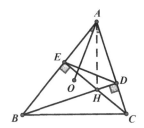

5.2 Connect BP, OP. It is easy to see that $\triangle PAB$ is a right angled isosceles triangle where $\angle APB = 90°$ and $\angle PAB = \angle PBA = 45°$.

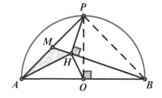

In the right angled triangle $\triangle PBM$, we have $\begin{cases} PM^2 = MH \cdot BM & (1) \\ PH^2 = MH \cdot BH & (2) \end{cases}$

Since $AM = PM$, (1) gives $AM^2 = MH \cdot BM$, or $\dfrac{AM}{MH} = \dfrac{BM}{AM}$.

It follows that $\triangle AHM \sim \triangle BAM$. Hence, $\angle MHA = \angle MAB = 45°$ and $\angle MAH = \angle MBA$.
On the other hand, since $\angle BHP = \angle BOP = 90°$, B, O, H, P are concyclic, which implies $\angle BHO = \angle BPO = 45° = \angle MHA$.

Now we have $\triangle AHM \sim \triangle BHO$ and hence, $\dfrac{MH}{AH} = \dfrac{OH}{BH}$, or

$AH \cdot OH = MH \cdot BH = PH^2$ by (2). This completes the proof.

Note: One sees the conclusion is essentially a property of the right angled isosceles triangle $\triangle PAB$ where only medians and perpendicular lines are introduced. Hence, one may solve it by brute force, i.e., calculating PH, AH and OH.

Let $AO = BO = OP = 1$. We have $PA = PB = \sqrt{2}$. Notice that $\triangle PBM$ is a right angled triangle whose sides are of the ratio $1 : 2 : \sqrt{5}$ (Pythagoras' Theorem). Hence, $PH = \dfrac{PM \cdot PB}{BM} = \dfrac{\sqrt{2}}{2} \cdot \dfrac{1 \times 2}{\sqrt{5}} = \dfrac{\sqrt{10}}{5}$.

Notice that $\dfrac{MH}{PH} = \dfrac{1}{2}$, i.e., $MH = \dfrac{\sqrt{10}}{10}$. Since $AM = \dfrac{\sqrt{2}}{2}$, we have $AH^2 = AM^2 + MH^2 - 2AM \cdot MH \cos\angle AMB$ by Cosine Rule, where $\cos\angle AMB = -\cos\angle PMB = -\dfrac{1}{\sqrt{5}}$.

Hence, $AH^2 = \left(\dfrac{\sqrt{2}}{2}\right)^2 + \left(\dfrac{\sqrt{10}}{10}\right)^2 + 2 \cdot \dfrac{1}{\sqrt{5}} \cdot \dfrac{\sqrt{2}}{2} \cdot \dfrac{\sqrt{10}}{10} = \dfrac{4}{5}$.

Notice that OH is a median of $\triangle ABH$, where $BH = 2PH = \dfrac{2\sqrt{10}}{5}$.

Hence, $OH^2 = \dfrac{1}{2} AH^2 + \dfrac{1}{2} BH^2 - \dfrac{1}{4} AB^2 = \dfrac{1}{2} \cdot \dfrac{4}{5} + \dfrac{1}{2} \left(\dfrac{2\sqrt{10}}{5}\right)^2 - \dfrac{1}{4} \times 2^2$
$= \dfrac{1}{5}$. It follows that $AH \cdot OH = \dfrac{2}{\sqrt{5}} \cdot \dfrac{1}{\sqrt{5}} = \dfrac{2}{5} = PH^2$.

5.3 Recall that $\angle AIB = 90° + \dfrac{1}{2}\angle C$. Since $CD = CE$, we have $\angle CDE = \dfrac{1}{2}(180° - \angle C)$ and hence, $\angle BDP = 90° + \dfrac{1}{2}\angle C = \angle AIB$. Since $\angle ABI = \angle PBC$, we must have $\triangle ABI \sim \triangle PBD$.

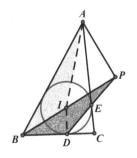

Hence, $\dfrac{AB}{BP} = \dfrac{IB}{BD}$ and we conclude that $\triangle ABP \sim \triangle IBD$.

It follows that $\angle APB = \angle IDB = 90°$.

Note: One may also show that $\angle API = \angle AEI = 90°$. In fact, once we obtain $\angle AIP = 90° - \frac{1}{2}\angle C = \angle CED = \angle AEP$, one immediately sees that A, I, E, P are concyclic and hence the conclusion.

5.4 Notice that in the right angled triangle $\triangle BCR$, $BD = CD = DR = \frac{1}{2}BC$.

Similarly, $DQ = \frac{1}{2}BC = DR$.

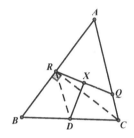

It follows that DX is the perpendicular bisector of QR. Similarly, EY, FZ are the perpendicular bisectors of PR, PQ respectively. Hence, DX, EY, FZ are concurrent at the circumcenter of $\triangle PQR$.

5.5 Let AA' be a diameter of $\odot O$. By Example 3.4.4, $A'BHC$ is a parallelogram and N is the midpoint of $A'H$. Notice that A', H, Q are collinear.
Let CC' be a diameter of $\odot O$. Similarly, we have C', H, P collinear and M is the midpoint of $C'H$.

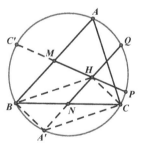

By the Intersecting Chords Theorem, $A'H \cdot HQ = C'H \cdot HP$, which implies $MH \cdot HQ = NH \cdot HP$ since $A'H = 2MH$ and $C'H = 2NH$. It follows that M, N, P, Q are concyclic.

Note: One may also notice that $MN // A'C'$ (Midpoint Theorem) and hence, A', C', Q, P concyclic implies M, N, Q, P are concyclic (Example 3.1.7).

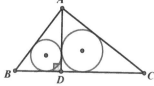

5.6 Notice $\triangle ABC$ is a right angled triangle. We have

$$r = \frac{1}{2}(AB + AC - BC).$$

Similarly, $r_1 = \frac{1}{2}(AD + BD - AB)$ and $r_2 = \frac{1}{2}(AD + CD - AC)$.

It follows that $r + r_1 + r_2 = AD + \frac{1}{2}(BD + CD - BC) = AD$.

5.7 Let C' be the reflection of C about BD. Draw $C'H \perp BC$ at H. One sees that $CP + PQ = C'P + PQ \geq C'Q \geq C'H$. Refer to the left diagram below.

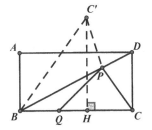 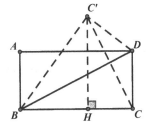

It is easy to see that $CC' \perp BD$ and hence, $\triangle CC'H \sim \triangle DBC$.

We have $\dfrac{C'H}{CC'} = \dfrac{BC}{BD} = \dfrac{2}{\sqrt{5}}$. Refer to the right diagram above.

Notice that $BC \cdot CD = 2[\triangle BCD] = BD \cdot \frac{1}{2}CC'$. Hence, $CC' = 2\dfrac{BC \cdot CD}{BD}$

$= \dfrac{4}{\sqrt{5}}$. It follows that $C'H = \dfrac{2}{\sqrt{5}}CC' = \dfrac{8}{5}$.

In conclusion, the smallest value of $CP + PQ$ is $\dfrac{8}{5}$, where $C'Q \perp BC$ at Q and $C'Q$ intersects BD at P.

5.8 Extend HM to D such that $HM = DM$. Clearly, $BDCH$ is a parallelogram where $\angle DBH = 180° - \angle BHC = \angle A$ because H is the orthocenter. (1)

Let CE be a height. Notice that $\angle APQ = 90° - \angle PHE = 90° - \angle CHQ = \angle CHM = \angle BDH$. (2)

(1) and (2) imply that $\triangle APQ \sim \triangle BDH$. Since $\angle HBM = 90° - \angle C = \angle CAH$, we conclude that H and M are corresponding points, i.e., $\dfrac{PH}{QH} = \dfrac{DM}{HM} = 1$. This completes the proof.

Note:

(1) One may also see that $\triangle APH \sim \triangle CHM$ and $\triangle AQH \sim \triangle BHM$.

Now $\dfrac{PH}{AH} = \dfrac{MH}{CM} = \dfrac{MH}{BM} = \dfrac{QH}{AH}$ leads to the conclusion.

(2) Notice that the diagram of this question is similar to Exercise 5.4. However, the techniques used are entirely different. In fact, this question is more closely related to Example 5.2.6. Can you see that $\triangle APQ$ and $\triangle BCH$ are related in a similar way as $\triangle ABC$ and $\triangle EFO$ in that example?

5.9 Refer to the diagram on the right. Since PB is tangent to $\odot O$, we have $\angle PBD = \angle BAD$. Since B, C, D, P are concyclic, we have $\angle ACD = \angle BPD$. It follows that $\triangle ACD \sim \triangle BPD$.

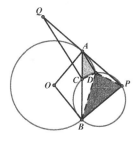

Hence, $\dfrac{BP}{AC} = \dfrac{BD}{AD}$. (1)

Since OA is the perpendicular bisector of PQ, we have $AQ = AP = BP$.

By (1), $\dfrac{AQ}{AC} = \dfrac{BD}{AD}$. Refer to the diagram on the right. Notice that $\angle BAQ = \angle ADB$

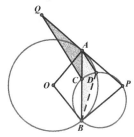

(because AQ is tangent to $\odot O$). We conclude that $\triangle ACQ \sim \triangle DAB$. Now $\angle DAB = \angle ACQ$ and we must have $AD \mathbin{/\mkern-5mu/} CQ$.

5.10 Let H be the orthocenter of $\triangle ABC$ and AD extended intersect the circumcircle of $\triangle ABC$ at F. It is easy to see that $DH = DF$ and $\angle BFH = \angle BHF = \angle AHQ$.

On the other hand, we have $\angle QAH = \angle PBF$ (angles in the same arc). It follows that $\triangle AHQ \sim \triangle BFP$. Refer to the left diagram below. Since $\angle CAF = \angle CBF$, D and E are corresponding points and $\dfrac{EQ}{EH} = \dfrac{PD}{DF}$. (1)

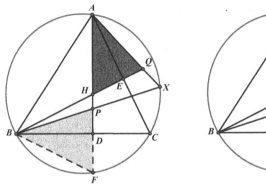 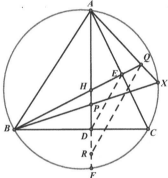

Draw $QR \mathbin{/\mkern-5mu/} DE$, intersecting AF at R. Refer to the right diagram above. Now $\dfrac{EQ}{EH} = \dfrac{DR}{DH}$. (2)

Since $DH = DF$, (1) and (2) imply $PD = DR$. By the Intercept Theorem, DE must pass through the midpoint of PQ.

Note: This is not an easy problem. Recognizing similar triangles $\triangle AHQ$, $\triangle BFP$ and $\triangle BHR$ is the key step, even though $\triangle BHR$ is not drawn explicitly in the proof. Indeed, one may see this problem as an extension of Example 3.4.3.

5.11 Since $\angle A = 90°$, one sees that BC is a diameter of Γ. Let DE intersect PH at G. Let M be the midpoint of PE. Let PH extended intersect Γ at Q.

We have $\angle PED = \angle D = \angle Q$ (angles in the same arc). It follows that $\triangle PGE \sim \triangle PEQ$.

Clearly, H is the midpoint of PQ. Since M is the midpoint of PE, we have $\triangle PGM \sim \triangle PEH$. Now $\dfrac{PG}{PM} = \dfrac{PE}{PH} = 1$, i.e., $PG = PM$.

It follows that $MG \mathbin{/\mkern-5mu/} EH$, and hence, G is the midpoint of PH.

Note:
(1) We introduced the midpoint M of PE instead of explicitly drawing a perpendicular from the center of Γ to the chord PE. Nevertheless, the motivation still comes from this technique.
(2) One may also connect MH and see that $MH \mathbin{/\mkern-5mu/} EQ$. Since $PE = PH$, we have $\triangle PMH \sim \triangle PEQ$ and $\triangle PMH \cong \triangle PGE$ (A.A.S.). Now $EHGM$ is an isosceles trapezium and the conclusion follows.
(3) There is an alternative solution by the Intersecting Chords Theorem. Refer to the diagram on the right. Draw the circumcircle of $\triangle DEH$.

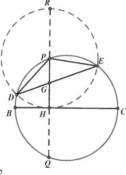

Since $PD = PE = PH$, P is the circumcenter of $\triangle DEH$. It is easy to see that $PH = HQ = PR$. Let $PH = r$, $PG = a$ and $GH = r - a$.

By the Intersecting Chords Theorem,
$PG \cdot GQ = DG \cdot GE = GH \cdot GR$.
Hence, $a(2r - a) = (r - a)(r + a)$, i.e., $2ra = r^2$.

It follows that $a = \dfrac{r}{2}$ and G is the midpoint of PH.

5.12 By considering the isosceles triangle $\triangle CDQ$, one sees that

$\angle CDQ = \angle DCQ = \angle CAD$ and hence, $\angle CQD = 180° - 2\angle CAD$.

Draw $\odot Q$ with radius CQ. Since AB is a diameter, $PD \perp AD$ and hence, $\angle CPD = 90° - \angle CAD = \dfrac{1}{2}\angle CQD$. It follows that P lies on $\odot Q$.

Hence, $PQ = CQ$ and we have $\angle CPQ = \angle PCQ$.
Since $AC \perp BC$, we have $\angle PCQ = 90° - \angle BCQ = 90° - \angle BAC$. This implies that $\angle CPQ + \angle BAC = 90°$, i.e., $AB \perp PQ$.
Since $PD \perp AE$, B must be the orthocenter of $\triangle AEP$. Now $BE \perp AC$, which implies B, C, E are collinear.

Note: One sees from the proof that $PQ = CQ = DQ = EQ$, i.e., Q is the midpoint of PE. Indeed, one may find an alternative solution as follows. Suppose the lines AD and BC intersect at E' where Q' is the midpoint of PE'. Connect $Q'C$ and $Q'D$. Refer to the diagram below.

Since $\triangle CPE'$ and $\triangle DE'P$ are right angled triangles sharing a common hypotenuse, we have $Q'C = \dfrac{1}{2}PE' = Q'D$.

Now it suffices to show that $Q'C$ and $Q'D$ are tangent to $\odot O$, or equivalently, $OC \perp Q'C$.

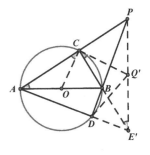

Since $AC \perp BC$, it suffices to show $\angle ACO = \angle BCQ'$. Notice that B is the orthocenter of $\triangle APE'$, i.e., $AB \perp PE'$. We have $\angle BCQ' = \angle CE'Q' = 90° - \angle APE' = \angle CAO = \angle ACO$. This implies $OC \perp Q'C$ and similarly, $OD \perp Q'D$. Hence, Q' coincides with Q and we conclude that E' coincides with E. This completes the proof because E' lies on BC.

5.13 Since I is the incenter of $\triangle ABC$, we know that $DC = DI$ and $EC = EI$. Hence, DE is the

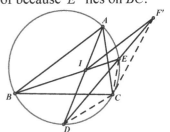

perpendicular bisector of CI. Let the line DE and ℓ_1 intersect at F'. It is easy to see that $\triangle EF'I \cong \triangle EF'C$ (S.S.S.) and hence, $\angle ECF' = \angle EIF'$.

Since $IF' /\!/ AB$, we have $\angle EIF' = \angle ABI = \angle CBI$. It follows that $\angle CBI = \angle ECF'$, which implies CF' is tangent to $\odot O$ at C. In conclusion, F and F' coincide. This completes the proof.

5.14 Draw $DF \perp BC$ at F and $EG \perp BC$ at G. Since BD bisects $\angle ABC$ and $\angle BAD = \angle BFD = 90°$, we must have $\triangle ABD \cong \triangle FBD$ (A.A.S.). Hence, $AB = BF$ and $AD = DF$.

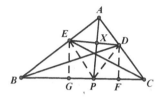

Similarly, $AC = CG$ and $AE = EG$. We claim that $FP = GP$.

Let AP and DE intersect at X. By applying Pythagoras' Theorem repeatedly, we have $FP^2 = PD^2 - DF^2 = PD^2 - AD^2$

$$= \left(PX^2 + DX^2\right) - \left(DX^2 + AX^2\right) = PX^2 - AX^2. \quad (1)$$

Similarly, $GP^2 = PE^2 - EG^2 = PE^2 - AE^2$

$$= \left(PX^2 + EX^2\right) - \left(EX^2 + AX^2\right) = PX^2 - AX^2. \quad (2)$$

(1) and (2) imply that $FP = GP$.
Now $AB - AC = BF - CG = \left(BP + FP\right) - \left(CP + GP\right) = BP - CP$.

5.15 Since $\angle ACD = \angle BCP$, one sees that $\angle PCD = \angle ACB = \angle CAD$ because $AD /\!/ BC$. Since $AB /\!/ CD$, we have $\angle PDC = \angle ABD = \angle AED$ (angles in the same arc). It follows that $\triangle ADE \sim \triangle CPD$ and hence, $\dfrac{PD}{CD} = \dfrac{DE}{AE}$.

Since $CD = AB$, we have $\dfrac{PD}{AB} = \dfrac{DE}{AE}$.

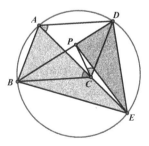

Clearly, $\angle BAE = \angle PDE$ (angles in the same arc).
We conclude that $\triangle ABE \sim \triangle DPE$. It follows that $\angle AEB = \angle DEP$, or equivalently, $\angle AED = \angle BEP$.

Note: We applied the technique of similar triangles sharing a common vertex to show $\triangle ABE \sim \triangle DPE$, where the "common" vertex is not E, but A and D: although these are different points, the corresponding angles at the vertices are the same due to the concyclicity.

5.16 Let A' be the point symmetric to A about O. Let the lines CA' and BD intersect at E'. Since $A'A$ is a diameter of $\odot O$, we must have $AC \perp CE'$. Now it suffices to show that P, O, E' are collinear.

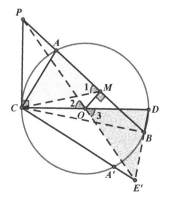

Connect BC. Notice that $\angle ABC = \angle A'CD$ because they correspond to equal arcs (i.e., $\overparen{AC} = \overparen{A'D}$ by symmetry).

Clearly, $\angle CAB = \angle CDB$. We must have $\triangle ACB \sim \triangle DE'C$. Draw $OM \perp AB$ at M.
We see that M is the midpoint of AB. Connect CM, OE'. Since O is the midpoint of CD, we have $\triangle ACM \sim \triangle DE'O$.
It follows that $\angle 1 = \angle 3$. Connect OP. Since $\angle OMP = \angle OCP = 90°$, P, C, O, M are concyclic and we have $\angle 1 = \angle 2$. Now $\angle 2 = \angle 3$, which implies P, O, E' are collinear.

Note: One may also re-write the proof in a direct approach: upon drawing $OM \perp AB$ at M, we show that $\triangle ACM \sim \triangle DEO$ and hence, $\triangle ACB \sim \triangle DEC$. Now the angles extended by $\overparen{AC}$ and $\overparen{A'D}$ on $\odot O$ are the same (where CE intersects $\odot O$ at A'). Hence, A and A' are symmetric about O. We conclude that AA' is a diameter of $\odot O$ and hence, $AC \perp CE$.

5.17 Notice that $\angle DAE = \angle BAD - \angle BAE$ where $\angle BAD = 180° - \angle C$ and $\angle BAE = \angle BFC$ (because A,B,C,D and A,B,F,E are concyclic). It follows that $\angle DAE = 180° - \angle C - \angle BFC = \angle CBF$. (1)
Refer to the left diagram below.

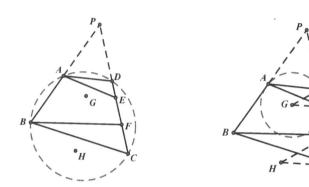

Since G is the circumcenter of $\triangle ADE$, we have $\angle DGE = 2\angle DAE$. Similarly, $\angle CHF = 2\angle CBF$. Refer to the right diagram above.
Now (1) implies $\angle DGE = \angle CHF$ and hence, the isosceles triangles $\triangle DEG$ and $\triangle CFH$ are similar. In particular, $DG \,/\!/\, FH$.

Consider $\triangle APE$.

Sine Rule gives $\dfrac{PD}{AP} = \dfrac{\sin \angle PAD}{\sin \angle PDA}$ and $\dfrac{DE}{AE} = \dfrac{\sin \angle DAE}{\sin \angle ADE}$.

Since $\sin \angle PDA = \sin \angle ADE$, we have $\dfrac{PD}{DE} = \dfrac{AP \sin \angle PAD}{AE \sin \angle DAE}$.

By Sine Rule, $\dfrac{AP}{AE} = \dfrac{\sin \angle AEP}{\sin \angle P}$. Now $\dfrac{PD}{DE} = \dfrac{\sin \angle AEP \cdot \sin \angle PAD}{\sin \angle P \cdot \sin \angle DAE}$. (2)

A similar argument applies in $\triangle BPC$, which gives

$$\dfrac{PF}{CF} = \dfrac{\sin \angle BCP \cdot \sin \angle PBF}{\sin \angle P \cdot \sin \angle CBF}. \quad (3)$$

Notice that $\angle AEP = \angle PBF$ and $\angle BCP = \angle PAD$ by concyclicity. We also have $\angle DAE = \angle CBF$ by (1).

Now (2) and (3) implies that $\dfrac{PD}{DE} = \dfrac{PF}{CF}$. (4)

Since $\triangle DEG \sim \triangle CFH$, we have $\dfrac{DG}{DE} = \dfrac{FH}{CF}$. (5)

(4) and (5) give $\dfrac{PD}{DG} = \dfrac{PF}{FH}$ and hence, $\triangle PDG \sim \triangle PFH$.

Now $\angle DPG = \angle FPH$ and it follows that P, G, H are collinear.

Chapter 6

6.1 Draw $AH \perp BC$ at H. Let OA and DE intersect at F. It is well-known (Example 3.4.1) that $\angle OAD = \angle CAH$.
If B, C, E, D are concyclic, we must have $\angle ADE = \angle C$.
Now $\angle ADE + \angle OAD = \angle C + \angle CAH = 90°$, i.e., $DE \perp OA$.

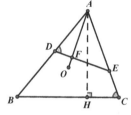

On the other hand, if $DE \perp OA$, we have $\angle ADE = 90° - \angle OAD$ $= 90° - \angle CAH = \angle C$ and hence, B, C, E, D are concyclic.

Note: Exercise 5.1 is a special case of this problem.

6.2 Let AM and BP intersect at D. It is easy to see that $\dfrac{AB}{BM} = 2$ and hence, in the right angled triangle $\triangle ABM$, $\dfrac{AD}{DM} = \left(\dfrac{AB}{BM}\right)^2 = 4$.

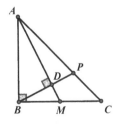

Apply Menelaus' Theorem when $\triangle ACM$ is intercepted by the line BP. We have $\dfrac{AD}{DM} \cdot \dfrac{BM}{BC} \cdot \dfrac{CP}{AP} = 1$. It follows that $\dfrac{CP}{AP} = \dfrac{1}{2}$.

Now $AP = 2\sqrt{2}$ and $AC = 3\sqrt{2}$. Hence, $AB = 3$.

Note: One may also draw $PE \perp BC$ at E. It is easy to see that $\triangle ABM \sim \triangle BEP$ and we have

$$PE = CE = \frac{1}{2}BE, \text{ i.e., } CE = \frac{1}{3}BC.$$

Since $CP = \sqrt{2}$, we have $CE = 1$ and $AB = 3$.

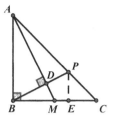

6.3 Refer to the diagram on the right. It is easy to see that $\triangle ABG \sim \triangle EDG$.

Hence, $\dfrac{AG}{EG} = \dfrac{BG}{DG} = \dfrac{AB}{DE} = \dfrac{1}{4}.$ (1)

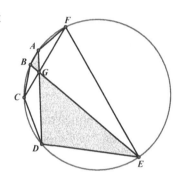

Similarly, $\triangle BCG \sim \triangle FEG$, which implies

$$\frac{BG}{FG} = \frac{CG}{EG} = \frac{BC}{EF} = \frac{2}{5}. \quad (2)$$

By (1) and (2), $AG = \dfrac{1}{4}EG$ and $CG = \dfrac{2}{5}EG = \dfrac{8}{5}AG$.

Since $\triangle CDG \sim \triangle AFG$, $\dfrac{AF}{CD} = \dfrac{AG}{CG} = \dfrac{5}{8}$. It follows that $AF = \dfrac{5}{8}CD = \dfrac{15}{8}$.

6.4 In the right angled triangle $\triangle BCD$, $DF = \dfrac{1}{2}BC$ because D is the midpoint of BC. Since $DE = DF$,

$DE = \dfrac{1}{2}BC$, which implies $\angle BEC = 90°$.

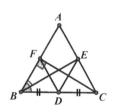

Since BE bisects $\angle ABC$, we have $\triangle ABE \cong \triangle CBE$ (A.A.S.), which implies $AB = BC$ and E is the midpoint of AC.

Hence, in the right angled triangle $\triangle ACF$, $EF = \dfrac{1}{2}AC$.

Since $DE = EF$, we have $AB = BC = 2DE = 2EF = AC$. This completes the proof.

6.5 Refer to the diagram on the right. Let AD, BE intersect at H, the orthocenter of $\triangle ABC$. It is easy to see that A, B, D, E are concyclic.

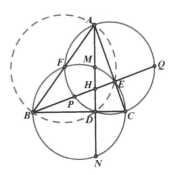

Apply the Intersecting Chords Theorem repeatedly: $PH \cdot QH = AH \cdot DH$
$= BH \cdot EH = MH \cdot NH$.
It follows that M, P, N, Q are concyclic.

Note:
(1) One may notice that M, P, N, Q lie on a circle centered at C. In fact, since BC is the perpendicular bisector of MN, we have $CM = CN$ and similarly, $CP = CQ$. We claim that $CM = CP$.
Since BC is a diameter, $\angle BMC = 90°$ and hence, $CM^2 = CD \cdot BC$ (Example 2.3.1). Similarly, we have $CP^2 = CE \cdot AC$. By the Tangent Secant Theorem, $CD \cdot BC = CE \cdot AC$. Hence, $CM = CP$ and M, P, N, Q lie on the circle centered at C with the radius CM.

(2) One may also draw $CF \perp AB$. at F. Since AC, BC are diameters, F lies on the circumcircles of $\triangle ACD$ and $\triangle BCE$. By the Intersecting Chords Theorem, $PH \cdot QH = CH \cdot FH = MH \cdot NH$ and hence the conclusion.

6.6 Refer to the diagram below. Apply Sine Rule to $\triangle ADX$ and $\triangle CDY$.

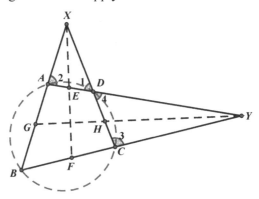

We have $\dfrac{AX}{DX} = \dfrac{\sin \angle 1}{\sin \angle 2}$ and $\dfrac{CY}{DY} = \dfrac{\sin \angle 4}{\sin \angle 3}$. Notice that $\angle 1 = \angle 4$ and

$\angle 2 + \angle 3 = 180°$, i.e., $\sin \angle 2 = \sin \angle 3$. It follows that $\dfrac{AX}{DX} = \dfrac{CY}{DY}$.

We have $\dfrac{AE}{DE} = \dfrac{AX}{DX}$ and $\dfrac{CH}{DH} = \dfrac{CY}{DY}$ by the Angle Bisector Theorem.

Hence, $\dfrac{AE}{DE} = \dfrac{CH}{DH}$, which implies $AC \mathbin{/\!/} EH$.

Similarly, $FG \mathbin{/\!/} AC$. (**Hint**: $\dfrac{AG}{BG} = \dfrac{AY}{BY} = \dfrac{CX}{BX} = \dfrac{CF}{BF}$.)

We conclude that $EH \mathbin{/\!/} FG$. Similarly $EG \mathbin{/\!/} BD \mathbin{/\!/} FH$. It follows that $EGFH$ is a parallelogram.

6.7 Let the circumcircle of $\triangle DEF$ intersect BC at D, D', AC at E, E' and AB at F, F'. Notice that the midpoints of BC and DD' coincide, i.e., D and D' are symmetric about the midpoint of BC.

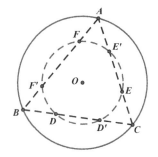

Let $\dfrac{BD}{CD} = \dfrac{CE}{AE} = \dfrac{AF}{BF} = k$.

We have $BD = \dfrac{k}{k+1} BC$ and $BD' = CD = \dfrac{1}{k+1} BC$.

Similarly, $BF = \dfrac{1}{k+1} AB$ and $BF' = AF = \dfrac{k}{k+1} AB$.

We have $BD \cdot BD' = \dfrac{k}{(k+1)^2} BC^2$ and $BF \cdot BF' = \dfrac{k}{(k+1)^2} AB^2$.

Since $BD \cdot BD' = BF \cdot BF'$ (Tangent Secant Theorem), we must have $AB^2 = BC^2$, i.e., $AB = AC$.
Similarly, $BC = AC$ and the conclusion follows.

6.8 Refer to the left diagram below. Since $OD = OE$ and OP bisects $\angle DOE$, we must have $PD = PE$ (because $\triangle OPD \cong \triangle OPE$). Clearly, $AD \neq AE$ because $\triangle ABC$ is non-isosceles. Since AP bisects $\angle A$, we must have A, D, P, E concyclic (Example 3.1.11). It follows that $\angle AED = \angle APD$.

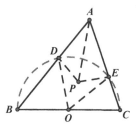 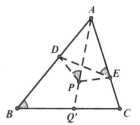

Refer to the right diagram above. Let AP extended intersect BC at Q'. Since B, C, D, E are concyclic, we must have $\angle B = \angle AED = \angle APD$. Hence, B, D, P, Q' are concyclic. Similarly, C, E, P, Q' are concyclic. It follows that the circumcircles of $\triangle BPD$ and $\triangle CPE$ intersect at P and Q', i.e., Q and Q' coincide. This completes the proof.

Note:
(1) It is easy to see that BE, CD are the heights of $\triangle ABC$, but this is not important when solving this problem.
(2) Recognizing A, D, P, E concyclic is the key step. This is the conclusion of Example 3.1.11, a commonly used fact.

6.9 Apply Menelaus' Theorem to $\triangle AMD$ intersected by BP, $\triangle AMF$ intersected by BC and $\triangle ADF$ intersected by PN:

$$\frac{AB}{MB} \cdot \frac{DP}{AP} \cdot \frac{ME}{DE} = 1 \quad (1)$$

$$\frac{MB}{AB} \cdot \frac{FD}{MD} \cdot \frac{AC}{FC} = 1 \quad (2)$$

$$\frac{AP}{DP} \cdot \frac{DN}{FN} \cdot \frac{FC}{AC} = 1 \quad (3)$$

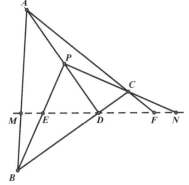

Multiplying (1), (2), (3) gives $\dfrac{ME \cdot FD \cdot DN}{DE \cdot MD \cdot FN} = 1$.

Since $DE = DF$, we have $\dfrac{DM}{EM} = \dfrac{DN}{FN}$, i.e., $\dfrac{DE}{EM} + 1 = \dfrac{DF}{FN} + 1$.

It follows that $EM = FN$ and hence, $DM = DN$.

Note: Multiplying (1), (2), (3) is a quick way to cancel out the terms. Of course, one may also manipulate each equation by moving the desired terms (DE, DF, MD, ME, etc.) to one side and the rest to the other side. This is a basic technique when applying Menelaus' Theorem.

6.10 Let $\odot O_2$ and $\odot O_3$ intersect at P and H. We have $PH \perp O_2O_3$ and $O_2O_3 /\!/ BC$ (Midpoint Theorem).

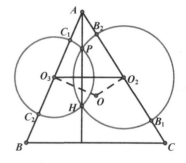

Hence, $PH \perp BC$, which implies A, P, H are collinear. By the Tangent Secant Theorem, we have

$$AC_1 \cdot AC_2 = AP \cdot AH = AB_1 \cdot AB_2.$$

Hence, B_1, B_2, C_1, C_2 are concyclic.

Let the perpendicular bisectors of BB_1, CC_1 intersect at O. Notice that OO_2, OO_3 are also the perpendicular bisectors of AC, AB respectively. Hence, O is the circumcenter of $\triangle ABC$, i.e., B_1, B_2, C_1, C_2 lie on $\odot O$ whose radius is OB_1.

A similar argument gives that A_1, A_2, B_1, B_2 also lie on $\odot O$. It follows that $A_1, A_2, B_1, B_2, C_1, C_2$ are concyclic on $\odot O$.

6.11 Let BM and CN intersect at X. Since $\angle C = \angle PAB$, we have $\triangle ABC \sim \triangle PBA$. Similarly, $\triangle ABC \sim \triangle QAC$.

Hence, $\angle 1 = \angle 2 = \angle BAC$ and we also have $\angle BPM = \angle NQC$.

Consider $\triangle BPM$ and $\triangle NQC$. Since P is the midpoint of AM, we have $\dfrac{BP}{PM} = \dfrac{BP}{AP}$ $= \dfrac{AB}{AC}$ because $\triangle ABC \sim \triangle PBA.$

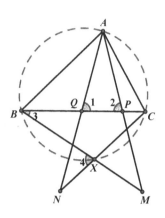

Similarly, $\dfrac{NQ}{CQ} = \dfrac{AQ}{CQ} = \dfrac{AB}{AC}.$

It follows that $\dfrac{BP}{PM} = \dfrac{NQ}{CQ}$ and hence, $\triangle BPM \sim \triangle NQC.$

Now $\angle 3 = \angle N$ and we must have $\angle 4 = \angle BQN = \angle 1 = \angle BAC$. It follows that A, B, X, C are concyclic.

6.12   Since $BE /\!/ CF$, we have $\angle BFC = \angle EBF = \angle CAD$ (angles in the same arc), which implies A, D, C, F are concyclic, say on $\odot O_1$.
Since M is the midpoint of BC, by the Tangent Secant Theorem, $AM \cdot DM = BM^2 = CM^2$, which implies BC is tangent to $\odot O_1$.
Since $\triangle BEG \sim \triangle CFG$, it follows that O and O_1 are corresponding points.

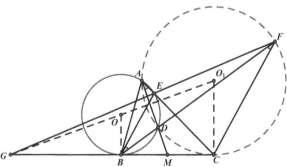

It follows that $\angle BGO = \angle CGO_1$ because they are corresponding angles in $\triangle BEG$ and $\triangle CFG$ respectively. This implies G lies on the line OO_1. Since OO_1 is the perpendicular bisector AD, we have $AG = DG$.

Note: One may also show G, O, O_1 collinear via $\triangle OBE \sim \triangle O_1CF$ and hence, $\dfrac{OB}{O_1C} = \dfrac{BE}{CF} = \dfrac{BG}{CG}$. Now $\triangle OBG \sim \triangle O_1CG$ and $\angle BGO = \angle CGO_1$.

6.13 Recall that $J_2J_3 \,/\!/\, EF$ because both are perpendicular to AI (Exercise 1.5). Similarly, $J_1J_2 \,/\!/\, DE$ and $J_1J_3 \,/\!/\, DF$. It follows that $\triangle DEF \sim \triangle J_1J_2J_3$.

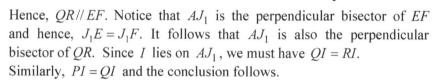

Now $\dfrac{J_1Q}{FQ} = \dfrac{DF}{J_1J_3}$ (since $J_1J_3 \,/\!/\, DF$)

$\quad = \dfrac{DE}{J_1J_2}$ (since $\triangle DEF \sim \triangle J_1J_2J_3$)

$\quad = \dfrac{J_1R}{ER}$ (since $DE \,/\!/\, J_1J_2$)

Hence, $QR \,/\!/\, EF$. Notice that AJ_1 is the perpendicular bisector of EF and hence, $J_1E = J_1F$. It follows that AJ_1 is also the perpendicular bisector of QR. Since I lies on AJ_1, we must have $QI = RI$. Similarly, $PI = QI$ and the conclusion follows.

6.14 Refer to the left diagram below. Extend AE to G such that $BC = EG$. Since $AB = BC + AE$, we have $AB = AG$.

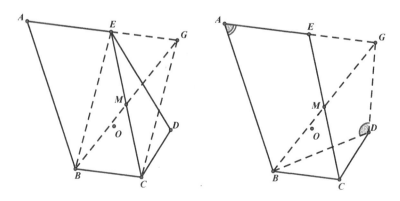

Now $\angle ABG = \angle AGB = \angle CBG$ (because $AE \,/\!/\, BC$), i.e., BG bisects $\angle ABC$. It is also easy to see that $BCGE$ is a parallelogram where M is the center.

We claim that A, B, D, G are concyclic. (1)

Notice that (1) would imply that $\angle ADB = \angle AGB$, which leads to the conclusion because $\angle AGB = \angle CBG = \dfrac{1}{2}\angle ABC = \dfrac{1}{2}\angle CDE$.

Refer to the right diagram above. It suffices to show that $\angle BDG = 180° - \angle A$, where $180° - \angle A = \angle ABC = \angle CDE$. Hence, it suffices to show $\angle BDG = \angle CDE$, or $\angle BDC = \angle EDG$. (2)

Let D' be the reflection of D about OM. Refer to the diagram on the right. Since $OD = OD'$, D' must lie on $\odot O$ whose radius is OD. Notice that $\odot O$ is exactly the circumcircle of $\triangle BCD$, i.e., B, C, D, D' are concyclic.

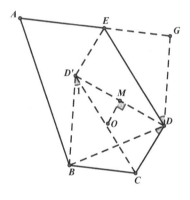

Now $\angle BDC = \angle 1$. (3)

On the other hand, one sees that $CDED'$ is a parallelogram because DD' and CE bisect each other at M.

It follows that $CD' = DE$ and $CD' /\!/ DE$. Now it is easy to see that $\triangle BCD' \cong \triangle GED$ (S.A.S.). We conclude that $\angle EDG = \angle 1$. (4).

(3) and (4) imply (2), which completes the proof.

Printed in the United States
By Bookmasters